The Character Map in Non-abelian Cohomology

Twisted, Differential, and Generalized

The Character Map in Non-abelian Cohomology

Twisted, Differential, and Generalized

Domenico Fiorenza
Sapienza Università di Roma, Italy

Hisham Sati
New York University Abu Dhabi, UAE

Urs Schreiber
New York University Abu Dhabi, UAE

World Scientific

NEW JERSEY • LONDON • SINGAPORE • BEIJING • SHANGHAI • HONG KONG • TAIPEI • CHENNAI • TOKYO

Published by

World Scientific Publishing Co. Pte. Ltd.
5 Toh Tuck Link, Singapore 596224
USA office: 27 Warren Street, Suite 401-402, Hackensack, NJ 07601
UK office: 57 Shelton Street, Covent Garden, London WC2H 9HE

Library of Congress Control Number: 2023033984

British Library Cataloguing-in-Publication Data
A catalogue record for this book is available from the British Library.

THE CHARACTER MAP IN NON-ABELIAN COHOMOLOGY
Twisted, Differential, and Generalized

ISBN 978-981-127-669-9 (hardcover)
ISBN 978-981-127-670-5 (ebook for institutions)
ISBN 978-981-127-671-2 (ebook for individuals)

For any available supplementary material, please visit
https://www.worldscientific.com/worldscibooks/10.1142/13422#t=suppl

Typeset by Stallion Press
Email: enquiries@stallionpress.com

Preface

This book is, first, a streamlined review of L_∞-algebraic rational homotopy theory in the modern guise of model category theory, and, second, the observation that the resulting rationalization Quillen adjunction, when understood as operating on generalized classifying spaces, is the general form of the "character map" in twisted, differential and non-abelian generalized cohomology theories. Finally it is an exposition of a zoo of examples and applications of this generalized character map, ranging all the way from the classical Chern character and the Chern-Weil homomorphism, over the Chern-Dold character in extraordinary cohomology theories such as topological modular forms and iterated higher K-theories, all the way to novel higher non-abelian examples such as unstable and twistorial Cohomotopy theory.

The Chern character is a famous construction in K-theory, generalizing, in its equivariant form, the classical character theory of group representations from vector spaces to vector bundles. And yet, the Chern character is just the first instance of a general and fundamental notion of character maps on generalized cohomology theories in a broad and encompassing sense. Part of this generalization is explicit in existing literature:

In the algebraic topology of stable homotopy theory, Chern-Dold character maps on Whitehead-generalized cohomology theories serve to approximate these *extraordinary* theories (such as elliptic cohomology or Cobordism cohomology) by ordinary rational cohomology, detecting their non-torsion information. In passage to differential topology, the extension of scalars of these Chern-Dold character to the real numbers is the key ingredient in the construction of *differential* generalized cohomology theories: commonly defined as the homotopy fiber products, formed in smooth higher stacks, of the Chern-Dold character map with the de Rham isomorphism.

In pure mathematics, differential cohomology was originally motivated through the Beilinson conjectures in arithmetic geometry, where Beilinson regulators have, more recently, been understood as character maps on algebraic K-theory. On the other hand, in modern mathematical physics, differential generalized cohomology theories encode "charge quantization" laws in fundamental high energy physics and, more recently, in topological phases of solid state physics. Here the Chern Dold character is interpreted as extracting the observable flux densities of higher gauge fields, and as such is at the heart of mathematical modelling of physical reality. In particular, the observation that differential form expressions for D-brane charges in string theory look like Chern character images led to the famous conjecture that D-brane charge is quantized in topological K-theory, a conjecture which has been and still is driving much of the activity in generalized cohomology theory, even though there are several indications that the conjecture needs to be refined in order to hold true.

Indeed, in all these applications the notion of "generalized cohomology" in the classical sense of Whitehead is not actually general enough: Besides the generalization to differential cohomology, one needs to consider twisted and equivariant and, crucially, also non-abelian (i.e.: unstable) generalizations of Whitehead-generalized cohomology theory, as we discuss below.

However, a general notion, certainly a general account, of what "character map" is to mean in the full generality of twisted equivariant differential non-abelian cohomology has been missing: Existing constructions of generalized variants of the Chern character tend to be ad-hoc, reliant on intuition and ingenuity; while required generalizations of more general Chern-Dold characters have hardly found any attention.

In this book we mean to fill this gap by laying out systematic foundations for a fully general notion of the character map in twisted, differential and non-abelian generalized cohomology theory (we relegate analogous discussion of the equivariant aspect to a followup), reviewing all relevant theory and pertinent examples as we go along.

The basic idea of general character maps which we promote here is simple and elegant, following the time-honored paradigm of representability: We **1.** observe that a twisted (differential) non-abelian generalized cohomology theory is exactly that which is *represented*, in a suitable slice homotopy category, by a local coefficient bundle of classifying spaces (of moduli ∞-stacks) and **2.** define the character map as that cohomology operation which is represented by the *rationalization reflection* of that local coefficient bundle, hence by its universal approximation by a bundle in rational homotopy theory. While shadows of this general abstract definition may easily be recognized throughout the existing literature, notably where ad-hoc constructions of character maps are justified *a posteriori* by checking that they become isomorphisms under tensoring with the rational numbers, a general and systematic discussion from this universal vantage point has been missing.

Based on this general notion, we highlight that there is a canonical tool for connecting general character maps to concrete computations and examples: This is provided by the *fundamental theorem of dg-algebraic rational homotopy theory* in the form obtained by Bousfield & Kan, following Quillen and Sullivan. We review this here in modernized and streamlined form, utilizing the full power of model category theory (surveyed in an appendix) and using the understanding of cofibrant dgc-algebras as formal duals of nilpotent L_∞-algebras. Combined with the above perspective on non-abelian cohomology, this reveals that the classical fundamental theorem of dg-algebraic rational homotopy theory may be re-cast as a *twisted non-abelian de Rham theorem* on twisted non-abelian generalized cohomology. We lay this out in some detail and review in a wealth of examples how this captures familiar and more exotic differential-form expressions for generalized character maps.

In view of the fact mentioned before, that differential generalized cohomology is essentially the homotopy fiber product of the general character map with the de Rham isomorphism, formed in smooth higher stacks, this allows us define and construct the full combination of twisted differential non-abelian generalized cohomology. We spell out how, besides the abelian Chern-Dold characters, classical non-abelian constructions like the Chern-Weil homomorphism on non-abelian cohomology in degree 1 (represented by principal bundles) and its refinement to Cheeger-Simons secondary characteristic classes, find their natural home in this general framework, which hence generalizes all this to non-abelian cohomology in higher degree, represented by higher non-abelian gerbes.

Finally, as a fundamental example that goes beyond the scope of existing literature, we discuss the twisted non-abelian character map on twistorial Cohomotopy theory, which, over 10-manifolds, may be viewed as a twisted non-abelian enhancement of topological modular forms (tmf) in degree 4. We review how this turns out to exhibit a list of subtle topological relations that in high energy physics are thought to govern the non-linear charge quantization of Yang-Mills instantons and of branes in non-perturbative string theory ("M-theory"), beyond what can be seen in K-theory or in any other Whitehead-generalized (hence abelian) cohomology theory.

This suggests ("Hypothesis H") that the non-abelian twisted differential character map constructed here, specifically on unstable Cohomotopy, is the fundamental object of interest at least for applications in high energy physics, to which various Chern- and Chern-Dold-character maps are but partial approximations.

Acknowledgements. We thank Tatiana Ezubova, John Lind, Chris Rogers, Carlos Simpson, Danny Stevenson, and Mathai Varghese for comments on an earlier version of this text. Particular thanks go to an anonymous referee for their detailed comments. Authors H. S. and U. S. acknowledge the support by *Tamkeen* under the NYU Abu Dhabi Research Institute grant CG008.

Domenico Fiorenza, *Dipartimento di Matematica, Sapienza Universitá di Roma, Piazzale Aldo Moro 2, 00185 Rome, Italy.*
fiorenza@mat.uniroma1.it

Hisham Sati, *Mathematics, Division of Science, and Center for Quantum and Topological Systems (CQTS), NYUAD Research Institute, New York University Abu Dhabi, UAE.*
hsati@nyu.edu

Urs Schreiber, *Mathematics, Division of Science, and Center for Quantum and Topological Systems (CQTS), NYUAD Research Institute, New York University Abu Dhabi, UAE.*
us13@nyu.edu

Contents

PART I
Introduction

Generalized cohomology theories [Whitehead (1962)][Adams (1974)] – such as K-theory, elliptic cohomology, stable Cobordism and stable Cohomotopy – are rich. This makes them fascinating but also intricate to deal with. In algebraic topology it has become commonplace to apply filtrations by iterative *localizations* [Bousfield (1979)] (review in [Elmendorf *et al.* (1997), §V][Bauer (2014)]) that allow generalized cohomology to be approximated in consecutive stages; a famous example of current interest is the chromatic filtration on complex oriented cohomology theories ([Mahowald and Ravenel (1987)], review in [Ravenel (1986)][Lurie (2010)]).

The Chern-Dold character. The primary approximation stage of generalized cohomology theories is their *rationalization* (e.g., [Hilton (1971)][Bauer (2014)]) to ordinary cohomology (e.g., singular cohomology) with rational coefficients or real coefficients (see Rem. 5.2). This goes back to [Dold (1972)]; and since on topological K-theory (Ex. 7.2) it reduces to the Chern character map [Hilton (1971), Thm. 5.8], this has been called the *Chern-Dold character* [Buhštaber (1970)]:

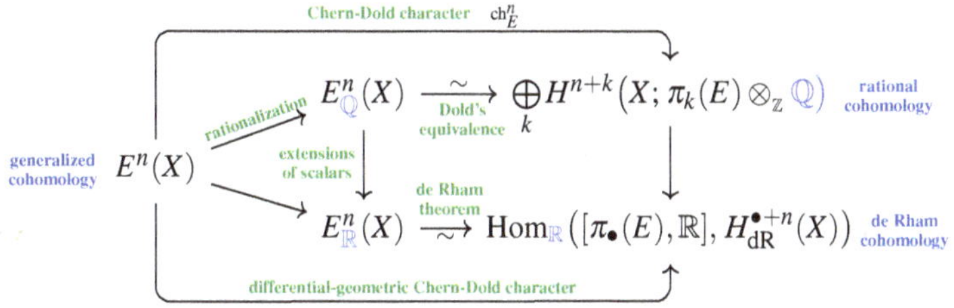

$$
\begin{array}{c}
\text{Chern-Dold character } \mathrm{ch}_E^n \\[4pt]
\hline
\end{array}
$$

$$
\begin{array}{ccc}
& E_{\mathbb{Q}}^n(X) \xrightarrow[\text{equivalence}]{\underset{\text{Dold's}}{\sim}} \bigoplus_k H^{n+k}\!\left(X; \pi_k(E) \otimes_{\mathbb{Z}} \mathbb{Q}\right) & \text{rational cohomology} \\
E^n(X) \;\overset{\text{rationalization}}{\nearrow}\;\; \Big\downarrow{\scriptstyle \text{extensions of scalars}} & & \Big\downarrow \\
\underset{\text{differential-geometric Chern-Dold character}}{\searrow}\; E_{\mathbb{R}}^n(X) \xrightarrow[\text{theorem}]{\overset{\text{de Rham}}{\sim}} \mathrm{Hom}_{\mathbb{R}}\!\left([\pi_\bullet(E), \mathbb{R}], H_{\mathrm{dR}}^{\bullet+n}(X)\right) & \text{de Rham cohomology}
\end{array}
\tag{0.1}
$$

(generalized cohomology $E^n(X)$)

That the left map in (0.1) is indeed the rationalization approximation on coefficient spectra is left somewhat implicit in [Buhštaber (1970)] (rationalization was properly formulated only in [Bousfield and Kan (1972b)]); a fully explicit statement is in [Lind *et al.* (2020), §2.1]. The equivalence on the top of (0.1) serves to make explicit how the result of that rationalization operation indeed lands in ordinary cohomology, and this was Dold's original observation [Dold (1972), Cor. 4] (see Prop. 7.2).

At the heart of differential cohomology. While rationalization is the coarsest of the localization approximations, it stands out in that it connects, via the de Rham theorem, to *differential geometric* data – when the base space X has the structure of a smooth manifold, and the coefficients are taken to be $\mathbb{R}$ instead of $\mathbb{Q}$. Indeed, this "differential-geometric Chern-Dold character" shown on the bottom of (0.1), underlies (usually without attribution to either Dold or Buchstaber) the pullback-construction of differential generalized cohomology theories [Hopkins and Singer (2005), §4.8] (see [Bunke and Nikolaus (2019), p. 17][Grady and Sati (2017), Def. 7][Grady and Sati (2021b), Def. 17][Grady and Sati (2019c), Def. 1], recalled as Def. 9.3 and Ex. 9.1 below).

At the heart of non perturbative field theory. It is in this differential-geometric form that the Chern-Dold character plays a pivotal role in high energy physics. Here closed differential forms encode *flux densities* $F_p \in \Omega_{\mathrm{dR}}^p(X)$ of generalized electromagnetic fields on spacetime manifolds X; and the condition that these lift through (i.e., are in the image of) the differential-geometric Chern-Dold character (0.1) for E-cohomology theory encodes a *charge quantization* condition in E-theory (see [Freed (2000)][Sati (2010)][Grady and Sati (2019b)]), generalizing Dirac's charge quantization of the ordinary electromagnetic field in ordinary cohomology [Dirac (1931)] (review in [Alvarez (1985), §2][Frankel (1997), 16.4e]):

$$\begin{array}{ccc} E^n(X) & \xrightarrow[\mathrm{ch}^n_E]{\text{differential-geometric Chern-Dold character}} & \mathrm{Hom}_{\mathbb{R}}\Big(\big[\pi_\bullet(E), \mathbb{R}\big], H^{n+\bullet}_{\mathrm{dR}}(X) \Big) \\ \underset{\text{class in } E\text{-cohomology}}{[c]} & \longmapsto & \underset{\text{charge-quantized flux densities}}{\Big[\big\{ F^{(a)}_{r_a} \in \Omega^{r_a}_{\mathrm{dR}}(X) \big\}_{1 \le a \le \dim[\pi_\bullet(E), \mathbb{R}]} \,\Big|\, d F^{(a)}_{r_a} = 0 \Big]} \end{array} \tag{0.2}$$

This idea of charge quantization in a generalized cohomology theory has become famous for the case of topological K-theory – $E = \mathrm{KU}, \mathrm{KO}$ – where it is argued to capture aspects of the expected nature of the Ramond-Ramond (RR) fields in type II/I string theory (see [Freed and Hopkins (2000)][Freed (2000)][Evslin (2006)][Grady and Sati (2019b)][Grady and Sati (2021b)]).

Need for non-abelian generalization. However, various further topological conditions (recalled in [Fiorenza *et al.* (2020b), Table 1][Fiorenza *et al.* (2021b), p. 2][Sati and Schreiber (2021b), Table 3][Fiorenza *et al.* (2022), p. 2], see Rem. 12.2 below), in non-perturbative type IIA string theory ("M-theory") are not captured by charge-quantization (0.2) in K-theory, nor in any Whitehead-generalized cohomology theory, since they involve *non-linear* functions (12.10) in the fluxes.

In order to systematically discuss the rich but under-appreciated area of non-abelian charge quantization, we introduce and explore, in Part IV and Part V, the natural non-linear/non-abelian generalization of the character map. This is based on classical constructions of dg-algebraic rational homotopy theory which we recall and develop in Part III.

The non-abelian character map. Indeed, despite their established name, generalized cohomology theories in the traditional sense of Whitehead [Whitehead (1962)][Adams (1974)] are not general enough for many purposes:

(i) Already the time-honored non-abelian cohomology that classifies principal bundles (Ex. 2.2 below), being the domain of the Chern-Weil homomorphism [Chern (1952)] (recalled as Def. 8.3, Prop. 8.5 below), falls outside the scope of Whitehead-generalized cohomology. Its *flat* sector alone, observed by secondary Cheeger-Simons invariants (re-derived as Thm. 9.9 below), is controlled by the classical *Maurer-Cartan equation* (e.g. [Nakahara (2003), §5.6.4][Rudolph and Schmidt (2017), Prop. 1.4.9]) on a Lie algebra valued differential form A_1:

$$d A_1^{(c)} = f^c_{ab} A_1^{(b)} \wedge A_1^{(a)} \quad \in \Omega^2_{\mathrm{dR}}(-) \tag{0.3}$$

(for f^c_{ab} the structure constants, recalled as Ex. 6.1 below) whose importance in large areas of mathematics and mathematical physics is hard to overstate (the "master equation",

e.g. [Markl (2012), Rem. 3.12][Chuang and Lazarev (2013)]), but whose cohomological content is not captured by Whitehead-generalized abelian cohomology theory.
(ii) Similarly outside the scope of Whitehead-generalized cohomology is the non-abelian cohomology classifying gerbes [Giraud (1971)] (see Ex. 2.5 below). In its flat sector this serves to adjoin to (0.3) the higher-degree condition

$$d B_2 \;=\; \mu_{abc} A_1^{(a)} \wedge A_1^{(b)} \wedge A_1^{(a)} \quad \in \Omega^3_{\mathrm{dR}}(-) \tag{0.4}$$

(for some differential 2-form, see Ex. 6.3) which has come to be recognized as a deep stringy refinement of the classical Maurer-Cartan equation (0.3) (see [Fiorenza *et al.* (2014b), App.] for pointers).

However – and this is our topic here – these two items are just the first two stages within a truly general concept of *higher non-abelian cohomology* (Def. 2.1 below), that classifies higher bundles/higher gerbes (Ex. 2.6 below), whose non-abelian character map (Def. IV.2 below) takes values in flat L_∞-algebra valued differential forms (Def. 6.1 below) satisfying non-linear polynomial differential relations (L_∞-algebraic Maurer-Cartan equations, e.g. [Doubek *et al.* (2007), (31)][Lazarev (2013), Def. 5.1][Manetti (2022), Def. 10.5.1]) which in string-theoretic applications (see (0.9) and Chapter 12 below) are identified with higher *Bianchi identities* on flux densities:

$$\begin{array}{ccc} \underset{\text{cohomology}}{\overset{\text{non-abelian}}{}} A(X) & \xrightarrow[\mathrm{ch}_A]{\text{non-abelian character}} & \overset{\text{non-abelian de Rham cohomology}}{\underset{}{}} H_{\mathrm{dR}}\big(X;\, \overset{\text{Whitehead } L_\infty\text{-algebra}}{\mathfrak{l}A}\big) \\ \underset{\substack{\text{class in}\\ A\text{-cohomology}}}{[c]} & \longmapsto & \Big[\underbrace{\big\{ F^{(a)}_{r_a} \in \Omega^{r_a}_{\mathrm{dR}}(X) \big\}_{1 \le a \le \dim[\pi_\bullet(A),\mathbb{R}]}}_{\text{charge-quantized flux densities}} \,\Big|\, \underbrace{d\,F^{(a)}_{r_a} \;=\; P_{r_a}\big(\{F^{(b)}_{r_b}\}_{b \le a}\big)}_{\text{higher Bianchi identities}} \Big] \end{array} \tag{0.5}$$

This generalizes (by Thm. 7.4 below) the Chern-Dold character (0.2) on Whitehead-generalized cohomology, which is subsumed as the abelian sector within the non-abelian theory (Ex. 2.10).

While the non-abelian character map (0.5) is built from mostly classical ingredients of dg-algebraic rational homotopy theory (recalled and developed in Chapter 5), its re-incarnation within non-abelian cohomology provides a new unifying perspective on mathematical phenomena expected to be relevant for non-perturbative physics:

Yang-Mills monopoles via higher non-abelian cohomology. In modern formulation, Dirac's charge quantization (e.g. [Alvarez (1985), §2])s of the electromagnetic field around a magnetic monopole with worldline $\mathbb{R}^{0,1} \hookrightarrow \mathbb{R}^{3,1}$, is the statement that the topological class of the field is encoded by a continuous map from the surrounding spacetime, which in the classical homotopy category (Ex. 1.14) is the 2-sphere $\mathbb{R}^{3,1} \setminus \mathbb{R}^{0,1} \simeq \mathbb{R}^3 \setminus \{0\} \simeq S^2 \in \mathrm{Ho}(\mathrm{TopSp}_{\mathrm{Qu}})$, to the classifying space of the circle group $B\mathrm{U}(1) \simeq K(\mathbb{Z},2)$ (2.6):

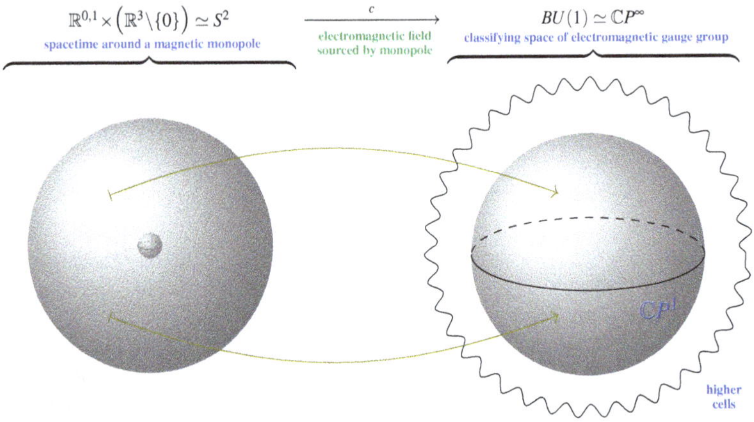

Since this is the classifying space for integral 2-cohomology (Ex. 2.1), one deduces, generally, that magnetic charge of the abelian U(1)-Yang-Mills field is measured in ordinary abelian cohomology $H^2(-;\mathbb{Z})$ (e.g. [Freed *et al.* (2007), p. 7]). But the minimal cell decomposition of this classifying space is by complex projective k-spaces for $k \in \mathbb{N}$:

$$BU(1) \simeq B^2\mathbb{Z} = K(\mathbb{Z},2) \simeq \overset{\substack{\text{infinite (stable, abelian)} \\ \text{complex projective space}}}{\mathbb{C}P^\infty} \simeq \overset{\substack{\text{direct} \\ \text{limit}}}{\varinjlim_{k\to\infty}} \mathbb{C}P^k \simeq \varinjlim_{k\to\infty} \overset{\substack{\text{finite (unstable, non-abelian)} \\ \text{complex projective } k\text{-space}}}{\frac{\mathrm{SU}(k+1)}{\mathrm{U}(k)}}. \tag{0.6}$$

While none of these finite-dimensional stages $\mathbb{C}P^k$ by themselves classify an abelian Whiehead-type cohomology theory, each of them classifies a *higher non-abelian cohomology theory* $H^1(-;\Omega\mathbb{C}P^k)$ (by Ex. 2.6 below).

We observe that this formal mathematical fact (Prop. 2.2 below) actually captures fine detail of the motivating physics, in that this higher non-abelian deformation of abelian cohomology measures magnetic charge of *non-abelian magnetic monopoles* in SU(k)-Yang-Mills theory (review in [Atiyah and Hitchin (1988)][Sutcliffe (1997)]) obtained by reduction from higher dimensional spacetimes $\mathbb{R}^{0,1} \times \mathbb{R}^3 \times X^d$ on a smooth fiber manifold X^d:

$$\begin{array}{ccccc}
\overset{\substack{\text{gauge-equivalence class of moduli} \\ \text{of } N \text{ magnetic monopoles on } \mathbb{R}^3 \\ \text{of } \mathrm{SU}(k+1)\text{-Yang-Mills theory} \\ \text{minimally broken to } \mathrm{U}(k)}}{\mathscr{M}\big(\mathrm{SU}(k+1)\big)_N} & \underset{\substack{\text{Don.-Jarvis} \\ \text{theorem}}}{\xrightarrow{\text{homeo}}} & \overset{\substack{\text{holomorphic maps of algebraic degree } N \\ \text{from Riemann sphere around monopoles} \\ \text{to the complex projective } k\text{-manifold}}}{\mathrm{Maps}_{\mathrm{hol}}\big(\mathbb{C}P^1, \mathbb{C}P^k\big)_{\deg = N}} & \underset{\substack{\text{Segal's} \\ \text{theorem}}}{\overset{\substack{N(2k-1)\text{-equiv.} \\ (1.14)}}{\hookrightarrow}} & \overset{\substack{\text{continuous maps of topological degree } N \\ \text{from 2-sphere around monopoles} \\ \text{to complex projective } k\text{-space}}}{\mathrm{Maps}\big(S^2, \mathbb{C}P^k\big)_{\deg = N}} \\[2em]
\overset{\substack{X^d\text{-parameterized deformation classes} \\ \text{of moduli of } N \text{ magnetic monopoles}}}{H\Big(X^d; \mathscr{M}\big(\mathrm{SU}(k+1)\big)_N\Big)} & & \overset{\text{for } d < N(2k-1)}{=\!=\!=\!=\!=} & & \overset{\text{higher non-abelian cohomology}}{H^1\big(X^d \times S^2; \Omega\mathbb{C}P^k\big)} \underset{\substack{\pi^2(S^2) \\ \text{Cohomotopy}}}{\times} \{N\}
\end{array} \tag{0.7}$$

This is a direct consequence (using Prop. 1.20 below) of classical theorems shown in the first line of (0.7): due to Donaldson ([Donaldson (1984)], for $k = 1$), Jarvis ([Jarvis (1998)]

[Jarvis (2000)] for general k, originally conjectured by Atiyah [Atiyah (1984), §5], review in [Ioannidou and Sutcliffe (1999)]) and Segal ([Segal (1979), Prop. 1.2], see also [Cohen *et al.* (1991)][Kamiyama (2007)]). Notice that the same moduli spaces of holomorphic maps out of $\mathbb{C}P^1$ (often regarded and referred to as *rational maps* out of $\mathbb{C}$), hence the same non-abelian cohomology sets (0.7), control numerous other aspects of non-abelian Yang-Mills theory, notably the topological field configurations known as *Skyrmions* (an observation due to [Houghton *et al.* (1998)][Manton and Piette (2001)] whose homotopy-theoretic implications through Segal's theorem (0.7) have been found in [Krusch (2003)][Krusch (2006)]), which are of deep relevance in non-perturbative quantum chromodynamics (hadrodynamics), not only theoretically but also experimentally (review in [Rho and Zahed (2016)], see [R. A. Battye and Sutcliffe (2010), p. 23] for the impact of Segal's theorem (0.7)). Moreover, the homotopy quotient of these spaces by the symmetries of $\mathbb{C}P^1$ (after compactification via adjoining of "stable maps" on degeneration limits of $\mathbb{C}P^1$) govern the Gromov-Witten invariants of $\mathbb{C}P^k$ (review in [Bertram (2004), §2][Katz (2006)]) and, for $k = 3$, the D-instantons of twistor string theory ([Witten (2004)]), the scattering amplitudes of $\mathcal{N} = 4$ super Yang-Mills theory [Roiban *et al.* (2004)] and those of $\mathcal{N} = 8$ supergravity ([Cachazo and Skinner (2013)][Adamo (2015)]), for mathematical review see [Atiyah *et al.* (2017), §7].

Non-abelian character of Yang-Mills monopoles. It turns out (Ex. IV.1) that the non-abelian character map (0.5) on these non-abelian magnetic monopole charges (0.7) extracts the underlying abelian magnetic flux density F_2 together with a non-linear differential relation:

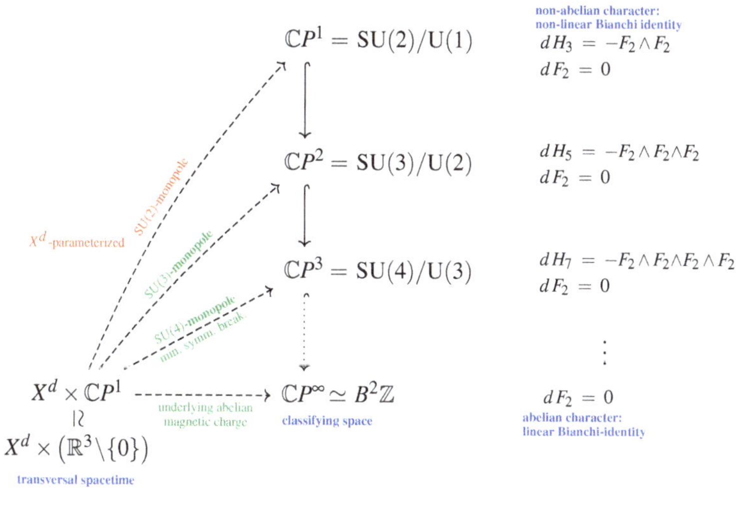

(0.8)

While the algebraic form of this non-abelian character data follows readily – once the non-abelian character map has been conceived in the first place, that is, according to our Def. IV.2 – from the well-known Sullivan model for complex projective spaces (Ex. 5.5), it is curious to observe [Fiorenza *et al.* (2022)][Sati and Schreiber (2020a)] and seems to have gone unnoticed before,[1] that its non-linear differential relations (0.8) on magnetic flux densities are those of important anomaly cancellation mechanisms in string theory. Notably the first non-linear relation in the list

$$\underset{\text{2-Cohomotopy}}{\pi^2(-)} := H^1\big(-;\Omega\mathbb{C}P^1\big) \xrightarrow[\mathrm{ch}_{\mathbb{C}P^1}]{\text{non-abelian character map}} H_{\mathrm{dR}}\big(-;\mathfrak{l}\mathbb{C}P^1\big)$$
$$[c] \longmapsto \left[\begin{array}{l|l} H_3 \in \Omega^3_{\mathrm{dR}}(-) & dH_3 = -F_2 \wedge F_2 \\ F_2 \in \Omega^2_{\mathrm{dR}}(-) & dF_2 = 0 \end{array}\right] \begin{array}{l}\text{Green-Schwarz-like Bianchi identity} \\ \text{ordinary abelian Bianchi identity}\end{array} \tag{0.9}$$

has the non-linear form of the Bianchi identity governing the *Green-Schwarz mechanism* [Green and Schwarz (1984), (4)-(6)][Candelas *et al.* (1985), [p. 49] (a mathematical account is in [Freed (2000), p. 40]) for anomaly cancellation in heterotic string theory (the "first superstring revolution" [Schwarz (2012)]), here for the case of *heterotic line bundles* (of phenomenological interest [Anderson *et al.* (2012)][Anderson *et al.* (2011)], see [Ashmore *et al.* (2021b), §4.2][Ashmore *et al.* (2021a), §2.2] for the case at hand), namely heterotic gauge bundles whose gauge group is reduced along the symmetry breaking $\mathrm{U}(1) \hookrightarrow \mathrm{SU}(2) \hookrightarrow E_8$ of the Yang-Mills monopole (0.8). Of course, the full Green-Schwarz mechanism is as in equation (0.9) but with a further contribution from a gravitational flux. This turns out to arise through tangential *twisting* of the non-abelian character, which is the main result of [Fiorenza *et al.* (2020b)][Fiorenza *et al.* (2021b)][Fiorenza *et al.* (2022)][Sati and Schreiber (2020a)] discussed in detail in Chapter 12 below, surveyed in a moment, in (0.14) below.

Cohomotopy theory as higher non-abelian cohomology. The higher non-abelian cohomology theory on the left of (0.9) is an example of a classical concept in homotopy theory, namely of *Cohomotopy* sets (Ex. 2.7) of homotopy classes of continuous maps into an n-sphere:

$$\underset{n\text{-Cohomotopy}}{\pi^n(-)} := \underset{\substack{\text{homotopy classes of contin.}\\ \text{maps into } n\text{-sphere}}}{\mathrm{Ho}(\mathrm{TopSp}_{\mathrm{Qu}})\big(-,S^n\big)} = \underset{\substack{\text{higher non-abelian}\\ S^n\text{-cohomology}}}{H^1\big(-;\Omega S^n\big)} \underset{\text{(on clsd. manifolds)}}{\overset{\text{Pontrjagin's theorem}}{=\!=\!=}} \underset{\substack{\text{cobordism classes of submanifolds}\\ \text{of codimension } n\\ \text{and normally framed}}}{\mathrm{Cob}^n_{\mathrm{Fr}}(-)}\,. \tag{0.10}$$

The stabilization of (0.10) to *stable Cohomotopy* (Ex. 2.13) is a widely recognized Whitehead-generalized cohomology theory, usually discussed in the context of the stable Pontrjagin-Thom theorem. But the original Pontrjagin theorem ([Pontryagin (1959)],

[1] Recently the string physics community is picking up some terminology of higher gauge theory in interpreting the Green-Schwarz mechanism, following [Sati *et al.* (2009)][Sati *et al.* (2012)][Fiorenza *et al.* (2014b)], identifying the Green-Schwarz-type Bianchi identity (0.9), Ex. 12.1, as reflecting 2-*group symmetry*, e.g. [Córdova *et al.* (2021), (1.18)][Del Zotto and Ohmori (2021), (3.3)]. To justify this terminology, one has to exhibit the GS-identity as the higher curvature invariant of a higher gauge bundle, hence as the non-abelian character of a higher non-abelian cohomology theory, foundations for which we mean to lay out here. Our results [Fiorenza *et al.* (2022)][Sati and Schreiber (2020a)] (see Chapter 12 below) indicate that to account for the fine structure of string/M-theory a 2-group is not sufficient, but a full ∞-group (Ex. 2.6, such as ΩS^4 (Rem. 2.1) equipped with twisting and equivariance, is required.

see [Sati and Schreiber (2021a)][Sati and Schreiber (2020b)] for review and further pointers) is decidedly unstable and as such says that the non-abelian Cohomotopy cohomology theory in (0.10) measures non-abelian charges carried by (normally framed) *sub*manifolds ("branes"), which generalize the monopole charges in (0.9) to higher codimension.

Non-abelian character of Cobordism. The non-abelian character (0.5) on unstable Cohomotopy/Cobordism (0.10) turns out (Ex. IV.1 below) to generalize the non-linear Green-Schwarz-type Bianchi identity (0.9) to higher even degrees. In degree 4 this yields the Bianchi identity of the C_3-field in $D = 11$ supergravity (due to [Sati (2018), §2.5][Fiorenza *et al.* (2017), §2], review in [Fiorenza *et al.* (2019), §7]), which merges with the monopole characters (0.8) to the mixed Bianchi identity[2] expected in Hořava-Witten's heterotic M-theory (due to [Fiorenza *et al.* (2022)][Sati and Schreiber (2020a)], see Chapter 12 below):

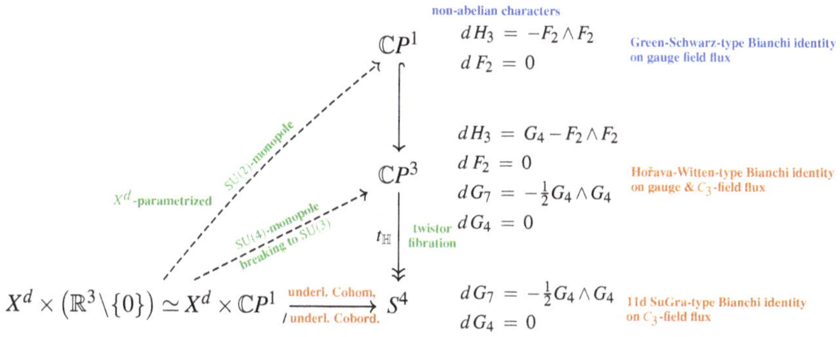

(0.11)

Twisted Cohomotopy as twisted non-abelian cohomology. Classical constructions in differential topology revolving around the Poincaré-Hopf theorem (e.g. [Bott and Tu (1982), §11]) involve deformation classes of non-vanishing vector fields on a smooth manifold X, hence of homotopy-classes of sections of the unit sphere bundle $S(TX)$ in the tangent bundle TX. Generally, for τ the class of a real vector bundle of rank $n+1$ over a paracompact Hausdoff space X, we may consider the homotopy-classes of sections of its unit sphere bundle $S(\tau)$ (with respect to any fiberwise metric) as the τ-*twisted* generalization (Ex. 3.8) of non-abelian n-Cohomotopy theory from (0.10):

$$\underset{\substack{\tau\text{-twisted} \\ \text{Cohomotopy}}}{\pi^{\tau}(-)} := \underset{\text{homotopy classes of cts. sections}}{\mathrm{Ho}\left(\mathrm{TopSp}_{\mathrm{Qu}}^{/X}\right)\left(-, S(\tau)\right)} = \underset{\substack{\tau\text{-twisted non-abelian} \\ S^n\text{-cohomology}}}{H^{\tau}(-; S^n)} \underset{\text{(for smth. } \tau \simeq \tau' \oplus 1\text{)}}{\overset{\substack{\text{twisted} \\ \text{Pontrjagin theorem}}}{=\!=\!=}} \underset{\substack{\text{cobordims classes of submanifolds} \\ \text{with normal } \tau'\text{-framing}}}{\mathrm{Cob}_{\mathrm{Fr}}^{\tau'}(-)}. \quad (0.12)$$

Indeed, when τ admits smooth structure and there is any section of $S(\tau)$ at all, then a twisted version of Pontrjagin's theorem still applies (e.g. [Cruickshank (2003), Lem. 5.2]) to show

[2]The physics-inclined reader may want to think of the broken SU(4) in (0.11) as a *flavor* symmetry group, along the lines of [Fiorenza *et al.* (2021a)].

that the twisted non-abelian cohomology theory which we may call *twisted Cohomotopy* [Fiorenza *et al.* (2020b), §2.1] still measures charges carried by cobordism classes of suitable submanifolds ("branes").

Non-abelian character of twisted Cohomotopy. In Part V below we construct the twisted generalization of the non-abelian character map (0.5), serving to extract differential form data underlying such twisted non-abelian cohomology classes. For instance, applied to the example of tangentially twisted Cohomotopy (0.12) on even-dimensional smooth manifolds, this deforms the Bianchi identity of ordinary odd-degree cohomology by the Euler form χ (8.8) of the manifold:

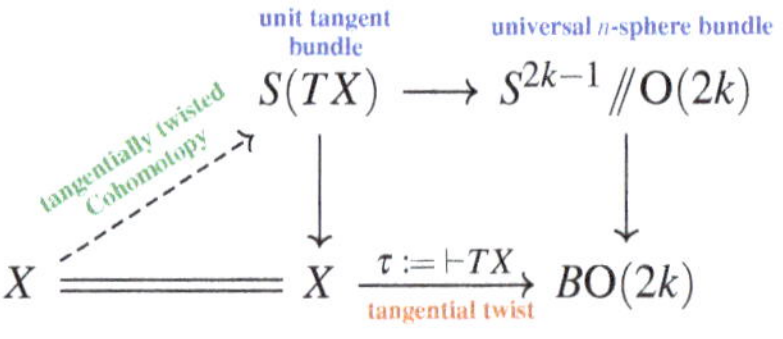

twisted non-abelian character

$$d\,\theta_{2k-1} = \chi_{2k}(\nabla) \quad \text{tangential de Rham twist} \tag{0.13}$$

Chern-Weil character of tangential twist

$$d\chi_{2k}(\nabla) = 0 \quad \text{Euler form}$$
$$d\,p_\bullet(\nabla) = 0 \quad \text{Pontrjagin form}$$

Hence the mere existence of the twisted non-abelian character on odd-degree Cohomotopy reflects part of the classical Poincaré-Hopf theorem (e.g. [Bott and Tu (1982), §11]), namely the vanishing of the Euler number of a manifold implied by the existence of a unit vector field. The further extension of this twisted non-abelian character to even-degree Cohomotopy yields a tangentially twisted enhancement of the classical Hopf invariant [Fiorenza *et al.* (2021b), §4].

Twisted non-abelian character of Yang-Mills monopoles. The exceptional isomorphism $\mathrm{Sp}(2) \simeq \mathrm{Spin}(5)$ between the quaternion unitary group (the compact "symplectic group") and the spin-group in 5 dimensions, together with the equivariance of the twistor fibration (0.11) under the canonical action of these groups implies a unification of all the above examples in a twisted non-abelian cohomology theory which we will call *twistorial Cohomotopy* (Ex. 3.11 below). The twisted non-abelian character on this theory is interesting in that it exhibits a variety of aspects expected of non-linear Bianchi identities in non-perturbative string theory (due to [Fiorenza *et al.* (2020b)][Fiorenza *et al.* (2021b)][Fiorenza *et al.* (2022)][Sati and Schreiber (2020a)], surveyed in Rem. 12.2 below) which cannot be explained by traditional twisted Whitehead-generalized cohomology theory (Ex. 3.5):

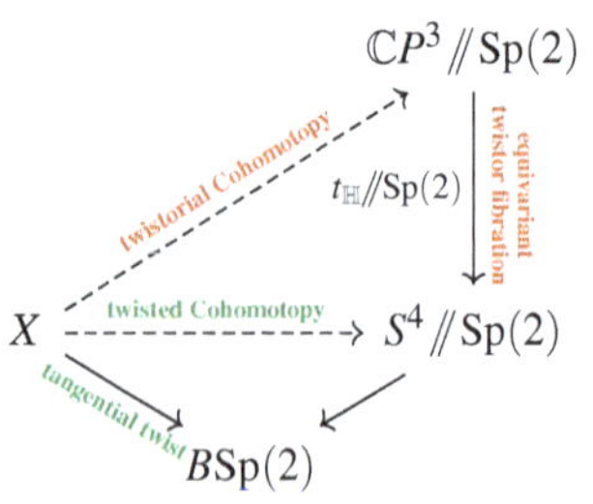

twisted non-abelian character

$$dH_3 = -F_2 \wedge F_2 + G_4 - \tfrac{1}{4} p_1(\nabla)$$

monopole char. tangential de Rham twist

$$dF_2 = 0 \tag{0.14}$$

cobordism char. tangential de Rham twist

$$d\,2G_7 = -G_4 \wedge G_4 + \tfrac{1}{4} p_1(\nabla) \wedge \tfrac{1}{4} p_1(\nabla) - \chi_8(\nabla)$$
$$dG_4 = 0$$

The desire to systematicall understand this rich example (see Chapter 12) as a non-abelian generalization of the traditional character map on twisted Whitehead-generalized cohomology originally motivated us to develop the theory of the twisted non-abelian character map presented here.

To fully bring out the unifying picture, we will discuss in detail how relevant examples of twisted Whitehead-generalized cohomology theories are subsumed by our twisted non-abelian character map (the "inverse Whitehead principle", Rem. 3.1).

Non-abelian cohomology via classifying spaces. As shown by these motivating examples, in higher non-abelian cohomology the very conceptualization of cohomology finds a beautiful culmination, as it is reduced to the pristine concept of homotopy types of mapping spaces (2.3), or rather, if geometric (differential, equivariant,...) structures are incorporated, of higher mapping stacks (Rem. 2.3 below). In particular, the concept of *twisted* non-abelian cohomology is most natural from this perspective (Def. 3.2 below) and naturally subsumes the traditional concept of twisted generalized cohomology theories (Prop. 3.5 below).

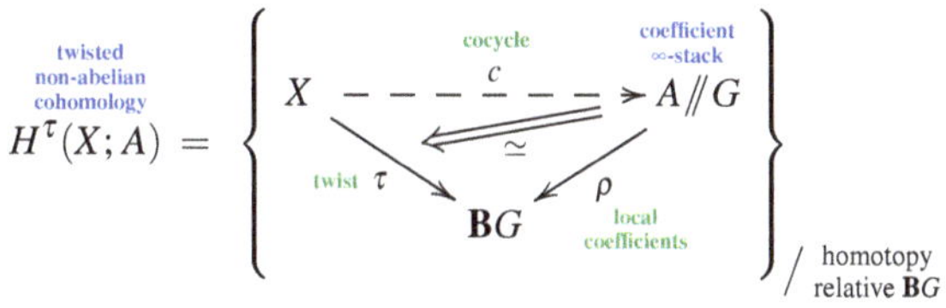

State of the literature. It is fair to say that this transparent fundamental nature of higher non-abelian cohomology is not easily recognized in much of the traditional literature on the topic, which is rife with unwieldy variants of cocycle conditions presented in combinatorial n-category-theoretic language. As a consequence, the development of non-abelian cohomology theory has seen little and slow progress, certainly as compared to the flourishing of Whitehead-generalized cohomology theory. In particular, the concepts of higher and of twisted non-abelian cohomology had tended to remain mysterious (see [Simpson (1997b), p. 1]). It is the more recently established homotopy-theoretic formulation of ∞-category theory (see Rem. 1.2) in its guise as ∞-topos theory (∞-*stacks*, recalled around Prop. 1.24 below) that provides the backdrop on which twisted higher non-abelian cohomology finds its true nature [Simpson (1997b)][Simpson (2002)][Toën (2002)][Lurie (2009a), §7.1][Sati *et al.* (2012)][Nikolaus *et al.* (2015a)][Nikolaus *et al.* (2015b)][Schreiber (2013)][Fiorenza *et al.* (2020b)][Sati and Schreiber (2020c)]; see Part II for details.

The non-abelian character map. From this homotopy-theoretic perspective, we observe, in Part IV and Part V, that the generalization of the Chern-Dold character (0.1) to twisted non-abelian cohomology naturally exists (Def. IV.2, Def. V.3), and that the non-abelian analogue of Dold's equivalence in (0.1) may neatly be understood as being, up to mild re-conceptualization, the fundamental theorem of dg-algebraic rational homotopy theory (recalled as Prop. 5.6 below). We highlight that this classical theorem is fruitfully recast as constituting a *non-abelian de Rham theorem* (Thm. 6.5 below) and, more generally, a *twisted non-abelian de Rham theorem* (Thm. 6.15 below). With this in hand, the notion of the (twisted) non-abelian character map appears naturally (Def. IV.2 and Def. V.3 below):

$$\underset{\substack{\text{twisted}\\\text{non-abelian}\\\text{character map}\\\text{(Def. V.3)}}}{\mathrm{ch}_\rho} : \overset{\substack{\text{twisted}\\\text{non-abelian cohomology}\\\text{(Def. 3.2)}}}{H^{\tau}(X;A)} \xrightarrow[\substack{\mathbb{R}\text{-rationalization}\\\text{(Def. IV.1)}}]{(\eta_\rho^{\mathbb{R}})_*} \overset{\substack{\text{twisted non-abelian}\\\text{real cohomology}\\\text{(Def. 5.19)}}}{H^{L_{\mathbb{R}}\tau}(X;L_{\mathbb{R}}A)} \xrightarrow[\substack{\text{twisted non-abelian}\\\text{de Rham theorem}\\\text{(Thm. 6.15)}}]{\simeq} \overset{\substack{\text{twisted non-abelian}\\\text{de Rham cohomology}\\\text{(Def. 6.9)}}}{H^{\tau_{\mathrm{dR}}}_{\mathrm{dR}}(X;\mathfrak{l}A)}\,. \tag{0.15}$$

Twisted differential non-abelian cohomology. Moreover, with the (twisted) non-abelian character in hand, the notion of (twisted) differential non-abelian cohomology appears naturally (Def. 9.3, Def. 11.2) together with the expected natural diagrams of twisted differential non-abelian cohomology operations:

$$\begin{array}{ccc}
\overset{\substack{\text{differential}\\\text{non-abelian cohomology}\\\text{(Def. 9.3)}}}{\widehat{H}(X;A)} & \xrightarrow{\substack{\text{curvature}\\\text{(9.13)}}} & \overset{\substack{\text{flat } L_\infty\text{-algebra valued}\\\text{differential forms}\\\text{(Def. 6.1)}}}{\Omega_{\mathrm{dR}}(X;\mathfrak{l}A)} \\
{\scriptstyle\substack{\text{characteristic}\\\text{class}\\\text{(9.12)}}}\Big\downarrow & \searrow^{\substack{\text{differential}\\\text{non-abelian character}\\\text{(9.14)}}} & \Big\downarrow \\
\underset{\substack{\text{non-abelian cohomology}\\\text{(Def. 2.1)}}}{H(X;A)} & \xrightarrow[\substack{\text{non-abelian character}\\\text{(Def. IV.2)}}]{\mathrm{ch}_A} & \underset{\substack{\text{non-abelian}\\\text{de Rham cohomology}\\\text{(Def. 6.3)}}}{H_{\mathrm{dR}}(X;\mathfrak{l}A)}
\end{array}$$

$$\begin{array}{ccc}
\overset{\substack{\text{twisted differential}\\\text{non-abelian cohomology}\\\text{(Def. 11.2)}}}{\widehat{H}^{\tau_{\mathrm{diff}}}(X;A)} & \xrightarrow{\substack{\text{twisted}\\\text{curvature}\\\text{(11.6)}}} & \overset{\substack{\text{twisted flat}\\L_\infty\text{-algebra valued}\\\text{differential forms}\\\text{(Def. 6.7)}}}{\Omega^{\tau_{\mathrm{dR}}}_{\mathrm{dR}}(X;\mathfrak{l}A)} \\
{\scriptstyle\substack{\text{twisted}\\\text{characteristic}\\\text{class}\\\text{(11.5)}}}\Big\downarrow & \searrow^{\substack{\text{twisted differential}\\\text{non-abelian character}\\\text{(11.7)}}} & \Big\downarrow \\
\underset{\substack{\text{twisted}\\\text{non-abelian cohomology}\\\text{(Def. 3.2)}}}{H^{\tau}(X;A)} & \xrightarrow[\substack{\text{twisted}\\\text{non-abelian character}\\\text{(Def. V.3)}}]{\mathrm{ch}^{\tau}_A} & \underset{\substack{\text{twisted non-abelian}\\\text{de Rham cohomology}\\\text{(Def. 6.9)}}}{H^{\tau_{\mathrm{dR}}}_{\mathrm{dR}}(X;\mathfrak{l}A)}
\end{array} \tag{0.16}$$

Unifying Chern-Dold, Chern-Weil and Cheeger-Simons. In order to show that this generalization of (twisted) character maps and (twisted) differential cohomology to higher non-abelian cohomology is sound, we proceed to prove that the non-abelian character map (Def. IV.2) specializes to:

the Chern-Dold character on generalized cohomology (Thm. 7.4),

the Chern-Weil homomorphism on degree-1 non-abelian cohomology (Thm. 8.6),

the Cheeger-Simons homomorphism on degree-1 differential non-abelian cohomology (Thm. 9.9).

All these classical invariants are thus seen as different low-degree aspects of the higher non-abelian character map.

Examples of twisted higher character maps. To illustrate the mechanism, we make explicit several examples of the (twisted) non-abelian character map on cohomology theories of relevance in high energy physics:

the Chern character on complex differential K-theory	(Ex. 7.2, 9.2),
the Pontrjagin character on real K-theory	(Ex. 7.3),
the Chern character on twisted differential K-theory	(Ex. 10.1, 11.3),
the MMS-character on cohomotopy-twisted K-theory	(Ex. 10.1),
the LSW-character on twisted higher K-theory	(Ex. 10.2),
the character on integral Morava K-theory	(Ex. 7.6),
the character on topological modular forms, tmf	(Ex. 7.4).

Once incarnated this way within the more general context of non-abelian cohomology theory, we may ask for non-abelian enhancements (Ex. 2.19) of these abelian character maps:

Non-abelian enhancement of the tmf**-character – the cohomotopical character.** Our culminating example, in Chapter 12, is the character map on twistorial Cohomotopy theory [Fiorenza *et al.* (2020b)][Fiorenza *et al.* (2022)], over 8-manifolds X^8 equipped with tangential $\mathrm{Sp}(2)$-structure τ (3.33). This may be understood (Rem. 7.4) as an enhancement of the tmf-character (Ex. 7.4) from traditional generalized cohomology to twisted differential non-abelian cohomology:

$$\overset{\substack{\text{tmf-cohomology}\\ \text{in degree 4}\\ \text{(Ex. 7.4)}}}{\mathrm{tmf}^4(X^8)} \underset{\substack{\text{tmf approximates}\\ \text{sphere spectrum}\\ \text{(Ex. 7.5)}}}{\simeq} \overset{\substack{\text{stable Cohomotopy}\\ \text{in degree 4}\\ \text{(Ex. 2.13)}}}{\mathbb{S}^4(X^8)} \underset{\substack{\text{non-abelian}\\ \text{enhancement}\\ \text{(Ex. 2.20)}}}{\rightsquigarrow} \overset{\substack{\text{unstable/non-abelian}\\ \text{4-Cohomotopy}\\ \text{(Ex. 2.7)}}}{\pi^4(X^8)} \underset{\substack{\text{twisting by}\\ \text{J-homomorphism}\\ \text{(Def. 3.2)}}}{\rightsquigarrow} \overset{\substack{\text{twisted non-abelian}\\ \text{4-Cohomotopy}\\ \text{(Ex. 3.8)}}}{\pi^{\tau^4}(X^8)} \underset{\substack{\text{lift through}\\ \text{twisted cohomology operation}\\ \text{induced by twistor fibration}\\ \text{(Ex. 3.11)}}}{\rightsquigarrow} \overset{\substack{\text{twistorial}\\ \text{Cohomotopy}\\ \text{(Ex. 3.11)}}}{\mathcal{T}^{\tau^4}(X^8)} \underset{\substack{\text{differential}\\ \text{enhancement}\\ \text{(Def. 11.2)}}}{\rightsquigarrow} \overset{\substack{\text{differential}\\ \text{twistorial}\\ \text{Cohomotopy}\\ \text{(Ex. 12.2)}}}{\widehat{\mathcal{T}}^{\tau^4}(X^8)}.$$

The non-abelian character map on twistorial Cohomotopy has the striking property (Prop. 12.1, the proof of which is the content of the companion physics article [Fiorenza *et al.* (2022), Prop. 3.9]) that the corresponding non-abelian version of Dirac's charge quantization (0.2) implies Hořava-Witten's Green-Schwarz mechanism in heterotic M-theory for heterotic line bundles F_2 (see [Fiorenza *et al.* (2022), §1]) and other subtle effects expected in non-perturbative high energy physics; these are discussed in Rem. 12.2 below.

Quadratic character functions from Whitehead brackets in non-abelian coefficient spaces. In summary, the crucial appearance of *quadratic functions* in the character map (12.10) is brought about by the intrinsic nature of (twisted) non-abelian cohomology theory, here specifically of Cohomotopy theory. These non-linearities originate in non-trivial Whitehead brackets (Rem. 5.4) on the non-abelian coefficient spaces, such as S^4 (Ex. 5.3) and $\mathbb{C}P^3$ (Ex. 6.8). Generally, the non-abelian character map (0.15) involves also higher monomial terms of any order (cubic, quartic, ...), originating in higher order Whitehead brackets on the non-abelian coefficient space (Rem. 5.4).

The desire to conceptually grasp character-like but quadratic functions appearing in M-theory had been the original motivation for developing differential generalized cohomology, in [Hopkins and Singer (2005)]. Here, in differential non-abelian cohomology, they appear intrinsically.

Outline.

- In Part I we motivate the need for (twisted) non-abelian characters and survey our key results.
- In Part II we recall (twisted) non-abelian cohomology theory with many examples.
- In Part III we review L_∞-algebras and dgc-algebraic rational homotopy theory in modernized form and re-cast the fundamental theorem as a de Rham theorem in non-abelian L_∞-algebraic de Rham cohomology.
- In Part IV we use this to construct the general character map in non-abelian cohomology and discuss applications.
- In Part V we generalize to twisted non-abelian cohomology and discuss further applications.

PART II

Non-abelian cohomology

Chapter 1

Model category theory

To set the scene, we begin here by reviewing basics of homotopy theory via model category theory [Quillen (1967)] (review in [Hovey (1999)][Hirschhorn (2003)][Lurie (2009a), A.2]) and of homotopy topos theory [Rezk (2010)] via model categories of simplicial presheaves [Brown (1973)][Jardine (1987)][Dugger (2001)] (review in [Dugger (1998)][Lurie (2009a), §A.3.3][Jardine (2015)]). The reader may want to skip this chapter and refer back to it as need be.

Topology. By

$$\mathrm{TopSp} \in \mathrm{Cats} \tag{1.1}$$

we denote a *convenient* [Steenrod (1967)] (in particular: cartesian closed) category of topological spaces such as compactly-generated spaces [Strickland (2009)] or Δ-generated spaces [Dugger (2003)], equivalently known as: numerically-generated spaces [Shimakawa *et al.* (2018)] or D-topological spaces [Sati and Schreiber (2020c), Prop. 2.4].

Categories. Let $\mathscr{C}$ be a category.
(i) For $X, A \in \mathscr{C}$ a pair of objects, we write

$$\mathscr{C}(X,A) := \mathrm{Hom}_{\mathscr{C}}(X,A)$$

for the set of *morphisms* from X to A.
(ii) For $\mathscr{C}, \mathscr{D}$ two categories, we denote a pair of *adjoint functors* between them by

$$\mathscr{D} \underset{R}{\overset{L}{\underset{\longrightarrow}{\overset{\longleftarrow}{\bot}}}} \mathscr{C} \quad \Leftrightarrow \quad \mathscr{D}\big(L(-), -\big) \underset{\sim}{\overset{\widetilde{(-)}}{\longleftrightarrow}} \mathscr{C}\big(-, R(-)\big), \tag{1.2}$$

and the corresponding *adjunction unit* and *adjunction counit* transformations by, respectively:

$$\eta^{RL}_C : C \xrightarrow{\widetilde{\mathrm{id}_{LC}}} R \circ L(C), \qquad \varepsilon^{LR}_D : L \circ R(D) \xrightarrow{\widetilde{\mathrm{id}_{RD}}} D \tag{1.3}$$

Notice/recall that this means that adjunct morphisms $f \leftrightarrow \widetilde{f}$ (1.2) and (co-)units (1.3) are related as follows:

$$L(c) \xrightarrow{\;\;f\;\;} d \quad \leftrightarrow \quad \overbrace{c \xrightarrow{\;\;\eta_c\;\;} R \circ L(c) \xrightarrow{\;\;R(f)\;\;} R(d)}^{\widetilde{f}}, \tag{1.4}$$

$$\overbrace{L(c) \xrightarrow{\;\;L(\widetilde{f})\;\;} L \circ R(d) \xrightarrow{\;\;\varepsilon_d\;\;} d}^{f} \quad \leftrightarrow \quad c \xrightarrow{\;\;\widetilde{f}\;\;} R(d)\,.$$

(iii) A Cartesian square in $\mathscr{C}$ we indicate by *pullback* notation $f^*(-)$ and/or by the symbol "(pb)":

$$\begin{array}{ccc} f^*A & \longrightarrow & A \\ {\scriptstyle f^*p}\big\downarrow & {\scriptstyle (\mathrm{pb})} & \big\downarrow{\scriptstyle p} \\ B_1 & \underset{f}{\longrightarrow} & B_2\,. \end{array} \tag{1.5}$$

Dually, a co-Cartesian square in $\mathscr{C}$ we indicate by *pushout* notation $f_*(-)$ and/or by the symbol "(po)":

$$\begin{array}{ccc} A_1 & \overset{f}{\longrightarrow} & A_2 \\ {\scriptstyle q}\big\downarrow & {\scriptstyle (\mathrm{po})} & \big\downarrow{\scriptstyle f_*q} \\ B & \longrightarrow & f_*B\,. \end{array} \tag{1.6}$$

Model categories.

Definition 1.1 (Weak equivalences). A *category with weak equivalences* is a category $\mathscr{C}$ equipped with a sub-class $\mathrm{W} \subset \mathrm{Mor}(\mathscr{C})$ of its morphisms, to be called the class of *weak equivalences*, such that

(i) W contains the class of isomorphisms;

(ii) W satisfies the cancellation property ("2-out-of-3"): if in any commuting triangle in $\mathscr{C}$

$$\begin{array}{ccccc} & & Y & & \\ & {\scriptstyle f}\nearrow & & \searrow{\scriptstyle g} & \\ X & & \underset{g\circ f}{\longrightarrow} & & Z \end{array} \tag{1.7}$$

two morphisms are in W, then so is the third.

Definition 1.2 (Weak factorization system). A *weak factorization system* in a category $\mathscr{C}$ is a pair of sub-classes of morphisms $\mathrm{Proj}, \mathrm{Inj} \subset \mathrm{Mor}(\mathscr{C})$ such that

(i) every morphisms $X \xrightarrow{f} Y$ in $\mathscr{C}$ may be factored through a morphism in Proj followed by one in Inj:

$$f : X \xrightarrow{\in \mathrm{Proj}} Z \xrightarrow{\in \mathrm{Inj}} Y \tag{1.8}$$

(ii) For every commuting square in $\mathscr{C}$ with left morphism in Proj and right morphism in Inj, there exists a lift, namely a dashed morphism

$$\begin{array}{ccc} X & \longrightarrow & A \\ {\scriptstyle \in \mathrm{Proj}}\downarrow & \overset{\exists}{\nearrow}\!\!\cdot\cdot & \downarrow{\scriptstyle \in \mathrm{Inj}} \\ Y & \longrightarrow & B \end{array} \tag{1.9}$$

making the resulting triangles commute.

(iii) Given Inj (resp. Proj), the class Proj (resp. Inj) is the largest class for which (1.9) holds.

Definition 1.3 (Model category [Joyal (2008c), Def. E.1.2][Riehl (2009)]**).** A *model category* is a category **C** that has all small limits and colimits, equipped with three subclasses of its class of morphisms, to be denoted

W – *weak equivalences*
Fib – *fibrations*
Cof – *cofibrations*

such that

(i) The class W makes **C** a category with weak equivalences (Def. 1.1);
(ii) The pairs $\big(\mathrm{Fib}, \mathrm{Cof}\cap \mathrm{W}\big)$ and $\big(\mathrm{Fib}\cap \mathrm{W}, \mathrm{Cof}\big)$ are weak factorization systems (Def. 1.2).

Remark 1.1 (Minimal assumptions). By item (iii) in Def. 1.2 a model category structure is specified already by the classes W and Fib, or alternatively by the classes W and Cof. Moreover, it follows from Def. 1.3 that also the class W is stable under retracts [Joyal (2008c), Prop. E.1.3][Riehl (2009), Lem. 2.4]: Given a commuting diagram in the model category **C** of the form on the left here

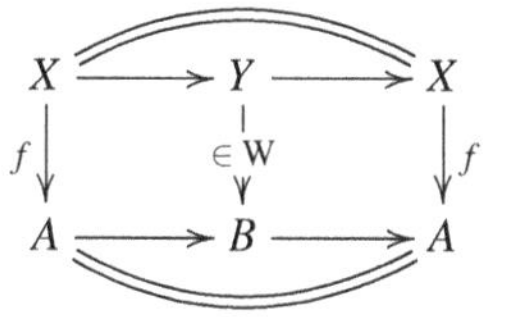

$\Rightarrow \quad f \in \mathrm{W}$ (1.10)

with the middle morphism a weak equivalence, then also f is a weak equivalence.

Definition 1.4 (Proper model category). A model category **C** Def. 1.3 is called

(i) *right proper*, if pullback along fibrations preserves weak equivalences:

$$\begin{array}{ccc} X & \xrightarrow{\;p^* f\;} & A \\ \downarrow & {\scriptstyle (\mathrm{pb})} & {\scriptstyle p}\downarrow{\scriptstyle \in \mathrm{Fib}} \\ Y & \xrightarrow[\;f \in \mathrm{W}\;]{} & B \end{array} \qquad \Rightarrow \qquad p^* f \in \mathrm{W} \tag{1.11}$$

(ii) *left proper*, if pushout along cofibrations preserves weak equivalences, hence if the opposite model category (Ex. 1.4) is right proper.

Notation 1.1 (Fibrant and cofibrant objects). Let **C** be a model category (Def. 1.3)

(i) We write $* \in \mathbf{C}$ for the terminal object and $\varnothing \in \mathbf{C}$ for the initial object.
(ii) An object $X \in \mathbf{C}$ is called:

(a) *fibrant* if the unique morphism to the terminal object is a fibration, $X \xrightarrow{\in \mathrm{Fib}} *$;
(b) *cofibrant* if the unique morphism from the initial object is a cofibration, $\varnothing \xrightarrow{\in \mathrm{Cof}} X$.

We write $\mathbf{C}_{\mathrm{fib}}, \mathbf{C}^{\mathrm{cof}}, \mathbf{C}^{\mathrm{cof}}_{\mathrm{fib}} \subset \mathbf{C}$ for the full subcategories on, respectively, fibrant objects, or cofibrant objects or objects that are both fibrant and cofibrant.
(iii) Given an object $X \in \mathbf{C}$

(a) A *fibrant replacement* is a factorization (1.8) of the terminal morphism as

$$X \xrightarrow[\in \mathrm{Cof}\cap\mathrm{W}]{j_X} PX \xrightarrow[\in \mathrm{Fib}]{q_X} * \,. \tag{1.12}$$

(b) A *cofibrant replacement* is a factorization (1.8) of the initial morphism as

$$\varnothing \xrightarrow[\in \mathrm{Cof}]{i_X} QX \xrightarrow[\in \mathrm{Fib}\cap\mathrm{W}]{p_X} X \,. \tag{1.13}$$

Recall that a continuous function f between topological spaces X, Y induces homomorphism on all homotopy groups $\pi_k(f,x) : \pi_k(X,x) \to \pi_k\big(Y, f(x)\big)$ and is called (e.g. [tom Dieck (2008), p. 144])

$$f \text{ is} \begin{cases} \text{an } n\text{-}\textit{equivalence} & \text{if} \quad \begin{array}{l}\pi_{\bullet<n}(f,x) \text{ is an isomorphism and} \\ \pi_n(f,x) \text{ is an epimorphism}\end{array} \\ \text{a } \textit{weak homotopy equivalence} & \text{if} \quad \pi_\bullet(f,x) \text{ is an isomorphism} \end{cases} \tag{1.14}$$

for all $x \in X$.

Example 1.1 (Classical model structure on topological spaces [Quillen (1967), §II.3][Hirschhorn (2019)]**).** The category TopSp (1.1) carries a model category structure whose

(i) W – weak equivalences are the weak homotopy equivalences (1.14);
(ii) Fib – fibrations are the Serre fibrations (e.g. [tom Dieck (2008), §5.5, 6.3]).

We denote this model category by

$$\mathrm{TopSp}_{\mathrm{Qu}} \in \mathrm{ModelCategories}\,.$$

Example 1.2 (Classical model structure on simplicial sets [Quillen (1967), §II.3][Gelfand and Manin (1996), §V.1-2][Goerss and Jardine (1999), §I.11]**).** The category of ΔSets of simplicial sets (e.g. [May (1967)][Curtis (1971)] exposition in [Friedman (2012)]) carries a model category structure whose

(i) W – weak equivalences are those whose geometric realization is a weak homotopy equivalence;

(ii) Cof – cofibrations are the monomorphisms (degreewise injections).

(iii) Fib – fibrations are the Kan fibrations.

We denote this model category by

$$\Delta\mathrm{Sets}_{\mathrm{Qu}} \in \mathrm{ModelCategories}.$$

Every simplicial set is cofibrant in the classical model structure (Ntn. 1.1):

$$\big(\Delta\mathrm{Sets}_{\mathrm{Qu}}\big)_{\mathrm{cof}} = \Delta\mathrm{Sets}, \tag{1.15}$$

while the fibrant simplicial sets are exactly the *Kan complexes* (e.g. [Goerss and Jardine (1999), §I.3], exposition in [Friedman (2012), §7])

$$\big(\Delta\mathrm{Sets}_{\mathrm{Qu}}\big)^{\mathrm{fib}} = \mathrm{KanComplexes}, \tag{1.16}$$

which we may think of as ∞-*groupoids* [Lurie (2009a), §1.1.2].

Example 1.3 (Simplicial nerves of groupoids).
(i) Let

$$\mathscr{G} = \Big(\mathscr{G}_1 \,{}_t\!\times_s \mathscr{G}_1 \xrightarrow{\;\circ\;} \mathscr{G}_1 \underset{t}{\overset{s}{\rightrightarrows}} \mathscr{G}_0\Big)$$

be a groupoid (exposition in [Weinstein (1996)]) then its *nerve* $N(\mathscr{G}) \in \Delta\mathrm{Sets}^{\mathrm{fib}}$ ([Segal (1968), §2]) is the Kan complex (1.16) whose k-cells are the sequences of k composable morphisms in $\mathscr{G}$.

$$N(\mathscr{G}) : [k] \mapsto \mathscr{G}_1 \,{}_t\!\times_s \mathscr{G}_1 \,{}_t\!\times_s \cdots \,{}_t\!\times_s \mathscr{G}_1 . \tag{1.17}$$

(ii) For $S \in \mathrm{Sets}$ any set, consider its *pair groupoid* $\mathrm{Pair}(S) := \big(S \times S \rightrightarrows S\big)$ whose objects are the elements of S and which has exactly one morphism $s_0 \xrightarrow{\exists !} s_1$ between any pair of elements. Its nerve (1.17) is contractible, in that it is weakly equivalent in the classical model category (Def. 1.2) to the point (the terminal simplicial set, which is constant on the singleton set):

$$N\big(\mathrm{Pair}(S)\big) \xrightarrow{\in \mathrm{W} \cap \mathrm{Fib}} * . \tag{1.18}$$

Example 1.4 (Opposite model category [Hirschhorn (2003), §7.1.8]**).** If $\mathbf{C}$ is a model category (Def. 1.3) then the opposite underlying category becomes a model category $\mathbf{C}^{\mathrm{op}}$ with the same weak equivalences (up to reversal) and with fibrations (resp. cofibrations) the cofibrations (resp. fibrations) of $\mathbf{C}$, up to reversal.

Example 1.5 (Slice model categories [Hirschhorn (2003), §7.6.4][May and Ponto (2012), Thm. 165.3.6]**).** Let $\mathbf{C}$ be a model category (Def. 1.3)

(i) For $X \in \mathbf{C}$ any object, the slice category $\mathbf{C}^{/X}$, whose objects are morphisms to X and whose morphisms are commuting triangles in $\mathbf{C}$ over X

$$\mathbf{C}^{/X}(a,b) \;:=\; \left\{ \begin{array}{ccc} A & \overset{f}{\dashrightarrow} & B \\ & {\scriptstyle a}\searrow \quad \swarrow{\scriptstyle b} & \\ & X & \end{array} \right\}$$

becomes itself a model category, whose weak equivalence, fibrations and cofibrations are those morphims whose underlying morphisms f are such in $\mathbf{C}$. This means in particular that:

$$a \in \left(\mathbf{C}^{/X}\right)^{\mathrm{cof}} \;\Leftrightarrow\; A \in \mathbf{C}^{\mathrm{cof}} \qquad \text{and} \qquad b \in \left(\mathbf{C}^{/X}\right)_{\mathrm{fib}} \;\Leftrightarrow\; b \in \mathbf{C}^{\mathrm{fib}}. \tag{1.19}$$

(ii) Dually there is the coslice model category $\mathbf{C}^{X/} := \left((\mathbf{C}^{\mathrm{op}})^{/X}\right)^{\mathrm{op}}$, being the opposite model category (Ex. 1.4) of the slice category of the opposite of $\mathbf{C}$:

$$\mathbf{C}^{X/}(a,b) \;:=\; \left\{ \begin{array}{ccc} & X & \\ & {\scriptstyle a}\swarrow \quad \searrow{\scriptstyle b} & \\ A & \underset{f}{\dashrightarrow} & B \end{array} \right\}.$$

Homotopy categories.

Definition 1.5 (Cylinder objects and Path space objects [Quillen (1967), Def. I.4]**).** Let $\mathbf{C}$ be a model category (Def. 1.3).

(a) With $A \in \mathbf{C}_{\mathrm{fib}}$ a fibrant object (Ntn. 1.1), a *path space object* for A is a factorization of the diagonal morphism Δ_A through a weak equivalence followed by a fibration:

$$A \xrightarrow{\;\in \mathrm{W}\;} \mathrm{Paths}(A) \xrightarrow{\;(p_0,p_1)\in \mathrm{Fib}\;} A \times A\,, \qquad \Delta_A : A \to A \times A \tag{1.20}$$

(b) With $X \in \mathbf{C}_{\mathrm{cof}}$ a cofibrant object (Ntn. 1.1), a *cylinder object* for X is a factorization of the co-diagonal morphism ∇_A through cofibration followed by a weak equivalence:

$$X \sqcup X \xrightarrow{\;(i_0,i_1)\in \mathrm{Cof}\;} \mathrm{Cyl}(X) \xrightarrow{\;\in \mathrm{W}\;} X\,, \qquad \nabla_X : X \sqcup X \to X \tag{1.21}$$

Example 1.6 (Standard cylinder object in simplicial sets). For $X \in \Delta\mathrm{Sets}_{\mathrm{Q}}$ (Ex. 1.2) a cylinder object (Def. 1.5) is evidently given by Cartesian product $X \times \Delta[1]$ with the 1-simplex, with (i_0,i_1) being the two endpoint inclusions.

Definition 1.6 (Homotopy). Let $\mathbf{C}$ be a model category (Def. 1.3), $X \in \mathbf{C}^{\mathrm{cof}}$ a cofibrant object, $A \in \mathbf{C}_{\mathrm{fib}}$ a fibrant object (Ntn. 1.1). Then a *homotopy* between a pair of morphisms $f, g \in \mathbf{C}(X, A)$, to be denoted

$$\phi : f \Rightarrow g \qquad \text{or} \qquad X \underset{g}{\overset{f}{\rightrightarrows}} A \quad (\Downarrow \phi)$$

is a morphism $\phi_l \in \mathbf{C}\big(\mathrm{Cyl}(X), A\big)$ out of a cylinder object for X *or* a morphism $\phi_r \in \mathbf{C}\big(X, \mathrm{Paths}(A)\big)$ to a path space object for A $\mathrm{Paths}(A)$ (Def. 1.5) which make either of these diagrams commute:

$$\begin{array}{ccc} X & \xrightarrow{\;f\;} & \\ {\scriptstyle i_0}\downarrow & & \\ \mathrm{Cyl}(X) & \xrightarrow{\;\phi_l\;} & A, \\ {\scriptstyle i_1}\uparrow & & \\ X & \xrightarrow{\;g\;} & \end{array} \qquad\qquad \begin{array}{ccc} & \xrightarrow{\;f\;} & A \\ & & \uparrow{\scriptstyle p_0} \\ X & \xrightarrow{\;\phi_r\;} & \mathrm{Paths}(A) \\ & & \downarrow{\scriptstyle p_1} \\ & \xrightarrow{\;g\;} & A \end{array}$$

Proposition 1.7 (Homotopy classes). *Let $\mathbf{C}$ be a model category, $X \in \mathbf{C}^{\mathrm{cof}}$ and $A \in \mathbf{C}_{\mathrm{fib}}$ (Ntn. 1.1). Then homotopy (Def. 1.6) is an equivalence relation $\sim$ on the hom-set $\mathbf{C}(X, A)$. We write*

$$\mathbf{C}(X, A)_{/\sim} \;\in\; \mathrm{Sets} \tag{1.22}$$

for the corresponding set of homotopy classes of morphisms from X to A.

Definition 1.8 (Homotopy category of a model category). For $\mathbf{C}$ a model category (Def. 1.3),

(i) we write

$$\mathrm{Ho}(\mathbf{C}) \;:=\; \big(\mathbf{C}^{\mathrm{cof}}_{\mathrm{fib}}\big)_{/\sim_r} \;\in\; \mathrm{Cats} \tag{1.23}$$

for the category whose objects are those objects of $\mathbf{C}$ that are both fibrant and cofibrant (Ntn. 1.1), and whose morphisms are the right homotopy classes of morphisms in $\mathbf{C}$ (Prop. 1.7):

$$X, A \in \mathbf{C}^{\mathrm{cof}}_{\mathrm{fib}} \quad \Rightarrow \quad \mathrm{Ho}(\mathbf{C})(X, A) \;:=\; \mathbf{C}(X, A)_{/\sim_r}$$

and composition of morphisms is induced from composition of representatives in $\mathbf{C}$.
(ii) Given a choice of fibrant replacement P and of cofibrant replacement Q for each object of $\mathbf{C}$ (Ntn. 1.1) we obtain a functor

$$\mathbf{C} \xrightarrow{\;\gamma_{\mathbf{C}}\;} \mathrm{Ho}(\mathbf{C}), \tag{1.24}$$

which **(a)** sends any object $X \in \mathbf{C}$ to PQX and sends **(b)** any morphism $X \xrightarrow{f} A$ to the right homotopy class (1.22) of any lift (1.9) PQf obtained from any lift Qf in the following diagrams:

$$
\begin{array}{ccc}
\varnothing & \longrightarrow & QY \\
{\scriptstyle i_X}\downarrow & \overset{Qf}{\nearrow} & \downarrow{\scriptstyle p_Y} \\
QX & \underset{f \circ p_x}{\longrightarrow} & Y
\end{array}
\quad \rightsquigarrow \quad
\begin{array}{ccc}
QX & \overset{j_{QY} \circ Qf}{\longrightarrow} & PQY \\
{\scriptstyle j_{QX}}\downarrow & \overset{PQf}{\nearrow} & \downarrow{\scriptstyle q_{QY}} \\
PQX & \longrightarrow & *
\end{array}
$$

Proposition 1.9 (Homotopy category is localization). *For a model category* $\mathbf{C}$ *(Def. 1.3) the functor* $\mathbf{C} \xrightarrow{\gamma_{\mathbf{C}}} \mathrm{Ho}(\mathbf{C})$ (1.24) *from Def. 1.8 exhibits the homotopy category as the localization of the model category at its class of weak equivalences:* $\gamma_{\mathbf{C}}$ *sends all weak equivalences in* $\mathbf{C}$ *to isomorphisms, and is the universal functor with this property.*

The restriction to fibrant-and-cofibrant objects in Def. 1.8 is convenient for defining composition of morphisms, but for computing hom-sets in the homotopy category it is sufficient that the domain object is cofibrant, and the codomain fibrant:

Proposition 1.10 ([Quillen (1967), §I.1 Cor. 7]). *Let* $\mathscr{C}$ *be a model category (Def. 1.3). For* $X \in \mathscr{C}^{\mathrm{cof}}$ *a cofibrant object and* $A \in \mathscr{C}_{\mathrm{fib}}$ *a fibrant object, any choice of fibrant replacement* PX *and cofibrant replacement* QA *(Ntn. 1.1). induces a bijection between the set of homotopy classes (Def. 1.6) and the hom-set in the homotopy category (Def. 1.8) between* X *and* A*:*

$$
\mathscr{C}(X,A)_{/\sim_r} \xrightarrow[\simeq]{\mathscr{C}(j_X, p_A)} \mathrm{Ho}(\mathbf{C})(X,A)\,.
$$

While the hom-functor of a homotopy category preserves almost no homotopy (co)limits, we do have:

Proposition 1.11 (Hom-functor of homotopy category respects (co)products). *The hom-functor of a homotopy category (Def. 1.8) respects coproducts in the first argument and products in its second argument, in that there are natural bijections of the following form:*

$$
\mathrm{Ho}(\mathbf{C})\Big(\coprod_{i \in I} X_i, \prod_{j \in J} A_j\Big) \;\simeq\; \prod_{\substack{i \in I \\ j \in J}} \mathrm{Ho}(\mathbf{C})(X_i, A_j)
$$

Proof. Noticing that coprodcts preserve cofibrancy and products preserve fibrancy, evidently, this follows from Prop. 1.10. □

Quillen adjunctions.

Definition 1.12 (Quillen adjunction). Let $\mathbf{D}, \mathbf{C}$ be model categories (Def. 1.3). Then a pair of adjoint functors $(L \dashv R)$ (1.2) between their underlying categories is called a *Quillen adjunction*, to be denoted

$$
\mathbf{D} \underset{R}{\overset{L}{\leftrightarrows}}{\scriptstyle \perp_{\mathrm{Qu}}} \mathbf{C} \tag{1.25}
$$

if the following equivalent conditions hold:

- L preserves Cof, and R preserves Fib;
- L preserves Cof and Cof $\cap$ W ("left Quillen functor");
- R preserves Fib and Fib $\cap$ W ("right Quillen functor").

Example 1.7 (Base change Quillen adjunction). Let $\mathbf{C}$ be a model category (Def. 1.3), $B_1, B_2 \in \mathbf{C}_{\text{fib}}$ a pair of fibrant objects (Ntn. 1.1) and

$$B_1 \xrightarrow{\ f\ } B_2 \quad \in \mathbf{C} \tag{1.26}$$

a morphism. Then we have a Quillen adjunction (Def. 1.12)

$$\mathbf{C}^{/B_2} \underset{f^*}{\overset{f_!}{\underset{\longrightarrow}{\overset{\longleftarrow}{\bot_{\text{Qu}}}}}} \mathbf{C}^{/B_1} \tag{1.27}$$

between the slice model categories (Ex. 1.5), where:

(i) The left adjoint functor $f_!$ is given by postcomposition in $\mathbf{C}$ with f (1.26):

$$f_* \;:\; \begin{array}{ccc} X & \xrightarrow{\ c\ } & A \\ & {\scriptstyle\tau}\searrow \quad \swarrow{\scriptstyle p} & \\ & B_1 & \end{array} \quad \longmapsto \quad \begin{array}{ccc} X & \xrightarrow{\ c\ } & A \\ & {\scriptstyle\tau}\searrow \quad \swarrow{\scriptstyle p} & \\ & B_1 & \\ & \downarrow{\scriptstyle f} & \\ & B_2 & \end{array} \tag{1.28}$$

(ii) The right adjoint functor f^* is given by pullback (1.5) along f (1.26).
That these functors indeed form an adjunction $f_! \dashv f^*$ follows from the defining universal property of the pullback (1.5):

$$\begin{array}{ccc} \mathbf{C}^{/B_2}(f_*\tau,\ \rho) & \simeq & \mathbf{C}^{/B_1}(\tau,\ f^*\rho) \\ \begin{array}{ccc} X & \overset{c}{\dashrightarrow} & A \\ {\scriptstyle\tau}\searrow & & \downarrow{\scriptstyle\rho} \\ B_1 & \xrightarrow{\ f\ } & B_2 \end{array} & \leftrightarrow & \begin{array}{ccccc} X & \overset{\tilde{c}}{\dashrightarrow} & f^*A & \longrightarrow & A \\ & {\scriptstyle\tau}\searrow & \downarrow{\scriptstyle f^*\rho} & \text{(pb)} & \downarrow{\scriptstyle\rho} \\ & & B_1 & \xrightarrow{\ f\ } & B_2 \end{array} \end{array} \tag{1.29}$$

That this adjunction is a Quillen adjunction (Def. 1.12) follows since $f_!$ (1.28) evidently preserves each of W and Cof (even Fib) separately, by Ex. 1.5.

Example 1.8 (Sliced Quillen adjunction). Given a Quillen adjunction $L \dashv_{\mathrm{Qu}} R$ (Def. 1.12) and an object of either of the two model categories, there is an induced *sliced* Quillen adjunctions (Def. 1.12) between slice model categories (Ex. 1.5) as follows:

$$\mathbf{D} \underset{R}{\overset{L}{\underset{\longrightarrow}{\overset{\longleftarrow}{\bot_{\mathrm{Qu}}}}}} \mathbf{C} \;\Rightarrow\; \left(\underset{c \in \mathbf{C}}{\forall} \;\; \mathbf{D}^{/L(c)} \underset{R^{/c}}{\overset{L^{/c}}{\underset{\longrightarrow}{\overset{\longleftarrow}{\bot_{\mathrm{Qu}}}}}} \mathbf{C}^{/c} \right) \text{ and } \left(\underset{d \in \mathbf{D}}{\forall} \;\; \mathbf{D}^{/d} \underset{R^{/d}}{\overset{L^{/d}}{\underset{\longrightarrow}{\overset{\longleftarrow}{\bot_{\mathrm{Qu}}}}}} \mathbf{C}^{/R(d)} \right), \tag{1.30}$$

where:

(i) $L^{/c}$ and $R^{/d}$ are given directly by applying L or R, respectively, to the triangular diagram that defines a morphism in the slice;

(ii) $R^{/c}$ and $L^{/d}$ are given by this direct application followed by right/left base change (1.7) along the adjunction unit/counit (1.3), respectively:

$$\begin{aligned} R^{/c} : \mathbf{D}^{/L(c)} \xrightarrow{\;R\;} \mathbf{C}^{/R \circ L(c)} \xrightarrow{(\eta_c)^*} \mathbf{D}^{/c}, \\ \mathbf{D}^{/d} \xleftarrow{(\varepsilon_d)_!} \mathbf{C}^{/L \circ R(d)} \xleftarrow{\;L\;} \mathbf{D}^{/R(d)} : L^{/d} \end{aligned} \tag{1.31}$$

In particular, this means that $L^{/d}$ sends a slicing morphism τ to its adjunct $\widetilde{\tau}$ (1.4), in that:

$$L^{/d} \left(\begin{array}{c} c \\ \downarrow{\scriptstyle \tau} \\ R(d) \end{array} \right) = \left(\begin{array}{c} L(c) \\ \downarrow{\scriptstyle \widetilde{\tau}} \\ d \end{array} \right) \in \mathbf{D}^{/d}. \tag{1.32}$$

Aspects of this statement appear in [Lurie (2009a), Prop. 5.2.5.1][Li (2016), Prop. 2.5(2)]. Since it is key to the proof of the twisted non-abelian de Rham theorem (Thm. 6.15) we spell it out:

Proof. It is clear that if we have adjunctions as claimed in (1.30), then $L^{/c}/R^{/d}$ are left/right Quillen functors, respectively, since these two act as the left/right Quillen functors *L*/*R* on underlying morphisms (by item (i) above), where the classes of slice morphisms are created, by Ex. 1.5.

To see that we have adjunctions as claimed, we may check their hom-isomorphisms (1.2) (for readability we now denote the object being sliced over by "*b*", in both cases, with "*c*" and "*d*" now being the variables in the hom-isomorphism):

(1) For the first case, consider the following transformations of slice hom-sets:

$$
\begin{array}{c}
L(c) \dashrightarrow^{f} d \\
\searrow \;\; \swarrow p \\
L(b)
\end{array}
\qquad \longmapsto \qquad
\begin{array}{c}
\overset{\widetilde{f}}{\overbrace{\qquad\qquad\qquad\qquad}} \\
c \xrightarrow{\eta_c} R \circ L(c) \dashrightarrow^{R(f)} R(d) \\
\searrow \qquad\qquad \swarrow R(p) \\
b \xrightarrow{\eta_b} R \circ L(b)
\end{array}
$$

$$\updownarrow \qquad\qquad\qquad\qquad \updownarrow$$

$$
\begin{array}{c}
\overset{f}{\overbrace{\qquad\qquad\qquad\qquad\qquad}} \\
L(c) \xrightarrow{L(\widetilde{f})} L \circ R(d) \xrightarrow{\varepsilon_d} d \\
\searrow \quad \swarrow \qquad\qquad \swarrow p \\
L(b) \underset{L(\eta_b)}{\rightarrow} L \circ R \circ L(b) \underset{\varepsilon_{L(b)}}{\rightarrow} L(b) \\
\underbrace{\qquad\qquad\qquad\qquad}_{\text{id}}
\end{array}
\qquad \longleftarrow \qquad
\begin{array}{c}
\overset{\widetilde{f}}{\overbrace{\qquad\qquad\qquad\qquad}} \\
c \dashrightarrow \eta_b^*(R(d)) \longrightarrow R(d) \\
\searrow \quad \swarrow \quad {}_{\text{(pb)}} \quad \swarrow R(p) \\
b \xrightarrow{\eta_b} R \circ L(b).
\end{array}
$$

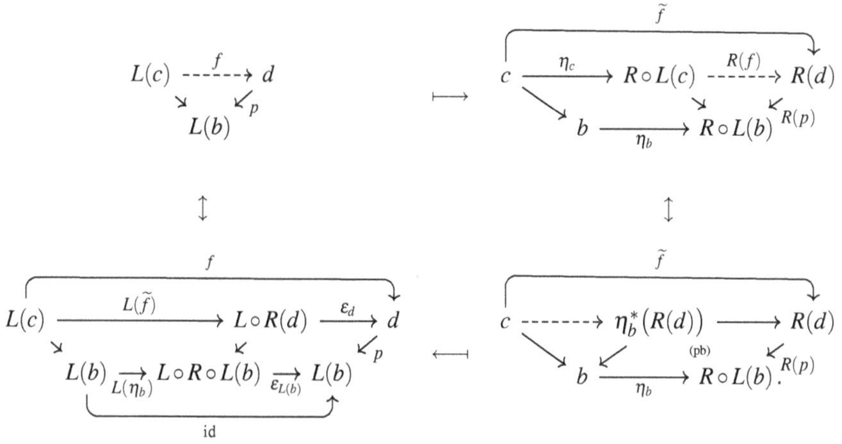

Here the horizontal transformations are given by applying the functors and then (post-)-composing with (co-)units, while the left vertical bijection is the formula (1.4) for adjuncts and the right vertical bijection is the universal property of the pullback. Evidently these operations commute in both possible ways, showing that also the horizontal operations are bijections (and they are natural by the naturality of the underlying hom-isomorphism).
(2) The second case follows analogously, but more directly as no pullback is involved here:

$$
\begin{array}{c}
c \dashrightarrow^{f} R(d) \\
\uparrow\searrow \quad \swarrow \\
R(b)
\end{array}
\longmapsto
\begin{array}{c}
\overset{\widetilde{f}}{\overbrace{\qquad\qquad\qquad}} \\
L(c) \overset{L(f)}{\dashrightarrow} L \circ R(d) \xrightarrow{\varepsilon_d} d \\
\searrow \quad \swarrow \quad \swarrow \\
L \circ R(b) \xrightarrow{\varepsilon_b} b
\end{array}
\longmapsto
\begin{array}{c}
\overset{f}{\overbrace{\qquad\qquad\qquad\qquad}} \\
c \xrightarrow{\eta_c} R \circ L(c) \xrightarrow{R(\widetilde{f})} R(d) \\
\searrow \qquad\qquad \swarrow \\
R(b) \underset{\eta_{R(b)}}{\rightarrow} R \circ L \circ R(b) \underset{R(\varepsilon_b)}{\rightarrow} R(b) \\
\underbrace{\qquad\qquad\qquad\qquad}_{\text{id}}
\end{array}
\qquad \square
$$

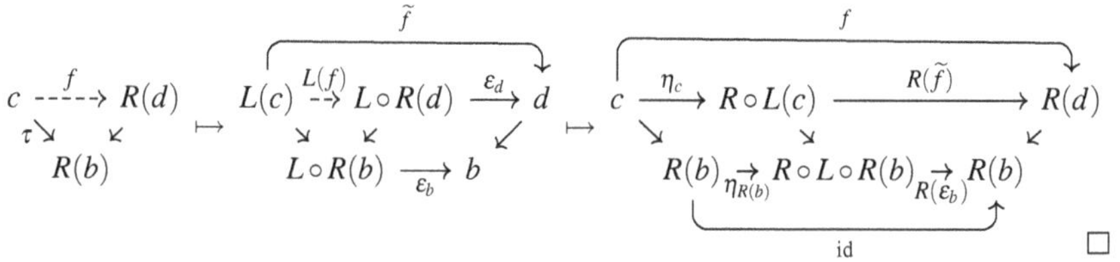

Example 1.9 (Induced Quillen adjunction on pointed objects). Given a Quillen adjunction $L \dashv_{\mathrm{Qu}} R$ (Def. 1.12) there is an induced Quillen adjunction of model categories of pointed objects, hence of coslice model structures (Ex. 1.5) under the terminal object

$$
\mathbf{D} \underset{R}{\overset{L}{\underset{\longrightarrow}{\overset{\longleftarrow}{\perp_{\mathrm{Qu}}}}}} \mathbf{C}
\;\Rightarrow\;
\left(\mathbf{D}^{*/} \underset{R^{*/}}{\overset{L^{*/}}{\underset{\longrightarrow}{\overset{\longleftarrow}{\perp_{\mathrm{Qu}}}}}} \mathbf{C}^{*/} \right)
=
\left((\mathbf{D}^{\mathrm{op}}_{/*})^{\mathrm{op}} \underset{\left(R^{\mathrm{op}}_{/*}\right)^{\mathrm{op}}}{\overset{\left(L^{\mathrm{op}}_{/*}\right)^{\mathrm{op}}}{\underset{\longrightarrow}{\overset{\longleftarrow}{\perp_{\mathrm{Qu}}}}}} (\mathbf{C}^{\mathrm{op}}_{/R(*)})^{\mathrm{op}} \right),
\qquad (1.33)
$$

where

(i) $R^{*/}$ is given directly by applying R to the underling triangular diagrams in $\mathbf{D}$;

(ii) $L^{*/}$ given by that direct application of L followed (using that the right adjoint R preserves the terminal object) by pushout along the adjunction counit, which is of the form $L(*) \simeq L \circ R(*) \xrightarrow{\varepsilon_*} *$:

$$L^{*/} \;:\; \mathbf{C}^{*/} \xrightarrow{L} \mathbf{D}^{L(*)/} \simeq \mathbf{D}^{L(*)/} \xrightarrow{(-)\underset{L\circ R(*)}{\sqcup} *} \mathbf{D}^{*/} . \tag{1.34}$$

This may be checked directly (e.g. [Hovey (1999), Prop. 1.3.5]), but it is also a special case of Ex. 1.8, as shown on the right of (1.33), observing that pullbacks are pushouts in the opposite category.

Lemma 1.1 (Ken Brown's lemma [Hovey (1999), Lem. 1.1.12][Brown (1973)]**).** *Given a Quillen adjunction $L \dashv R$ (Def. 1.12),*

(i) *the right Quillen functor R preserves all weak equivalences between fibrant objects.*
(ii) *the left Quillen functor L preserves all weak equivalences between cofibrant objects.*

Proposition 1.13 (Derived functors). *Given a Quillen adjunction $(L \dashv_{\mathrm{Qu}} R)$ (Def. 1.12), there are adjoint functors $\mathbb{D}L \dashv \mathbb{D}R$*[1] (1.2) *between the homotopy categories (Def. 1.8)*

$$\mathrm{Ho}(\mathbf{D}) \underset{\mathbb{D}R}{\overset{\mathbb{D}L}{\underset{\perp}{\rightleftarrows}}} \mathrm{Ho}(\mathbf{C}) \tag{1.35}$$

whose composites with the localization functors (1.24) *make the following squares commute up to natural isomorphism:*

$$\begin{array}{ccc} \mathbf{D} & \xrightarrow{R} & \mathbf{C} \\ {\scriptstyle \gamma_{\mathbf{D}}}\downarrow & \swarrow\!\!\!\simeq & \downarrow{\scriptstyle \gamma_{\mathbf{C}}} \\ \mathrm{Ho}(\mathbf{D}) & \xrightarrow[\mathbb{D}R]{} & \mathrm{Ho}(\mathbf{C}) \end{array} \qquad \begin{array}{ccc} \mathbf{D} & \xleftarrow{L} & \mathbf{C} \\ {\scriptstyle \gamma_{\mathbf{D}}}\downarrow & \simeq\!\!\searrow & \downarrow{\scriptstyle \gamma_{\mathbf{C}}} \\ \mathrm{Ho}(\mathbf{D}) & \xleftarrow[\mathbb{D}L]{} & \mathrm{Ho}(\mathbf{C}) . \end{array}$$

These are unique up to natural isomorphism, and are called the left and right derived functors of L and R, respectively.

Example 1.10 (Derived functors via (co-)fibrant replacement). It is convenient to leave the localization functors γ (1.24) notationally implicit, and understand objects of $\mathbf{C}$ as objects of $\mathrm{Ho}(\mathbf{C})$, via γ. Then:

(i) The value of a left derived functor $\mathbb{D}L$ (Prop. 1.13) on an object $c \in \mathbf{C}$ is equivalently the value of L on a cofibrant replacement Qc (1.13):

$$\mathbb{D}L(c) \simeq L(Qc) \quad \in \mathrm{Ho}(\mathbf{D}). \tag{1.36}$$

(ii) The value of a right derived functor $\mathbb{D}R$ (Prop. 1.13) on an object $d \in \mathbf{D}$ is equivalently the value of R on a fibrant replacement Pd (1.12):

$$\mathbb{D}R(d) \simeq R(Pd) \quad \in \mathrm{Ho}(\mathbf{C}) . \tag{1.37}$$

[1] We avoid the common notation $\mathbb{L}L \dashv \mathbb{R}R$ for derived functors, since this clashes with the prominent role that "$\mathbb{R}$" plays as notation for the field of real numbers in the main text.

(iii) The derived unit $\mathbb{D}\eta$ (1.3) of the derived adjunction (1.35) is, on any cofibrant object $c \in \mathbf{C}^{\mathrm{cof}}$, given by

$$\mathbb{D}\eta_c \,:\, c \xrightarrow{\eta_c} R\big(L(c)\big) \xrightarrow{R(j_{L(c)})} R\big(PL(c)\big) \quad \in \mathrm{Ho}(\mathbf{C}) \tag{1.38}$$

where $L(c) \xrightarrow{j_{L(c)}} PL(c)$ is any fibrant replacement (1.12).

(iv) The derived co-unit $\mathbb{D}\varepsilon$ (1.3) of the derived adjunction (1.35), is, on any fibrant object $d \in \mathbf{D}_{\mathrm{fib}}$, given by

$$\mathbb{D}\varepsilon_d \,:\, L\big(QR(d)\big) \xrightarrow{L(p_{R(d)})} L\big(R(d)\big) \xrightarrow{\varepsilon_d} d \quad \in \mathrm{Ho}(\mathbf{D}) \tag{1.39}$$

where $QR(d) \xrightarrow{p_{R(d)}} R(d)$ is any cofibrant replacement (1.13).

Homotopy fibers and homotopy pullback.

Definition 1.14 (Homotopy fiber). Let $\mathbf{C}$ be a model category (Def. 1.3).

(i) For $A \xrightarrow{p} B$ a morphism in $\mathbf{C}$ with $B \in \mathbf{C}_{\mathrm{fib}} \subset \mathbf{C}$ a fibrant object (Ntn. 1.1), and for $* \xrightarrow{b} B$ a morphism from the terminal object (a "point in B"), the *homotopy fiber* of p over b is the image in the homotopy category (1.24) of the ordinary fiber over b, i.e. the pullback (1.5) along b in $\mathbf{C}$, of any fibration $\widetilde{p}$ weakly equivalent to p:

$$\begin{array}{ccc} \mathrm{hofib}_b(p) & \longrightarrow & A \\ & & \big\downarrow{\scriptstyle p} \\ & & B \end{array} \;:=\; \gamma_{\mathbf{C}}\left(\begin{array}{ccccc} \mathrm{fib}_b(\widetilde{p}) & \longrightarrow & \widetilde{A} & \xleftarrow{\in \mathrm{W}} & A \\ \big\downarrow & \text{(pb)} & {\scriptstyle \widetilde{p}}\big\downarrow{\scriptstyle \in\,\mathrm{Fib}} & \swarrow{\scriptstyle p} & \\ * & \xrightarrow[b]{} & B & & \end{array} \right) \;\in \mathrm{Ho}(\mathbf{C})\,. \tag{1.40}$$

This is well-defined in that $\mathrm{hofib}_b(p) \in \mathrm{Ho}(\mathbf{C})$ depends on the choice of fibration replacement $\widetilde{p}$ only up to isomorphism in the homotopy category.

(ii) Dually, homotopy co-fibers are homotopy fibers in the opposite model category (Def. 1.4).

More generally:

Definition 1.15 (Homotopy pullback). Given a model category $\mathbf{C}$ (Def. 1.3) and a pair of coincident morphisms

$$\begin{array}{ccc} & & A \\ & & \big\downarrow{\scriptstyle \rho} \\ X & \xrightarrow[\tau]{} & B \end{array}$$

between fibrant objects, the *homotopy pullback* of ρ along τ (or *homotopy fiber product of ρ with τ*) is the image of ρ, regarded as an object in the homotopy category (Def. 1.8) of the slice model category (Ex. 1.5) under the right derived functor (Prop. 1.13) of the right base change functor along τ (Ex. 1.7):

$$\mathrm{Ho}\big(\mathbf{C}^{/B}\big) \ni \begin{array}{c} A \\ \big\downarrow{\scriptstyle \rho} \\ B \end{array} \quad \overset{\text{homotopy pullback}}{\longmapsto} \quad \begin{array}{c} \mathbb{D}\tau^* A \\ \big\downarrow{\scriptstyle \mathbb{D}\tau^*\rho} \\ X \end{array} \quad := \quad \mathrm{Ho}\big(\mathbf{C}^{/X}\big). \tag{1.41}$$

By (1.29), the derived adjunction counit (1.39) on (1.41) gives a commuting square in (1.24) the homotopy category of $\mathbf{C}$

$$\begin{array}{ccc} \mathbb{D}\tau^* A & \longrightarrow & A \\ {\scriptstyle \mathbb{D}\tau^*\rho}\big\downarrow & \text{(hpb)} & \big\downarrow{\scriptstyle \rho} \\ X & \underset{\tau}{\longrightarrow} & B \end{array} \quad := \quad \gamma_{\mathbf{C}} \left(\begin{array}{ccccc} \tau^*\widetilde{A} & \longrightarrow & \widetilde{A} & \overset{\in \mathrm{W}}{\longleftarrow} & A \\ \big\downarrow & \text{(pb)} & {\scriptstyle \widetilde{\rho}}\big\downarrow{\scriptstyle \in \mathrm{Fib}} & \swarrow{\scriptstyle \rho} & \\ X & \underset{\tau}{\longrightarrow} & B & & \end{array} \right) \in \mathrm{Ho}(\mathbf{C})\,. \tag{1.42}$$

This square in the homotopy category, together with its pre-image pullback square in the model category, is the *homotopy pullback square* of ρ along τ.

Example 1.11 (Homotopy fiber is homotopy pullback to the point). Homotopy fibers (Def. 1.14) are the homotopy pullbacks (Def. 1.15) to the terminal object, by (1.37).

Lemma 1.2 (Factorization lemma [Brown (1973), p. 421]**).** *Let* $\mathbf{C}$ *be a model category (Def. 1.3) and* $A \overset{\rho}{\longrightarrow} B \in \mathbf{C}_{\mathrm{fib}}$ *a morphism between fibrant objects. Then for* $\mathrm{Paths}(B)$ *a path space object for* B *(Def. 1.5) the vertical composite in the following diagram*

$$\begin{array}{ccccc} A & \overset{\in \mathrm{W}}{\longrightarrow} & p_1^* A & \overset{\in \mathrm{W}}{\longrightarrow} & A \\ & \searrow & \big\downarrow & \text{(pb)} & \big\downarrow{\scriptstyle \rho} \\ & {\scriptstyle \rho} & \mathrm{Paths}(B) & \underset{p_1}{\longrightarrow} & B \\ & & \big\downarrow{\scriptstyle p_0} & & \\ & & B & & \end{array} \tag{1.43}$$

is a fibration, and in fact a fibration resolution of ρ*, in that it factors* ρ *through a weak equivalence.*

Example 1.12 (Homotopy pullback via triples). Given a model category $\mathbf{C}$ (Def. 1.3) and a pair of coincident morphisms

$$\begin{array}{ccc} & & A \\ & & \big\downarrow{\scriptstyle \rho} \\ X & \overset{\tau}{\longrightarrow} & B \end{array}$$

between fibrant objects, Lem. 1.2 says that the corresponding homotopy pullback (Def. 1.15) is computed by the following diagram

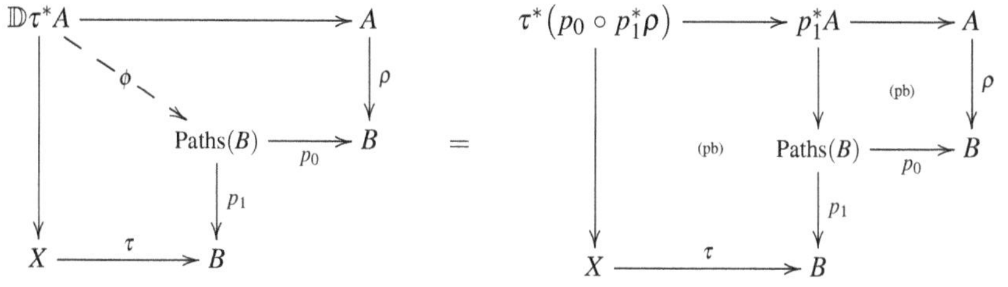

Here the right hand side exhibits the left hand side as a limit cone. This means that the homotopy pullback $\mathbb{D}\tau^*A$ is universally characterized by the fact that morphisms into it are triples (f, g, ϕ), consisting of a pair of morphisms f, g to A, X, respectively, and a right homotopy ϕ (Def. 1.6) between their composites with ρ and τ, respectively:

$$\mathbf{C}\big(-;\mathbb{D}\tau^*A\big) \simeq \left\{ (f,g,\phi) \;\middle|\; \begin{array}{ccc} & \overset{f}{\dashrightarrow} & A \\ g\big\downarrow & \overset{\phi}{\Leftarrow} & \big\downarrow\rho \\ X & \xrightarrow{\;\tau\;} & B \end{array} \right\}. \qquad (1.44)$$

Quillen equivalences.

Lemma 1.3 (Conditions characterizing Quillen equivalences). *Given a Quillen adjunction $L \dashv_{\mathrm{Qu}} R$ (Def. 1.12), the following conditions are equivalent:*

- *The left derived functor (Prop. 1.13) is an equivalence of homotopy categories (Def. 1.8)* $\mathrm{Ho}(\mathscr{D}) \underset{\simeq}{\overset{\mathbb{D}L}{\longleftarrow}} \mathrm{Ho}(\mathscr{C})$.
- *The right derived functor (Prop. 1.13) is an equivalence of homotopy categories (Def. 1.8)* $\mathrm{Ho}(\mathscr{D}) \underset{\simeq}{\overset{\mathbb{D}R}{\longrightarrow}} \mathrm{Ho}(\mathscr{C})$.
- *Both of the following two conditions hold:*
  - **(i)** *The derived adjunction unit $\mathbb{D}\eta$ (1.38) is a natural isomorphism, hence (1.38) is a weak equivalence in* $\mathbf{C}$*;*
  - **(ii)** *The derived adjunction counit $\mathbb{D}\varepsilon$ (1.39) is a natural isomorphism, hence (1.39) is a weak equivalence in* $\mathbf{D}$.
- *For $c \in \mathbf{C}^{\mathrm{cof}}$ and $d \in \mathbf{D}_{\mathrm{fib}}$, a morphism out of $L(c)$ is a weak equivalence precisely if its adjunct into $R(d)$ is:*

$$L(c) \xrightarrow[\in \mathrm{W}]{\;f\;} d \quad \Leftrightarrow \quad c \xrightarrow[\in \mathrm{W}]{\;\widetilde{f}\;} R(d)\,. \qquad (1.45)$$

Definition 1.16 (Quillen equivalence). If the equivalent conditions from Lem. 1.3 are met, a Quillen adjunction $L \dashv_{\mathrm{Qu}} R$ (Def. 1.12) is called a *Quillen equivalence*, which we denote as follows:

$$\mathscr{D} \underset{R}{\overset{L}{\underset{\simeq_{\mathrm{Qu}}}{\rightleftarrows}}} \mathscr{C}\,.$$

Hence:

Proposition 1.17 (Derived equivalence of homotopy categories). *The derived adjunction (Prop. 1.13) of a Quillen equivalence (Def. 1.16) is an adjoint equivalence of homotopy categories (Def. 1.8):*

$$\mathrm{Ho}(\mathbf{D}) \underset{\mathbb{D}R}{\overset{\mathbb{D}L}{\underset{\simeq}{\rightleftarrows}}} \mathrm{Ho}(\mathbf{C})\,. \tag{1.46}$$

Remark 1.2 (∞-Category theory). As each model category (Def. 1.3) provides a context of *homotopy theory* (with its own notion of homotopy-coherent universal constructions such as homotopy pullbacks, Def. 1.15, etc.), Prop. 1.17 is a first indication that Quillen equivalent (Def. 1.16) model categories represent the *same* context of homotopy theory, for a suitably homotopy-theoretic notion of sameness. This suggests that model categories regarded up to Quillen equivalence are but coordinate presentations of a more intrinsic notion of homotopy theories, now known as *∞-categories* [Joyal (2008c)][Joyal (2008b)][Joyal (2008a)][Lurie (2009a)][Cisinski (2019)][Riehl and Verity (2021)].

Lemma 1.4 (Quillen equivalence when left adjoint creates weak equivalences [Erdal and Güçlükan İlhan (2019), Lem. 3.3]**).** *Let $L \dashv_{\mathrm{Qu}} R$ be a Quillen adjunction (Def. 1.12) such that the left adjoint functor L creates weak equivalences, in that for all morphisms f in $\mathbf{C}$ we have*

$$f \in \mathrm{W}_{\mathbf{C}} \quad \Leftrightarrow \quad L(f) \in \mathrm{W}_{\mathbf{D}}\,. \tag{1.47}$$

Then $L \dashv_{\mathrm{Qu}} R$ is a Quillen equivalence (Def. 1.16) precisely if the adjunction co-unit ε_d is a weak equivalence on all fibrant objects $d \in \mathbf{C}_{\mathrm{fib}}$.

Proof. By Lem. 1.3, it is sufficient to check that the **(i)** derived unit and **(ii)** derived counit of the adjunction are weak equivalences precisely if the ordinary counit is a weak equivalence.
(ii) For the derived counit (1.39)

$$\mathbb{D}\varepsilon_c \,:\, L\big(QR(d)\big) \xrightarrow[\in \mathrm{W}]{L(p_{R(d)})} L\big(R(d)\big) \xrightarrow{\varepsilon_d} d$$

we have that $p_{R(d)}$ is a weak equivalence (1.13), and since L preserves this, by assumption, so is $L\big(p_{R(d)}\big)$. Therefore $\mathbb{D}\varepsilon_d$ is a weak equivalence precisely if ε_d is, by 2-out-of-3 (1.7).
(i) For the derived unit (1.38)

$$c \xrightarrow{\eta_c} R\big(L(c)\big) \xrightarrow{R(j_{L(c)})} R\big(PL(c)\big)$$

consider the composite of its image under L with the adjunction counit, as shown in the middle row of the following diagram:

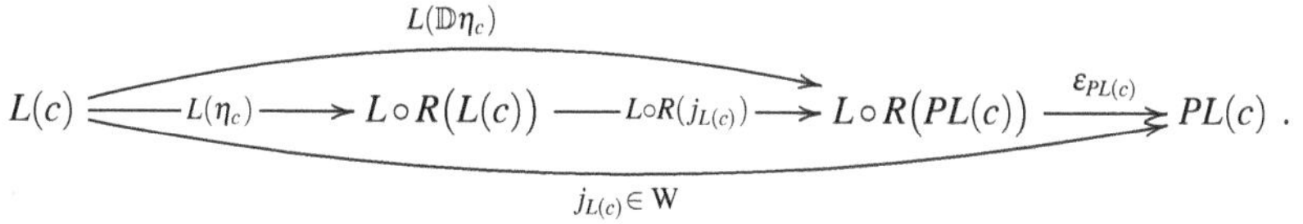

By the formula (1.4) for adjuncts, this composite equals the adjunct of the derived adjunction unit, hence $j_{L(c)}$, as shown by the bottom morphism, which is a weak equivalence (1.12). Now, since L creates weak equivalences by assumption, $L(\mathbb{D}\eta_c)$ is a weak equivalence precisely if $\mathbb{D}\eta_c$ is a weak equivalence. Therefore it follows, again by 2-out-of-3 (1.7), that this is the case precisely if the adjunction counit ε is a weak equivalence on the fibrant object $PL(c)$. □

Proposition 1.18 (Base change along weak equivalence in right proper model category). *Let* $\mathbf{C}$ *be a right proper model category (Def. 1.4). Then its base change Quillen adjunction (Ex. 1.7) along any weak equivalence*

$$B_1 \xrightarrow[\in \mathrm{W}]{f} B_2 \qquad \in \mathbf{C}$$

is a Quillen equivalence (Def. 1.16):

$$\mathbf{C}^{/B_2} \underset{f^*}{\overset{f_!}{\underset{\simeq_{\mathrm{Qu}}}{\rightleftarrows}}} \mathbf{C}^{/B_1} .$$

Proof. Observe that $B_2 \xrightarrow{\mathrm{id}} B_2$ is the terminal object of $\mathbf{C}^{/B_2}$, so that the fibrant objects of $\mathbf{C}^{/B_2}$ correspond to the fibrations in $\mathbf{C}$ over B_2. Therefore, the condition (1.45) says that for $f_! \dashv f^*$ to be a Quillen equivalence it is sufficient that in (1.29) c is a weak equivalence precisely if $\widetilde{c}$ is, assuming that ρ is a fibration:

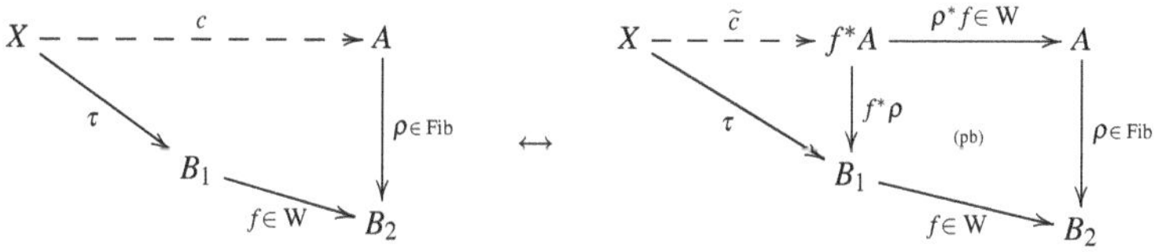

(1.48)

But under this assumption, right-properness implies that $\rho^* f$ is a weak equivalence (1.11), so that the statement follows by 2-out-of-3 (1.7).

Alternative Proof. The conclusion also follows with Lem. 1.4: The left adjoint functor $L = f_!$ clearly creates weak equivalences (1.47) (by the nature of the slice model structure, Ex. 1.5), so that Lem. 1.4 asserts that we have a Quillen equivalence as soon as the ordinary adjunction counit is a weak equivalence on all fibrant objects. By (1.29), the adjunction counit on a fibration $\rho \in \mathrm{Fib}$ is the dashed morphism $\rho^* f$ in the following diagram on the right:

$$
\begin{array}{ccccc}
f^*A & \overset{\mathrm{id}}{\dashrightarrow} & f^*A & \xrightarrow{\rho^* f \in \mathrm{W}} & A \\
 & \searrow^{f^*\rho} & \downarrow{\scriptstyle f^*\rho} & \text{(pb)} & \downarrow{\scriptstyle \rho \in \mathrm{Fib}} \\
 & & B_1 & \xrightarrow{f \in \mathrm{W}} & B_2
\end{array}
\quad \leftrightarrow \quad
\begin{array}{ccc}
f^*A & \overset{\rho^* f \in \mathrm{W}}{\dashrightarrow} & A \\
 \searrow^{f^*\rho} & & \downarrow{\scriptstyle \rho \in \mathrm{Fib}} \\
 B_1 & \xrightarrow{f \in \mathrm{W}} & B_2
\end{array}
\tag{1.49}
$$

Therefore this is a weak equivalence, again by right-properness. □

Example 1.13 (Quillen equivalence between topological spaces and simplicial sets [Quillen (1967)]). Forming simplicial sets constitutes a Quillen equivalence (Def. 1.16)

$$
\mathrm{TopSp}_{\mathrm{Qu}} \underset{\underset{\text{singular simplicial complex}}{\mathrm{Sing}}}{\overset{\overset{\text{geometric realization}}{|-|}}{\rightleftarrows}}_{\simeq_{\mathrm{Qu}}} \Delta\mathrm{Sets}_{\mathrm{Qu}}
\tag{1.50}
$$

between the classical model structure on topological spaces (Ex. 1.1) and the classical model structure on simplicial sets (Ex. 1.2).

Example 1.14 (Classical homotopy category). The derived adjunction (Prop. 1.13) of the $|-| \dashv \mathrm{Sing}$-adjunction (Ex. 1.13) is an equivalence between the homotopy categories (Def. 1.8) of the classical model category of topological spaces (Ex. 1.1) and the classical model category of simplicial sets (Ex. 1.2):

$$
\mathrm{Ho}\big(\mathrm{TopSp}_{\mathrm{Qu}}\big) \underset{\mathbb{D}\mathrm{Sing}}{\overset{\mathbb{D}|-|}{\rightleftarrows}}_{\simeq} \mathrm{Ho}\big(\Delta\mathrm{Sets}_{\mathrm{Qu}}\big).
\tag{1.51}
$$

Either of these is the *classical homotopy category*. We refer to its objects as *homotopy types*, to be distinguished from the actual topological spaces or simplicial sets that represent them.

Example 1.15 (Simplicial sets are weakly equivalent to singular simplicial sets of their realization). The characterization of Quillen equivalences (Lem. 1.3) implies, with Ex. 1.13, that for each $S \in \Delta\mathrm{Sets}$ the composite

$$
S \xrightarrow{\eta_S} \mathrm{Sing}(|S|) \xrightarrow{\mathrm{Sing}(|j_{|S|}|)} \mathrm{Sing}(P|S|)
$$

is a weak equivalence, where $j_{|S|}$ is a fibrant replacement for $|S|$. But since all topological spaces are fibrant (Ex. 1.1), it follows that the ordinary unit of the adjunction (1.50) is already a weak equivalence:

$$
S \xrightarrow[\in \mathrm{W}]{\eta_S} \mathrm{Sing}(|S|)\ .
\tag{1.52}
$$

Cell complexes.

Proposition 1.19 (Skeleta and truncation [May (1967), §II.8][Dwyer and Kan (1984), §1.2 (vi)]**).** *For each $n \in \mathbb{N}$ there is a pair of adjoint functors*

$$\Delta\mathrm{Sets} \underset{\mathrm{cosk}_n}{\overset{\mathrm{sk}_n}{\underset{\longrightarrow}{\overset{\longleftarrow}{\bot}}}} \Delta\mathrm{Sets}\,, \tag{1.53}$$

where $\mathrm{sk}_n(S)$ is the simplicial sub-set generated by the simplices in S of dimension $\leq n$ (hence including only all their degenerate higher simplices), and where

$$\mathrm{cosk}_n(S) \;:\; [k] \;\mapsto\; \Delta\mathrm{Sets}\big(\mathrm{sk}_n(\Delta[k])\,,\,S\big)\,.$$

One says that S is n-coskeletal if the comparison morphism $S \to \mathrm{cosk}_n(S)$ is an isomorphism.

Here cosk_{n+1} *preserves* ([Dwyer and Kan (1984), p. 141], for proofs see [Low (2013)][Deflorin (2019), Lem. 10.12]) *fibrant objects of the classical model structure (Ex. 1.2), hence preserves Kan complexes* (1.16)*, and models n-truncation, in that:*

$$\pi_k\big|\mathrm{cosk}_{n+1}(S)\big| \;=\; 0 \qquad \textit{for } k \geq n+1$$

and there are natural morphisms

$$S \xrightarrow{\;\;p_n\;\;} \mathrm{cosk}_n(S) \tag{1.54}$$

such that

$$\pi_k\big|S\big| \xrightarrow[\simeq]{\;\;\pi_k|p_n|\;\;} \pi_k\big|\mathrm{cosk}_{n+1}(S)\big| \qquad \textit{for } k \leq n.$$

For $A \in \mathrm{Ho}\big(\Delta\mathrm{Sets}_{\mathrm{Qu}}\big)$ we write

$$A(n) \;:=\; \big|\mathrm{cosk}_{n+1}\big(\mathrm{Sing}(A)\big)\big| \tag{1.55}$$

We say that A is *n-truncated* if it is equivalent to its n-truncation (1.55):

$$A \text{ is } n\text{-truncated} \quad \Leftrightarrow \quad A \simeq A(n)\,. \tag{1.56}$$

Example 1.16 (Homotopy types of manifolds via triangulations). For $X \in \mathrm{TopSp}$ equipped with the structure of a smooth n-manifold, there exists a *triangulation* of X (e.g. [Whitney (1957), §IV.B][Munkres (1966), Thm. 10.6], see also [Manolescu (2014)]), namely a simplicial set (in fact a simplicial complex) which is n-sleletal (Prop. 1.19)

$$\mathrm{Tr}(X) \in \Delta\mathrm{Sets}\,, \qquad \mathrm{sk}_n\big(\mathrm{Tr}(X)\big) \;=\; \mathrm{Tr}(X) \tag{1.57}$$

equipped with a homeomorphism to X out of its geometric realization (1.50)

$$\big|\mathrm{Tr}(X)\big| \xrightarrow[\text{homeo}]{\;\;p\;\;} X \tag{1.58}$$

which restricts in the interior of each simplex to a diffeomorphism onto its image. Since the inclusion

$$\mathrm{Tr}(X) \underset{\in \mathrm{W}}{\xhookrightarrow{\eta_{\mathrm{Tr}(X)}}} \mathrm{Sing}\big(|\mathrm{Tr}(X)|\big) \underset{\in \mathrm{Iso}}{\xrightarrow{\mathrm{Sing}(p)}} \mathrm{Sing}(X)\,, \tag{1.59}$$

is a weak equivalence (by Ex. 1.15), the triangulation represents the homotopy type (1.51) of the manifold.

Proposition 1.20 (Homotopy classes of maps out of n-manifolds). *Let $X \in \mathrm{TopSp}$ admit the structure of an n-manifold. Then for any $A \in \mathrm{Ho}(\Delta\mathrm{Sets}_{\mathrm{Qu}})$ (Ex. 1.14) the homotopy classes of maps $X \to A$ are in natural bijection to the homotopy classes into the n-truncation* (1.55) *of A:*

$$\mathrm{Ho}(\Delta\mathrm{Sets}_{\mathrm{Qu}})(X\,,A) \simeq \mathrm{Ho}(\Delta\mathrm{Sets}_{\mathrm{Qu}})(X\,,A(n)) \tag{1.60}$$

Proof. Consider the following sequence of natural isomorphisms

$$\begin{aligned}
&\mathrm{Ho}(\Delta\mathrm{Sets}_{\mathrm{Qu}})(X\,,A)\\
&\simeq \mathrm{Ho}(\Delta\mathrm{Sets}_{\mathrm{Qu}})(\mathrm{Sing}(X)\,,\mathrm{Sing}(A))\\
&\simeq \mathrm{Ho}(\Delta\mathrm{Sets}_{\mathrm{Qu}})(\mathrm{Tr}(X)\,,\mathrm{Sing}(A))\\
&\simeq \Delta\mathrm{Sets}\Big(\mathrm{Tr}(X)\,,\mathrm{Sing}(A)\Big)\Big/\Delta\mathrm{Sets}\Big(\mathrm{Tr}(X)\times\Delta[1]\,,\mathrm{Sing}(A)\Big)\\
&\simeq \Delta\mathrm{Sets}\Big(\mathrm{sk}_{n+1}(\mathrm{Tr}(X))\,,\mathrm{Sing}(A)\Big)\Big/\Delta\mathrm{Sets}\Big(\mathrm{sk}_{n+1}(\mathrm{Tr}(X)\times\Delta[1])\,,\mathrm{Sing}(A)\Big)\\
&\simeq \Delta\mathrm{Sets}\Big(\mathrm{Tr}(X)\,,\mathrm{cosk}_{n+1}(\mathrm{Sing}(A))\Big)\Big/\Delta\mathrm{Sets}\Big(\mathrm{Tr}(X)\times\Delta[1]\,,\mathrm{cosk}_{n+1}(\mathrm{Sing}(A))\Big)\\
&\simeq \mathrm{Ho}(\Delta\mathrm{Sets}_{\mathrm{Qu}})\Big(\mathrm{Tr}(X)\,,\mathrm{cosk}_{n+1}(\mathrm{Sing}(A))\Big)\\
&\simeq \mathrm{Ho}(\Delta\mathrm{Sets}_{\mathrm{Qu}})\Big(|\mathrm{Tr}(X)|\,,|\mathrm{cosk}_{n+1}(\mathrm{Sing}(A))|\Big)\\
&\simeq \mathrm{Ho}(\Delta\mathrm{Sets}_{\mathrm{Qu}})(X\,,A(n)).
\end{aligned}$$

Here the first step is (1.14), using, with Ex. 1.10, that all topological spaces are fibrant and all simplicial sets cofibrant. The second step uses (1.59). The third step uses Ex. 1.6 with Prop. 1.10 (observing that $\mathrm{Sing}(A)$ is fibrant as A is and Sing is right Quillen) to express the morphisms in the homotopy category as equivalence classes of simplicial maps under the relation that identifies those pairs of maps that extend to a map on the cylinder $\mathrm{Tr}(X)\times\Delta[1]$. The fourth step observes that with $\mathrm{Tr}(X)$ being n-skeletal (1.57), its cylinder is $(n+1)$-skeletal. The fifth step is thus the $\mathrm{sk}_{n+1} \dashv \mathrm{cosk}_{n+1}$-adjunction isomorphism (1.53). The sixth step applies again Prop. 1.10, using that cosk_{n+1} preserves fibrancy (Prop. 1.19). The seventh step is the reverse of the first step, with the same argument on (co-)fibrancy. The last step uses (1.58) in the first argument and (1.55) in the second. The composite of these isomorphisms is the desired (1.60). □

Proposition 1.21 (Postnikov tower [Goerss and Jardine (1999), Cor. 3.7]**).** *Let $X \in \mathrm{Ho}(\Delta\mathrm{Sets}_{\mathrm{Qu}})$ (Ex. 1.14). If X is connected, then its sequence of n-truncations* (1.55) *forms* *a system of maps with homotopy fibers (Def. 1.14) the Eilenberg-MacLane spaces* (2.6) *of the homotopy group in the given degree:*

$$
\begin{array}{ccc}
 & & \vdots \\
 & & \downarrow \\
K(\pi_3(X),3) & \xrightarrow{\mathrm{hfib}(p_3^X)} & X(3) \\
 & & \downarrow{\scriptstyle p_3^X} \\
K(\pi_2(X),2) & \xrightarrow{\mathrm{hfib}(p_2^X)} & X(2) \\
 & & \downarrow{\scriptstyle p_2^X} \\
K(\pi_1(X),1) & \xrightarrow{\mathrm{hfib}(p_1^X)} & X(1) \\
 & & \downarrow{\scriptstyle p_1^X} \\
 & & X(0)\,.
\end{array}
$$

If X is not connected then this applies to each of its connected components.

Stable model categories.

Example 1.17 (Looping/suspension-adjunction). On the category of pointed topological spaces, equipped with the coslice model structure under the point (Ex. 1.5) of the classical model structure (Ex. 1.1), the operation of forming based loop spaces $\Omega X := \mathrm{Maps}^{*/}(S^1, X)$ is the right adjoint in a Quillen adjunction (Def. 1.12)

$$
\mathrm{TopSp}^{*/}_{\mathrm{Qu}} \underset{\Omega}{\overset{\Sigma}{\underset{\longrightarrow}{\overset{\longleftarrow}{\bot_{\mathrm{Qu}}}}}} \mathrm{TopSp}^{*/}_{\mathrm{Qu}} \tag{1.61}
$$

whose left adjoint is the reduced suspension operation $\Sigma X := S^1 \wedge X := \big(S^1 \times X\big)/\big(S^1 \times \{*_X\} \sqcup \{*_{S^1}\} \times X\big)$.

Example 1.18 (Stable model category of sequential spectra [Bousfield and Friedlander (1978)][Goerss and Jardine (1999), §X.4]**).** There exists a model category (Def. 1.3) $\mathrm{SequentialSpectra}_{\mathrm{BF}}$ whose objects are sequences

$$
E := \Big\{ E_n \in \mathrm{TopSp},\ \Sigma E_n \xrightarrow{\sigma_n} E_{n+1} \Big\}_{n \in \mathbb{N}}
$$

of topological spaces E_n and continuous function σ_n from their suspension ΣE_n (Ex. 1.17) to the next space in the sequences; and whose morphisms $E \xrightarrow{f} F$ are sequences of component maps $E_n \xrightarrow{f_n} F_n$ that commute with the σs. Moreover:

W – *weak equivalences* are the morphisms that induce isomorphisms on all *stable homotopy groups* $\pi_\bullet(X) := \underset{\overrightarrow{n}}{\lim}\, \pi_{\bullet+k}(X_k)$ (where the colimit is formed using the σ's);

Cof – *cofibrations* are those morphisms $E \xrightarrow{f} F$ such that the maps

$$E_0 \xrightarrow[\in \mathrm{Cof}]{f_0} F_0 \quad \text{and} \quad \underset{n\in\mathbb{N}}{\forall} \quad E_{n+1} \underset{\Sigma E_n}{\sqcup} \Sigma F_n \xrightarrow[\in \mathrm{Cof}]{(f_{n+1}, \sigma_n^F)} F_{n+1}$$

are cofibrations in the classical model structure on topological spaces (Ex. 1.1).

Fib – *Fibrant objects* are the *Ω-spectra*, namely those sequences of spaces $\{E_n\}$ for which the $\Sigma \dashv \Omega$-adjunct (1.61) of each σ_n is a weak equivalence:

$$\left\{ E_n \in \mathrm{TopSp}^{*/}_{\mathrm{Qu}},\ E_n \xrightarrow[\in \mathrm{W}]{\tilde{\sigma}_n} \Omega E_{n+1} \right\}_{n\in\mathbb{N}} \tag{1.62}$$

Example 1.19 (Derived stabilization adjunction). The suspension/looping Quillen adjunction on pointed spaces (Ex. 1.17) extends to a commuting diagram of Quillen adjunctions (Def. 1.12) to and on the stable model category of spectra (Ex. 1.18)

$$\begin{array}{ccc} \mathrm{TopSp}^{*/}_{\mathrm{Qu}} & \underset{\Omega}{\overset{\Sigma}{\rightleftarrows}}\ (\bot_{\mathrm{Qu}}) & \mathrm{TopSp}^{*/}_{\mathrm{Qu}} \\ \Sigma^\infty \downarrow \dashv_{\mathrm{Qu}} \uparrow \Omega^\infty & & \Sigma^\infty \downarrow \dashv_{\mathrm{Qu}} \uparrow \Omega^\infty \\ \mathrm{SequentialSpectra}_{\mathrm{BF}} & \underset{\Omega}{\overset{\Sigma}{\rightleftarrows}}\ (\bot_{\mathrm{Qu}}) & \mathrm{SequentialSpectra}_{\mathrm{BF}} \end{array} \tag{1.63}$$

such that the bottom adjunction is a Quillen equivalence (Def. 1.16), hence such that under passage to derived adjunctions (Prop. 1.13)

$$\begin{array}{ccc} \mathrm{Ho}\big(\mathrm{TopSp}^{*/}_{\mathrm{Qu}}\big) & \underset{\mathbb{D}\Omega}{\overset{\mathbb{D}\Sigma}{\rightleftarrows}}\ (\bot) & \mathrm{Ho}\big(\mathrm{TopSp}^{*/}_{\mathrm{Qu}}\big) \\ \mathbb{D}\Sigma^\infty \downarrow \dashv \uparrow \mathbb{D}\Omega^\infty & & \mathbb{D}\Sigma^\infty \downarrow \dashv \uparrow \mathbb{D}\Omega^\infty \\ \mathrm{Ho}\big(\mathrm{SequentialSpectra}_{\mathrm{BF}}\big) & \underset{\mathbb{D}\Omega}{\overset{\mathbb{D}\Sigma}{\rightleftarrows}}\ (\simeq) & \mathrm{Ho}\big(\mathrm{SequentialSpectra}_{\mathrm{BF}}\big) \end{array} \tag{1.64}$$

the bottom adjunction is an equivalence, thus exhibiting the homotopy category of spectra as being *stable* under looping/suspension.
We say that

(i) $\mathrm{Ho}\big(\mathrm{SequentialSpectra}_{\mathrm{BF}}\big)$ is the *stable homotopy category* of spectra;
(ii) the vertical adjunction $(\mathbb{D}\Sigma^\infty \dashv \mathbb{D}\Omega^\infty)$ is the *stabilization adjunction* between homotopy types (1.51) and spectra.
(iii) the images of Σ^∞ are the *suspension spectra*.
(iv) For $E \in \mathrm{Ho}\big(\mathrm{SequentialSpectra}_{\mathrm{BF}}\big)$ and $n \in \mathbb{N}$ we write (for brevity and in view of (1.62))

$$E_n := \mathbb{D}\Omega^\infty\big((\mathbb{D}\Sigma)^n E\big) \quad \in \mathrm{Ho}\big(\Delta\mathrm{Sets}^{*/}_{\mathrm{Qu}}\big) \tag{1.65}$$

for the homotopy type of the nth component space of the spectrum.

Smooth ∞-stacks. We briefly highlight some basics of smooth ∞-stack theory, for more details and more exposition see [Fiorenza *et al.* (2013), §2][Fiorenza *et al.* (2015b)][Schreiber (2013)][Sati and Schreiber (2021b), §1].

Definition 1.22 (Simplicial presheaves over Cartesian spaces). We write

(i)

$$\mathrm{CartSp} \;:=\; \left\{ \mathbb{R}^{n_1} \xrightarrow{\text{smooth}} \mathbb{R}^{n_2} \right\}_{n_i \in \mathbb{N}} \hookrightarrow \mathrm{SmthMfds} \tag{1.66}$$

for the category whose objects are the Cartesian spaces $\mathbb{R}^n$, for $n \in \mathbb{N}$, and whose morphisms are the *smooth* functions between these (hence the full subcategory of SmthMfds on the Cartesian spaces).

(ii)

$$\mathrm{PSh}\big(\mathrm{CartSp}, \Delta\mathrm{Sets}\big) \;:=\; \mathrm{Functors}\big(\mathrm{CartSp}^{\mathrm{op}}, \Delta\mathrm{Sets}\big) \tag{1.67}$$

for the category of functors from the opposite of CartSp (1.66) to ΔSets (Ex. 1.2).

Example 1.20 (Model structure on simplicial presheaves over Cartesian spaces [Dugger (1998)][Dugger (2001)][Fiorenza *et al.* (2012), §A]). The category of simplicial presheaves over Cartesian spaces (Def. 1.22) carries the following model category structures (Def. 1.3):

(i) The global projective model structure

$$\mathrm{PSh}\big(\mathrm{CartSp}, \Delta\mathrm{Sets}_{\mathrm{Qu}}\big)_{\mathrm{proj}} \;\in\; \mathrm{ModelCategories} \tag{1.68}$$

whose

W – *weak equivalences* are the morphisms which over each $\mathbb{R}^n$ are weak equivalence in $\Delta\mathrm{Sets}_{\mathrm{Qu}}$ (Ex. 1.2),

Fib – *fibrations* are the morphisms which over each $\mathbb{R}^n$ are fibrations in $\Delta\mathrm{Sets}_{\mathrm{Qu}}$ (Ex. 1.2),

(ii) The local projective model structure

$$\mathrm{PSh}\big(\mathrm{CartSp}, \Delta\mathrm{Sets}_{\mathrm{Qu}}\big)_{\substack{\mathrm{proj}\\ \mathrm{loc}}} \;\in\; \mathrm{ModelCategories} \tag{1.69}$$

whose:

W – *weak equivalences* are the morphisms whose stalk at $0 \in \mathbb{R}^n$ is a weak equivalence in $\Delta\mathrm{Sets}_{\mathrm{Qu}}$ (Ex. 1.2), for all $n \in \mathbb{N}$;

Cof – *cofibrations* are the morphisms with the left lifting property (1.9) against the class of morphisms which over each $\mathbb{R}^n$ are in $\mathrm{Fib} \cap \mathrm{W}$ of $\Delta\mathrm{Sets}_{\mathrm{Qu}}$.

Example 1.21 (Smooth manifolds as simplicial presheaves). Consider a smooth manifold. $X \in \mathrm{SmthMfds}$.

(i) The manifold is incarnated as a simplicial presheaf (Def. 1.20) by the rule which assigns to a Cartesian space the set of smooth functions $\mathbb{R}^n \xrightarrow{\text{smooth}} X$, regarded as a simplicially constant simplicial set. This construction constitutes a full subcategory inclusion:

$$\begin{array}{ccc} \mathrm{SmthMfds} & \hookrightarrow & \mathrm{PSh}\big(\mathrm{CartSp}, \Delta\mathrm{Sets}\big) \\ X & \longmapsto & \Big(\mathbb{R}^n \mapsto \big([k] \mapsto \mathrm{SmthMfds}(\mathbb{R}^n, X)\big)\Big). \end{array} \tag{1.70}$$

(ii) For $p_n \in \mathbb{R}^n$ any point, the *stalk* of this presheaf is the set of *germs of smooth functions* from an open neighbourhood of p_n to X. This set depends, in general, on $n \in \mathbb{N}$, but does not depend on the choice of p_n.

Example 1.22 (Lie groupoids as simplicial presheaves). Consider a Lie groupoid $\mathscr{G} = \big(\mathscr{G}_1 \rightrightarrows \mathscr{G}_0\big)$ (review in [Mackenzie (1987)][Moerdijk and Mrčun (2003)] [Mackenzie (2005)]) hence a groupoid internal to smooth manifolds $\mathscr{G} \in \mathrm{Grpds}(\mathrm{SmthMfds})$. Notice that for each $\mathbb{R}^n \in \mathrm{CartSp}$ there is an induced bare groupoid of smooth functions into the component manifolds:

$$\mathbb{R}^n \longmapsto \mathscr{G}(\mathbb{R}^n) := \big(\mathrm{SmthMfds}(\mathbb{R}^n, \mathscr{G}_1) \rightrightarrows \mathrm{SmthMfds}(\mathbb{R}^n, \mathscr{G}_0)\big) \in \mathrm{Grpds}(\mathrm{Sets}).$$

The simplicial nerves (Ex. 1.3) of these mapping groupoids arrange into a simplicial presheaf (Ex. 1.20) and this construction is the inclusion of a full subcategory, extending the full inclusion of smooth manifolds (1.70):

$$\begin{array}{ccc} \mathrm{Grpds}\big(\mathrm{SmthMfds}\big) & \hookrightarrow & \mathrm{PSh}\big(\mathrm{CartSp}, \Delta\mathrm{Sets}\big) \\ \mathscr{G} & \longmapsto & \big(\mathbb{R}^n \mapsto N\big(\mathscr{G}(\mathbb{R}^n)\big)\big) \end{array}$$

Example 1.23 (Čech groupoids of open covers as simplicial presheaves). Let X be a smooth manifold equipped with a cover by a set of open subsets $\big\{U_i \overset{\text{open}}{\hookrightarrow} X\big\}_{i \in I}$.

(i) The *Čech nerve* of the open cover is the simplicial presheaf (Def. 1.20)

$$N\big(\{U_i\}_{i\in I}\big) \in \mathrm{PSh}\big(\mathrm{CartSp}, \Delta\mathrm{Sets}\big) \tag{1.71}$$

whose k-cells over any $\mathbb{R}^n$ are the smooth functions

$$N\big(\{U_i\}_{i\in I}\big) : \big(\mathbb{R}^n, [k]\big) \longmapsto \mathrm{SmthMfds}\Big(\mathbb{R}^n, \big(\sqcup_i U_i\big)^{\times_X^{(k+1)}}\Big) \tag{1.72}$$

into the $(k+1)$-fold intersections of the patches U_i in X:

$$\big(\sqcup_i U_i\big)^{\times_X^{k+1}} := \underbrace{\big(\sqcup_i U_i\big) \times_X \cdots \times_X \big(\sqcup_i U_i\big)}_{k+1 \text{ factors}} \overset{\iota_{k+1}}{\hookrightarrow} X. \tag{1.73}$$

This is, for each $\mathbb{R}^n$, a Kan complex (1.16) and as such is the nerve of the Lie groupoid (Ex. 1.22) which is the *smooth Čech-groupoid* of the open cover:

$$N\big(\{U_i\}_{i\in I}\big) = N\Big(\big(\sqcup_i U_i\big) \times_X \big(\sqcup_i U_i\big) \rightrightarrows \big(\sqcup_i U_i\big)\Big). \tag{1.74}$$

(ii) For any point $p_n \in \mathbb{R}^n$, the *stalk* of the Čech nerve (1.71) at p_n is the disjoint union over the germs of smooth functions $\mathbb{R}^n \xrightarrow{\phi} X$ of the nerves of the pair groupoids on the subset $I_{\phi(p_n)} \subset X$ of patches U_i that contain $\phi(p_n)$.
(iii) Postcomposition with the inclusions (1.73) yields a canonical morphism of simplicial presheaves from the cover's Čech nerve (1.72) to the presheaf incarnation (1.70) of the underlying manifold:

$$\begin{array}{ccc} N\big(\{U_i\}_{i\in I}\big) & \xrightarrow{\qquad \in \mathrm{W} \qquad} & X \\ \Big(\mathbb{R}^n \xrightarrow{\phi} \big(\sqcup_i U_i\big)^{\times_X^{k+1}}\Big) & \mapsto & \Big(\mathbb{R}^n \xrightarrow{\phi} \big(\sqcup_i U_i\big)^{\times_X^{k+1}} \xrightarrow{\iota_{k+1}} X\Big). \end{array} \qquad \in \mathrm{PSh}\big(\mathrm{CartSp}, \Delta\mathrm{Sets}_{\mathrm{Qu}}\big)_{\substack{\mathrm{proj}\\ \mathrm{loc}}} \tag{1.75}$$

On stalks, this map takes the nerve of the pair groupoid on the set of factorizations through the patches U_i of a the germ of a given smooth $\mathbb{R}^n \xrightarrow{\phi} X$ to that germ itself. Since nerves of pair groupoids are contractible (1.18), this means that (1.75) is a weak equivalence in the local model structure of Ex. 1.20.

This says that (Čech nerves of) open covers serve as resolutions of smooth manifolds in the local model structure Ex. 1.20; in fact as *cofibrant resolutions* if the cover is "good":

Proposition 1.23 (Dugger's cofibrancy recognition [Dugger (2001), Cor. 9.4]). *A sufficient condition for $\mathscr{X} \in \mathrm{PSh}(\mathrm{CartSp}, \Delta\mathrm{Sets})_{\substack{\mathrm{proj}\\ \mathrm{loc}}}$ (Ex. 1.20) to be cofibrant (Ntn. 1.1) is that in each simplicial degree k, the component presheaf X_k is*

(i) *a coproduct (as presheaves, using Ex. 1.21) of Cartesian spaces: $\mathscr{X}_k \simeq \coprod_{i_k} \mathbb{R}^{n_{i_k}}$;*

(ii) *whose degenerate cells split off as a disjoint summand.*

Example 1.24 (Good open covers are projectively cofibrant resolutions of smooth manifolds). Prop. 1.23 applied to Ex. 1.23 says that the Čech nerve of an open cover is a cofibrant resolution of the underlying manifold if the open cover is *good*, or rather: *differentiably good*, in that each non-empty intersection of a finite number of its patches is diffeomorphic to an open ball (namely, equivalently: to a Cartesian space):

$$\big\{U_i \overset{\mathrm{open}}{\hookrightarrow} X\big\}_{i\in I} \text{ is good} \quad \Rightarrow \quad \varnothing \underset{\in \mathrm{Cof}}{\rightarrow} N\big(\{U_i\}_{i\in I}\big) \underset{\in \mathrm{W}}{\overset{p_{\{U_i\}_i}}{\rightarrow}} X \quad \in \mathrm{PSh}\big(\mathrm{CartSp}, \Delta\mathrm{Sets}\big)_{\substack{\mathrm{proj}\\ \mathrm{loc}}}. \tag{1.76}$$

Notice that every smooth manifold admits a *differentiably* good open cover (which is a somewhat subtle point that is traditionally being glossed over, for details see [Fiorenza *et al.* (2012), Prop. A.1]).

Example 1.25 (Hom-complexes of simplicial presheaves).
(i) For $\mathscr{X} \in \mathrm{PSh}\big(\mathrm{CartSp}, \Delta\mathrm{Sets}\big)$ and (Def. 1.22) $S \in \Delta\mathrm{Sets}$ (Ex. 1.1) there is the *tensored* simplicial presheaf

$$\mathscr{X} \times S \;\in\; \mathrm{PSh}\big(\mathrm{CartSp}, \Delta\mathrm{Sets}\big)$$

given by value-wise Cartesian product of simplicial sets:

$$\mathscr{X} \times S \;:\; \mathbb{R}^n \;\mapsto\; \mathscr{X}(\mathbb{R}^n) \times S. \tag{1.77}$$

(ii) For $\mathscr{X}, \mathscr{A} \in \mathrm{PSh}(\mathrm{CartSp}, \Delta\mathrm{Sets})$ the *simplicial hom-complex* from $\mathscr{X}$ to $\mathscr{A}$ is the simplicial set of morphisms of simplicial presheaves

$$\underbrace{\mathrm{Maps}(\mathscr{X}, \mathscr{A})}_{\text{simplicial mapping complex}} := \mathrm{PSh}(\mathrm{CartSp}, \Delta\mathrm{Sets})\big(\mathscr{X} \times \Delta[\bullet], \mathscr{A}\big) \in \Delta\mathrm{Sets}. \tag{1.78}$$

into $\mathscr{A}$ out of the tensoring (1.77) of $\mathscr{X}$ with the simplicial simplices $\Delta[n] \in \Delta\mathrm{Sets}$, $n \in \mathbb{N}$. Its image in the classical homotopy category (Ex. 1.14) is the *mapping space*

$$\mathrm{Maps}(\mathscr{X}, \mathscr{A}) \in \mathrm{Ho}(\Delta\mathrm{Sets}_{\mathrm{Qu}}). \tag{1.79}$$

(iii) The (simplicially enriched) Yoneda lemma says that simplicial hom-complexes (1.78) out of a Cartesian space (1.66) regarded as a simplicial presheaf via Ex. 1.21:

$$\mathscr{X}(\mathbb{R}^n) \simeq \mathrm{Maps}(\mathbb{R}^n, \mathscr{X}). \tag{1.80}$$

Proposition 1.24 (Smooth ∞-Stacks). *The fibrant objects (Ntn. 1.1) in the local projective model structure* (1.69)*, are to be called the smooth ∞-stacks (or smooth ∞-groupoids)*

$$\mathrm{SmoothStacks}_\infty := \Big(\mathrm{PSh}(\mathrm{CartSp}, \Delta\mathrm{Sets}_{\mathrm{Qu}})_{\substack{\mathrm{proj}\\ \mathrm{loc}}}\Big)^{\mathrm{fib}}, \tag{1.81}$$

are precisely those simplicial presheaves which:

(i) *are presheaves of ∞-groupoids in that they take values in Kan complexes* (1.16)*;*
(ii) *respect gluing of patches of good open covers of Cartesian spaces ("satisfy descent") in that for each $n \in \mathbb{N}$ and each good open cover $\{U_i \hookrightarrow \mathbb{R}^n\}_{i \in I}$ (Ex. 1.24) the following map* (1.82) *of simplicial hom-complexes* (1.78) *– induced by precomposition with the comparison morphism* (1.76) *from the Čech nerve* (1.75) *– is a weak equivalence of simplicial sets (Ex. 1.2):*

$$\mathscr{X}(\mathbb{R}^n) \simeq \mathrm{Maps}(\mathbb{R}^n, \mathscr{X}) \xrightarrow[\in \mathrm{W}]{\mathrm{Maps}(p_{\{U_i\}_i}, \mathscr{X})} \mathrm{Maps}\Big(N(\{U_i\}_{i \in I}), \mathscr{X}\Big) \in \Delta\mathrm{Sets}_{\mathrm{Qu}}. \tag{1.82}$$

Proof. By the discussion in [Dugger (2001), §5.1] the claimed condition characterizes the fibrant objects in the left Bousfield localization of the global projective model category (1.68) at the Čech nerve projections (1.75) By [Dugger (1998), Prop. 3.4.8] this left Bousfield localization is the local model structure (1.69). □

Definition 1.25 (Homotopy category of smooth ∞-stacks). In view of Prop. 1.24, we write

$$\mathrm{Ho}(\mathrm{SmthStacks}_\infty) := \mathrm{Ho}\Big(\mathrm{PSh}(\mathrm{CartSp}, \Delta\mathrm{Sets})_{\substack{\mathrm{proj}\\ \mathrm{loc}}}\Big) \tag{1.83}$$

for the homotopy category (Def. 1.8) of the local projective model category of simplicial presheaves over CartSp (Ex. 1.20). We say that the objects of $\mathrm{Ho}(\mathrm{SmthStacks}_\infty)$ (1.83) are *smooth ∞-stacks*.

Example 1.26 (Truncated smooth ∞-stacks [Sati and Schreiber (2020c), Ex. 3.18]**).**
(i) Those smooth ∞-stacks (Prop. 1.24) which take values in 2-coskeletal, hence 1-truncated, Kan complexes (Prop. 1.19) are 1-groupoid valued, hence are *smooth 1-stacks* or just smooth *stacks* [Jardine (2001)][Hollander (2008)].
(ii) Those smooth ∞-stacks which are 0-truncated take values in sets and hence are *sheaves* on CartSp. We call these *smooth spaces*. The *concrete sheaves* among these are the *diffeological spaces* ([Souriau (1980)][Souriau (1984)][Iglesias-Zemmour (1985)], see [Baez and Hoffnung (2011)][Iglesias-Zemmour (2013)]).

$$\underset{\text{diffeological spaces}}{\mathrm{PSh}(\mathrm{CartSp},\mathrm{Sets})^{\mathrm{fib}}_{\mathrm{conc}}} \hookrightarrow \underset{\substack{\text{smooth sets}\\ \text{(smooth spaces)}}}{\mathrm{PSh}(\mathrm{CartSp},\mathrm{Sets})^{\mathrm{fib}}} \hookrightarrow \underset{\substack{\text{smooth groupoids}\\ \text{(smooth stacks)}}}{\mathrm{PSh}(\mathrm{CartSp},\Delta\mathrm{Sets}_{\mathrm{cosk}_2})^{\mathrm{fib}}} \hookrightarrow \underset{\substack{\text{smooth }\infty\text{-groupoids}\\ \text{(smooth }\infty\text{-stacks)}}}{\mathrm{PSh}(\mathrm{CartSp},\Delta\mathrm{Sets})^{\mathrm{fib}}}$$

Lemma 1.5 (∞-Stackification preserves finite homotopy limits). *The identity functors constitute a Quillen adjunction (Def. 1.12) between the local and the global projective model categories of Ex. 1.20:*

$$\mathrm{PSh}(\mathrm{CartSp},\Delta\mathrm{Sets})_{\substack{\mathrm{proj}\\ \mathrm{loc}}} \underset{\mathrm{id}}{\overset{\mathrm{id}}{\rightleftarrows}}\!{}_{\perp_{\mathrm{Qu}}} \mathrm{PSh}(\mathrm{CartSp},\Delta\mathrm{Sets})_{\mathrm{proj}}\,.$$

Moreover, this is such that the derived left adjoint functor (Prop. 1.13)

$$L^{\mathrm{loc}}:\ \mathrm{Ho}\Big(\mathrm{PSh}(\mathrm{CartSp},\Delta\mathrm{Sets})_{\mathrm{proj}}\Big) \xrightarrow{\ \mathbb{D}\mathrm{id}\ } \mathrm{Ho}(\mathrm{SmthStacks}_\infty) \tag{1.84}$$

(the ∞-stackification functor) preserves homotopy pullbacks (Def. 1.15).

Proposition 1.26 (Shape Quillen adjunction [Schreiber (2013), Prop. 4.4.8][Sati and Schreiber (2021b), Ex. 3.18]**).** *We have a Quillen adjunction (Def. 1.12)*

$$\mathrm{PSh}(\mathrm{CartSp},\Delta\mathrm{Sets})_{\substack{\mathrm{proj}\\ \mathrm{loc}}} \underset{\mathrm{Disc}}{\overset{\mathrm{Shp}}{\rightleftarrows}}\!{}_{\perp_{\mathrm{Qu}}} \Delta\mathrm{Sets}_{\mathrm{Qu}}$$

between the projective local model structure on simplicial presheaves over CartSp *(Ex. 1.20) and the classical model structure on simplicial sets (Ex. 1.2), hence a derived adjunction (Prop. 1.13) between homotopy category of ∞-stacks (Def. 1.25) and the classical homotopy category (Ex. 1.14)*

$$\mathrm{Ho}(\mathrm{SmthStacks}_\infty) \underset{\mathbb{D}\mathrm{Disc}}{\overset{\mathbb{D}\mathrm{Shp}}{\rightleftarrows}}\!{}_{\perp_{\mathrm{Qu}}} \mathrm{Ho}(\Delta\mathrm{Sets}_{\mathrm{Qu}})$$

whose (underived) right adjoint sends a simplicial set to the presheaf which is constant on that simplicial set:

$$\mathrm{Disc}(S) := \mathrm{const}(S) : \big(\mathbb{R}^n \mapsto S\big). \tag{1.85}$$

Homological algebra.

Example 1.27 (Projective model structure on connective chain complexes [Quillen (1967), §II.4 (5.)]**).** The category $\mathrm{ChainComplexes}_{\mathbb{Z}}^{\geq 0}$ of connective chain complexes of abelian groups (i.e. concentrated in non-negative degrees with differential of degree -1) carries a model category structure (Def. 1.3) whose

- W – weak equivalences are the quasi-isomorphisms (those inducing isomorphisms on all chain homology groups)
- Fib – fibrations are the morphisms that are surjections in each positive degree
- Cof – cofibrations are the morphisms with degreewise injective kernels.

We write $\big(\mathrm{ChainComplexes}_{\mathbb{Z}}^{\geq 0}\big)_{\mathrm{proj}}$ for this model category.

More generally:

Example 1.28 (Projective model structure on presheaves of connective chain complexes [Jardine (2003), p. 7]**).** The category of presheaves of connective chain complexes over CartSp (1.66) carries the structure of a model category whose weak equivalences and fibrations are objectwise those of $\big(\mathrm{ChainComplexes}_{\mathbb{Z}}^{\geq 0}\big)_{\mathrm{proj}}$ (Ex. 1.27). We write $\mathrm{PSh}\big(\mathrm{CartSp}, \mathrm{ChainComplexes}_{\mathbb{Z}}^{\geq 0}\big)_{\mathrm{proj}}$ for this model category.

Proposition 1.27 (Dold-Kan correspondence [Dold (1958), Thm. 1.9][Kan (1958)][Goerss and Jardine (1999), §III.2][Schwede and Shipley (2003a), §2.1]**).** *Given* $A_\bullet \in \Delta\mathrm{AbGrps}$, *its normalized chain complex*

$$N(A)_\bullet \in \mathrm{ChainComplexes}_{\mathbb{Z}}^{\geq 0}$$

is the connective chain complex of abelian groups (Ex. 1.27) which in degree $n \in \mathbb{N}$ *is the quotient of* A_n *by the degenerate cells and whose differential is the alternating sum of the face maps:*

$$N(A)_\bullet := \Big\{ N(A)_n := A_n/\sigma(A_{n+1}),\ \partial_n := \sum_{i=0}^{n} (-1)^i d_i \,:\, N(A)_n \longrightarrow N(A)_{n-1} \Big\}_{n\in\mathbb{N}}. \tag{1.86}$$

(i) *This construction constitutes an adjoint equivalence of categories*

$$\mathrm{ChainComplexes}_{\mathbb{Z}}^{\geq 0} \underset{\simeq}{\overset{N}{\rightleftarrows}} \Delta\mathrm{AbGrps} \tag{1.87}$$

(ii) *such that simplicial homotopy groups of* $A \in \Delta\mathrm{AbGrps} \to \mathrm{SimplicialSet}$ *are identified with chain homology groups of the normalized chain complex ([Goerss and Jardine (1999), Cor. III.2.5]):*

$$\pi_\bullet(A) \simeq H_\bullet(NA)\,. \tag{1.88}$$

Example 1.29 (Model structure on simplicial abelian groups [Quillen (1969), §III.2][Schwede and Shipley (2003a), §4.1]**).**   The category ΔAbGrps carries a model category structure (Def. 1.3) whose

W – weak equivalences are the morphisms which are weak equivaleces as morphisms in ΔSets$_{\text{Qu}}$ (Ex. 1.2)

Fib – fibrations are the morphisms which are fibrations as morphisms in ΔSets$_{\text{Qu}}$ (Ex. 1.2)

In other words, this is the transferred model structure along the free/forgetful adjunction, which thus becomes a Quillen adjunction (Def. 1.12):

$$\Delta\text{AbGrps}_{\text{proj}} \underset{\underset{\perp_{\text{Qu}}}{\xrightarrow{\hspace{2cm}}}}{\overset{\mathbb{Z}[-]}{\xleftarrow{\hspace{2cm}}}} \Delta\text{Sets}_{\text{Qu}} . \qquad (1.89)$$

Proposition 1.28 (Dold-Kan Quillen equivalence [Schwede and Shipley (2003a), §4.1][Jardine (2003), Lem. 1.5]**).** *With respect to the projective model structure on connective chain complexes (Ex. 1.27) and the projective model structure on simplicial abelian groups (Ex. 1.29) the Dold-Kan correspondence (Prop. 1.27) is a Quillen equivalence (Def. 1.16):*

$$\left(\text{ChainComplexes}_{\mathbb{Z}}^{\geq 0}\right)_{\text{proj}} \underset{\underset{\simeq_{\text{Qu}}}{\xrightarrow{\hspace{2cm}}}}{\overset{N}{\xleftarrow{\hspace{2cm}}}} \Delta\text{AbGrps}_{\text{proj}} , \qquad (1.90)$$

where both functors preserve all three classes of morphims, Fib, Cof *and* W, *separately.*

Example 1.30 (Dold-Kan construction [Fiorenza *et al.* (2012), §3.2.3][Fiorenza *et al.* (2013), §2.4]**).**   **(i)** We write DK for the total right adjoint in the composite of the free Quillen adjunction (1.89) and the Dold-Kan equivalence (1.90):

$$\left(\text{ChainComplexes}_{\mathbb{Z}}^{\geq 0}\right)_{\text{proj}} \underset{\underset{\simeq_{\text{Qu}}}{\xrightarrow{\hspace{1.5cm}}}}{\overset{N}{\xleftarrow{\hspace{1.5cm}}}} \Delta\text{AbGrps}_{\text{proj}} \underset{\underset{\perp_{\text{Qu}}}{\xrightarrow{\hspace{1.5cm}}}}{\overset{\mathbb{Z}[-]}{\xleftarrow{\hspace{1.5cm}}}} \Delta\text{Sets}_{\text{Qu}} . \quad (1.91)$$
$$\underset{\text{DK}}{\underbrace{\hspace{6cm}}}$$

(ii) This extends to a right Quillen functor on global projective model categories of presheaves (Ex. 1.20, Ex. 1.28). whose right derived functor (Prop. 1.13) $\mathbb{D}$DK composed with the ∞-stackification functor (1.84) is thus of the form

$$\text{Ho}\left(\text{PSh}\left(\text{CartSp}, \text{ChainComplexes}_{\mathbb{Z}}^{\geq 0}\right)_{\text{proj}}\right) \overset{\overset{\text{derived}}{\overset{\text{Dold-Kan construction}}{\mathbb{D}\text{DK}}}}{\xrightarrow{\hspace{2cm}}} \text{Ho}\left(\text{PSh}\left(\text{CartSp}, \Delta\text{Sets}\right)_{\text{proj}}\right)$$

$$\underset{\substack{\infty\text{-stackified} \\ \text{Dold-Kan construction}}}{\searrow} \qquad \qquad \downarrow L^{\text{loc}} \; {\scriptstyle \infty\text{-stackification}}$$

$$\text{Ho}(\text{SmthStacks}_{\infty})$$

and preserves homotopy pullbacks (by Lem. 1.5).

Example 1.31 (Projective model structure on unbounded chain complexes [Hovey (1999), Thm. 2.3.11]**).** The category $\mathrm{ChainComplexes}_{\mathbb{Z}}$ of unbounded chain complexes of abelian groups carries a model category structure (Def. 1.3) whose:

W – weak equivalences are the quasi-isomorphisms;

Fib – fibrations are the degreewise surjections.

We write $\big(\mathrm{ChainComplexes}_{\mathbb{Z}}\big)_{\mathrm{proj}}$ for this model category.

Proposition 1.29 (Stable Dold-Kan construction). *The Dold-Kan construction (Def. 1.30) lifts along the stabilization adjunction (Ex. 1.19) from connective to unbounded chain complexes (Ex. 1.31), such as to make the following diagram commute:*

$$
\begin{array}{ccccc}
\mathrm{Ho}\Big(\big(\mathrm{ChainComplexes}_{\mathbb{Z}}^{\geq 0}\big)_{\mathrm{proj}}\Big) & \xrightarrow{\simeq} & \mathrm{Ho}\big(\Delta\mathrm{AbGrps}_{\mathrm{proj}}\big) & \longrightarrow & \mathrm{Ho}\big(\Delta\mathrm{Sets}_{\mathrm{Qu}}\big) \\
\big\downarrow \dashv \big\uparrow{\scriptstyle \mathbb{D}\Omega^\infty} & & & & {\scriptstyle \mathbb{D}\Sigma^\infty}\big\downarrow \dashv \big\uparrow{\scriptstyle \mathbb{D}\Omega^\infty} \\
\mathrm{Ho}\Big(\big(\mathrm{ChainComplexes}_{\mathbb{Z}}\big)_{\mathrm{proj}}\Big) & \xrightarrow{\simeq} & \mathrm{Ho}\big((H\mathbb{Z})\mathrm{ModuleSpectra}\big) & \longrightarrow & \mathrm{Ho}\big(\mathrm{SequentialSpectra}_{\mathrm{BF}}\big)\,.
\end{array}
\tag{1.92}
$$

(top curved arrow: $\mathbb{D}\mathrm{DK}$, "Dold-Kan correspondence"; bottom curved arrow: $\mathbb{D}\mathrm{DK}_{\mathrm{st}}$, "stable Dold-Kan construction")

Here the right adjoint on chain complexes is the homological truncation from below:

$$
\mathbb{D}\Omega^\infty\Big(\cdots \xrightarrow{\partial_2} V_2 \xrightarrow{\partial_1} V_1 \xrightarrow{\partial_0} V_0 \xrightarrow{\partial_{-1}} V_{-1} \xrightarrow{\partial_{-2}} \cdots\Big) = \Big(\cdots \xrightarrow{\partial_2} V_2 \xrightarrow{\partial_1} V_1 \xrightarrow{\partial_0} \ker(\partial_{-1})\Big). \tag{1.93}
$$

Proof. **(i)** It is clear from inspection that the assignment (1.93) is right adjoint to the inclusion of connective chain complexes, so that we have a pair of adjoint functors

$$
\big(\mathrm{ChainComplexes}_{\mathbb{Z}}\big)_{\mathrm{proj}} \underset{\Omega^\infty}{\overset{\longleftarrow\!\!\!\!\supset}{\underset{\longrightarrow}{\perp_{\mathrm{Qu}}}}} \big(\mathrm{ChainComplexes}_{\mathbb{Z}}^{\geq 0}\big)_{\mathrm{proj}}\,. \tag{1.94}
$$

Moreover, it is immediate that this is a Quillen adjunction (Def. 1.12) between the projective model structure on connective chain complexes (Ex. 1.27) and that on unbounded chain complexes (Ex. 1.31): Ω^∞ clearly preserves fibrations (using that those between connective chain complexes need to be surjective only in positive degrees!) and clearly preserves all weak equivalences. Finally, since all chain complexes in the projective model structure are fibrant, we have that with Ω^∞ also $\mathbb{D}\Omega^\infty$ is given by (1.93), via Ex. 1.10.

(ii) A Quillen adjunction of the form

$$\underbrace{(\mathrm{ChainComplexes}_{\mathbb{Z}})_{\mathrm{proj}} \underset{H}{\overset{\longleftarrow}{\underset{\longrightarrow}{\bot_{\simeq}}}} (H\mathbb{Z})\mathrm{ModuleSpectra} \overset{\longleftarrow}{\underset{\longrightarrow}{\bot_{\mathrm{Qu}}}} \mathrm{SequentialSpectra}_{\mathrm{BF}}}_{\mathrm{DK}_{\mathrm{st}}} \tag{1.95}$$

is established in [Schwede and Shipley (2003b), §B.1], where:

(a) the first step is a Quillen equivalence (Def. 1.16) between the projective model structure on unbounded chain complexes (Ex. 1.31) and a model category of module spectra over the Eilenberg-MacLane spectrum $H\mathbb{Z}$ [Schwede and Shipley (2003b), §B.1.11];

(b) the second step is a Quillen adjunction [Schwede and Shipley (2003b), p. 37, item ii)] to the Bousfield-Friedlander model structure (Ex. 1.18) whose right adjoint assigns underlying sequential spectra; such that

(c) the composite right adjoint $\mathrm{DK}_{\mathrm{st}}$ (1.95) further composed with Ω^∞ on spectra (1.63) equals the composite of Ω^∞ on chain complexes (1.94) with the unstable Dold-Kan construction (1.91):

$$\Omega^\infty \circ \mathrm{DK}_{\mathrm{st}} \simeq \mathrm{DK} \circ \Omega^\infty$$

(by immediate inspection of the construction in [Schwede and Shipley (2003b), p. 38-39]).

(iii) By uniqueness of adjoints, this implies that the Quillen adjunction of the stable Dold-Kan construction (1.95) is intertwined by the Quillen adjunctions involving Ω^∞ with the Quillen adjunction of the unstable Dold-Kan construction (1.91), and hence the commuting diagram of derived functors (1.29) follows (Prop. 1.13). □

Chapter 2

Non-abelian cohomology theories

We make explicit the concept of general non-abelian cohomology (Def. 2.1 below) and of twisted non-abelian cohomology (Def. 3.2 below), following [Simpson (1997b)][Simpson (2002)][Toën (2002)][Sati *et al.* (2012)][Nikolaus *et al.* (2015a)][Nikolaus *et al.* (2015b)][Fiorenza *et al.* (2020b)][Sati and Schreiber (2020c)]; and we survey how this concept subsumes essentially every notion of cohomology known.

In the following, we make free use of the basic language of category theory and homotopy theory (for joint introduction see [Riehl (2014)][Richter (2020)]). For $\mathscr{C}$ a category and $X, A \in \mathscr{C}$ a pair of its objects, we write

$$\mathscr{C}(X, A) := \mathrm{Hom}_{\mathscr{C}}(X, A) \in \mathrm{Sets} \tag{2.1}$$

for the set of morphisms from X to A. These are, of course, contravariantly and covariantly functorial in their first and second argument, respectively:

$$\mathscr{C} \xrightarrow{\mathscr{C}(X,-)} \mathrm{Sets}\,, \qquad \mathscr{C}^{\mathrm{op}} \xrightarrow{\mathscr{C}(-,A)} \mathrm{Sets}\,. \tag{2.2}$$

Basic as this is, contravariant hom-functors are of paramount interest in the case where $\mathscr{C}$ is the *homotopy category* $\mathrm{Ho}(\mathbf{C})$ (Def. 1.8) of a model category (Def. 1.3), such as the classical homotopy category of topological spaces or, equivalently, of simplicial sets (Ex. 1.14).

Definition 2.1 (Non-abelian cohomology). For $X, A \in \mathrm{Ho}(\Delta\mathrm{Sets}_{\mathrm{Qu}})$ (Ex. 1.14) we say that their hom-set (2.1) is the *non-abelian cohomology* of X *with coefficients* in A, or the *non-abelian A-cohomology* of X, to be denoted:

$$\underset{\text{non-abelian cohomology}}{H(X;A)} := \mathrm{Ho}\big(\Delta\mathrm{Sets}_{\mathrm{Qu}}\big)(X, A) = \left\{ X \underset{\underset{\text{map = cocycle}}{c'}}{\overset{\overset{\text{map = cocycle}}{c}}{\Downarrow\,{\scriptstyle \text{homotopy = coboundary}}}} A \right\}_{\big/\text{homotopy}} \tag{2.3}$$

We also call the contravariant hom-functor (2.2)

$$H(-;A) \;:\; \mathrm{Ho}\big(\Delta\mathrm{Sets}_{\mathrm{Qu}}\big) \longrightarrow \mathrm{Sets} \tag{2.4}$$

the non-abelian *A-cohomology theory*.

Example 2.1 (Ordinary cohomology). For $n \in \mathbb{N}$ and A a discrete abelian group, the ordinary cohomology (e.g. singular cohomology) in degree n with coefficients in A is equivalently ([Eilenberg (1940), p. 243][Eilenberg and MacLane (1954), p. 520-521], review in [Steenrod (1972), §19][May (1999), §22][Aguilar *et al.* (2002), §7.1, Cor. 12.1.20]) non-abelian cohomology in the sense of Def. 2.1

$$\overset{\text{ordinary cohomology}}{H^n(-;A)} \;\simeq\; H\big(-;K(A,n)\big) \tag{2.5}$$

with coefficients in an *Eilenberg-MacLane space* [Eilenberg and MacLane (1953)][Eilenberg and MacLane (1954)]:

$$K(A,n) \in \mathrm{Ho}\big(\Delta\mathrm{Sets}_{\mathrm{Qu}}\big) \quad \text{such that} \quad \pi_k\big(K(A,n)\big) \;=\; \begin{cases} A \mid k=n \\ 0 \mid k\neq n\,. \end{cases} \tag{2.6}$$

Example 2.2 (Traditional non-abelian cohomology). For G a well-behaved[1] topological group, the traditional non-abelian cohomology $H^1(-;G)$ classifying G-principal bundles, is equivalently ([Steenrod (1951), §19.3][Roberts and Stevenson (2016), Thm. 1], review in [Addington (2007), §5]) non-abelian cohomology in the general sense of Def. 2.1

$$\overset{\text{classification of principal bundles}}{H^1(-;G)} \simeq H(-;BG) \tag{2.7}$$

with coefficients in the *classifying space BG* ([Milnor (1956)][Segal (1968)][Steenrod (1968)][Stasheff (1971)], review in [Kochman (1996), §1.3][May (1999), §23.1] [Aguilar *et al.* (2002), §8.3][Nikolaus *et al.* (2015b), §3.7.1]). The latter may be given as the homotopy colimit (in the classical model structure of $\mathrm{TopSp}_{\mathrm{Qu}}$, Ex. 1.1) over the nerve of the topological group G (e.g. [Nikolaus *et al.* (2015a), Rem. 2.23]):

$$BG \simeq \underset{\longrightarrow}{\mathrm{holim}} \left(\cdots \;\vdots\; G\times G \;\underset{\longleftarrow}{\overset{\longrightarrow}{\underset{\longleftarrow}{\xrightarrow{(-)\cdot(-)}}}}\; G \;\underset{\longrightarrow}{\overset{\longrightarrow}{\xleftarrow{e}}}\; * \right). \tag{2.8}$$

---

[1]The technical condition is that G be *well-pointed*, which means that the inclusion $* \xrightarrow{e} G$ of the neutral element is a *closed Hurewicz cofibration*, hence that $(G,\{e\})$ is an *NDR pair*, see [Baez and Stevenson (2009)] for pointers and [Roberts and Stevenson (2016)] for details. All Lie groups are well-pointed.

Example 2.3 (Group cohomology and Characteristic classes). Conversely, the ordinary cohomology (Ex. 2.1) *of* the classifying space BG (2.8) of a Lie group G with coefficients in a discrete group $A \in \mathrm{Grps}(\mathrm{Sets})$ (such as $A = \mathbb{Z}$) is, equivalently:[2]

(i) the group cohomology of G;
(ii) the universal characteristic classes of G-principal bundles:

$$\overset{\substack{\text{group}\\\text{cohomology}}}{H\big(BG; K(A,n)\big)} \;\simeq\; H^n(BG; A) \;\simeq\; H^n_{\mathrm{Grp}}(G; A)\,.$$

Example 2.4 (Non-abelian cohomology in degree 2). For a well-behaved topological 2-group, such as the *string 2-group* $\mathrm{String}(G)$ (of a connected, simply connected semi-simple Lie group G) [Baez *et al.* (2007)][Henriques (2008), Thm. 4.8][Nikolaus *et al.* (2013)], the non-abelian cohomology $H^1(-;\mathrm{String}(G))$ classifying principal 2-bundles [Nikolaus and Waldorf (2013)] with structure 2-group $\mathrm{String}(G)$ is, equivalently [Baez and Stevenson (2009)],

$$\overset{\substack{\text{classification of}\\\text{String-bundles}}}{H^1\big(-;\mathrm{String}(G)\big)} \;\simeq\; H\big(-; B\mathrm{String}(G)\big) \tag{2.9}$$

non-abelian cohomology in the general sense of Def. 2.1 with coefficients in the classifying space $B\mathrm{String}(G)$.

Example 2.5 (Non-abelian gerbes). For G a well-behaved topological group, a non-abelian *G-gerbe* [Giraud (1971)][Breen (2010)] is equivalently [Nikolaus *et al.* (2015a), §4.4] a fiber 2-bundle associated to principal 2-bundles with a certain topological structure 2-group $\mathrm{Aut}(\mathbf{B}G)$ (the automorphism 2-group of the moduli stack of G, see Rem. 2.3). Hence, as in Ex. 2.4, G-gerbes are classified by non-abelian cohomology with coefficients in $B\mathrm{Aut}(\mathbf{B}G)$ [Nikolaus *et al.* (2015a), Cor. 4.51]:

$$\overset{\substack{\text{classification of}\\\text{non-abelian gerbes}}}{G\mathrm{Gerbes}(X)_{/\sim}} \;\simeq\; H^1\big(X; \mathrm{Aut}(\mathbf{B}G)\big) \;\simeq\; H\big(X; B\mathrm{Aut}(\mathbf{B}G)\big)\,.$$

Example 2.6 (Non-abelian cohomology in unbounded degree). For any ∞-group $\mathcal{G}$ (see [Nikolaus *et al.* (2015a), §2.2][Nikolaus *et al.* (2015b), §3.5]), the non-abelian cohomology $H^1(-;\mathcal{G})$ classifying principal ∞-bundles [Glenn (1982)][Jardine and Luo (2006)][Nikolaus *et al.* (2015a)][Nikolaus *et al.* (2015b)] with structure ∞-group $\mathcal{G}$ is, equivalently [Wendt (2011)][Roberts and Stevenson (2016)],

$$\overset{\substack{\text{classification of}\\\text{non-abelian }\infty\text{-gerbes}}}{H^1(-;\mathcal{G})} \;\simeq\; H(-; B\mathcal{G}) \tag{2.10}$$

[2] If G is a topological or Lie group, then the appropriate (*continuous* or *smooth*, respectively) group cohomology of G is (by [Schreiber (2013), Thm. 4.4.36]) in general not that of the classifying space BG, but of the *universal moduli stack* $\mathbf{B}G$ (Rem. 2.3) with coefficients in the higher stack $\mathbf{B}^n A$. However, for *discrete* coefficients A this reduces (by [Schreiber (2013), Prop. 4.4.35]) to the cohomology of the geometric realization of $\mathbf{B}G$, which, at least for Lie groups G, coincides (by [Schreiber (2013), Prop. 4.4.30]) with that of the classifying space BG.

non-abelian cohomology in the general sense of Def. 2.1 with coefficients in the classifying space $B\mathscr{G}$ (see also [Stevenson (2012)]).

Ex. 2.6 is, in fact, universal:

Proposition 2.2 (Connected homotopy types are higher non-abelian classifying spaces [May (1972)][Lurie (2009a), 7.2.2.11], [Nikolaus *et al.* (2015a), Thm. 2.19][Nikolaus *et al.* (2015b), Thm. 3.30, Cor. 3.34]). *Every connected homotopy type* $A \in \mathrm{Ho}(\Delta\mathrm{Sets}_{\mathrm{Qu}})$ (1.51) *is the classifying space of a topological group, namely of its loop group*[3] ΩA

$$A \simeq B(\Omega A) \quad \in \mathrm{Ho}(\Delta\mathrm{Sets}_{\mathrm{Qu}}). \tag{2.11}$$

This allows to make precise the core nature of non-abelian cohomology:

Remark 2.1 (From non-abelian to abelian ∞-groups). For $A \simeq BG$ (2.11), the ∞-group structure on G is reflected by its weak homotopy equivalence $G \simeq \Omega BG$ with a based loop space.

- There is no commutativity of composition of loops in a generic loop space, and hence this exhibits G as a *non-abelian* ∞-group.
- But it may happen that A itself is already equivalent to a loop space, which by (2.11) means that $A \simeq B(BG) =: B^2G$ is a *double delooping*. In this case $G \simeq \Omega(\Omega A) =: \Omega^2 A$ is an *iterated loop space* [May (1972)], specifically a *double loop space*; hence a *braided ∞-group* ([Garzon and Miranda (1997), §1][Garzón and Miranda (2000), §6][Fiorenza *et al.* (2014b), Def. 4.28]). By the Eckmann-Hilton argument [Eckmann and Hilton (1961/62), Thm. 1.12][Schlank and Yanovski (2019)], this implies a first level of commutativity of the group operation in G. Indeed, in the special case that such G is also 0-truncated (1.55), it implies that G is an ordinary abelian group.
- Next, it may happen that $A \simeq B^3G$ is a 3-fold delooping, hence that $G \simeq \Omega^3 A$ is a 3-fold loop space, hence a *sylleptic ∞-group* (where the terminology follows [Day and Street (1997), §5][Crans (1998), §4] see [Gurski and Osorno (2013), §2.2] for relation to our context). This is one step "more abelian" than a braided ∞-group.
- In the limiting case that G is an n-fold loop space for any $n \in \mathbb{N}$, hence an *infinite loop space* [May (1977b)][Adams (1978)], it is as abelian as possible for an ∞-group. Such *symmetric* (in the monoidal ∞-category theoretic terminology of [Lurie (2009b)]) or, we may say, *abelian ∞-groups* are the coefficients of abelian cohomology theories, namely of generalized cohomology theories in the sense of Whitehead (Ex. 2.10).
- The fewer deloopings an ∞-group G admits, the "more non-abelian" is the cohomology theory represented by BG.

[3] A priori, the loop group is an A_∞-group, for which classifying spaces are defined as in [Nikolaus *et al.* (2015a), Rem. 2.23], but each such is weakly equivalent to an actual topological group, see [Nikolaus *et al.* (2015b), Prop. 3.35].

Coefficients		$H(X;BG)$	**Examples**
non-abelian ∞-group	$G \simeq \Omega\ B\ G$	non-abelian cohomology	$\pi^n(-)$ (Ex. 2.7)
braided ∞-group	$G \simeq \Omega^2 B^2 G$		$\pi^3(-)$
sylleptic ∞-group	$G \simeq \Omega^3 B^3 G$		
⋮	$G \simeq \Omega^n B^n G$		
abelian ∞-group	$G \simeq \Omega^\infty B^\infty G$	abelian cohomology	$E^n(-)$ (Ex. 2.10)

The most fundamental connected homotopy types are the n-spheres (all other are obtained by gluing n-spheres to each other):

Example 2.7 (Cohomotopy theory). The non-abelian cohomology theory (Def. 2.1) with coefficients in the homotopy types of n-spheres is (unstable) *Cohomotopy* theory [Borsuk (1936)][Spanier (1949)][Peterson (1956)][Taylor (2012)][Kirby *et al.* (2012)]:

$$\overset{\text{Cohomotopy}}{\pi^n(-)} = H(-;S^n) \simeq H^1(-;\Omega S^n) \quad \text{for } n \in \mathbb{N}_+ .$$

(i) By Prop. 2.2, Cohomotopy theory classifies principal ∞-bundles (Ex. 2.6) with structure ∞-group of the homotopy type of the ∞-group ΩS^n.
(ii) By Rem. 2.1, Cohomotopy theory is a *maximally non-abelian* cohomology theory, in that S^n does not admit deloopings, for general n (it admits a single delooping for $n = 3$ and arbitrary deloopings for $n = 0, 1$).

Example 2.8 (Bundle gerbes). The classifying space (2.8) of the circle group $\mathrm{U}(1)$ is an Eilenberg-MacLane space (2.6)

$$B\mathrm{U}(1) \simeq K(\mathbb{Z},2) \quad \in \mathrm{Ho}\big(\Delta\mathrm{Sets}_{\mathrm{Qu}}\big) .$$

Since $\mathrm{U}(1)$ is abelian, this space carries itself the structure of (the homotopy type of) a 2-group, and hence has a higher classifying space

$$B^2\mathrm{U}(1) := B(B\mathrm{U}(1)) \simeq K(\mathbb{Z},3) \quad \in \mathrm{Ho}\big(\Delta\mathrm{Sets}_{\mathrm{Qu}}\big) ,$$

in the sense of Ex. 2.4, which is an Eilenberg-MacLane space in one degree higher. The higher principal 2-bundles with topological structure 2-group $\mathbf{B}\mathrm{U}(1)$ are equivalently [Nikolaus *et al.* (2015a), Rem. 4.36] known as *bundle gerbes* [Murray (1996)][Schweigert and Waldorf (2011)]. Therefore, Ex. 2.6 combined with Ex. 2.1 gives the classification of bundle gerbes by ordinary integral cohomology in degree 3:

$$\overset{\substack{\text{classification of} \\ \text{bundle gerbes}}}{H^1\big(-;\mathbf{B}\mathrm{U}(1)\big) \simeq H\big(-;B^2\mathrm{U}(1)\big) \simeq H^3(-;\mathbb{Z})} .$$

Example 2.9 (Higher bundle gerbes). In fact, Prop. 2.2 implies that, for all $n \in \mathbb{N}$,

$$B^{n+1}\mathrm{U}(1) := B\big(B^n\mathrm{U}(1)\big) \simeq K(\mathbb{Z},n+2) \quad \in \mathrm{Ho}\big(\Delta\mathrm{Sets}_{\mathrm{Qu}}\big) , \tag{2.12}$$

in the sense of Ex. 2.6. The higher principal bundles with structure $(n+1)$-group $\mathbf{B}^n\mathrm{U}(1)$ [Gajer (1997)][Fiorenza *et al.* (2012), §3.2.3][Fiorenza *et al.* (2013), §2.6] are also known as *higher bundle gerbes* (for $n=2$ see [Carey *et al.* (1997)][Stevenson (2004)]). On these coefficients, Ex. 2.6 reduces to the classification of higher bundle gerbes by ordinary integral cohomology in higher degree:

$$\underbrace{H^1\big(-;\mathbf{B}^n\mathrm{U}(1)\big)}_{\text{classification of higher bundle gerbes}} \simeq H\big(-;B^{n+1}\mathrm{U}(1)\big) \simeq H^{n+2}(-;\mathbb{Z}).$$

More generally, the special case of Ex. 2.6 where the coefficient ∞-group happens to be abelian is "generalized cohomology" in the standard sense of algebraic topology (including cohomology theories such as K-theory, elliptic cohomology, stable Cobordism theory, stable Cohomotopy theory, etc.):

Example 2.10 (Whitehead-generalized cohomology). For E a generalized cohomology theory in the traditional sense of [Whitehead (1962)] (review in [Adams (1974)][Adams (1978)][Kono and Tamaki (2006)]), Brown's representability theorem ([Adams (1974), §III.6][Kochman (1996), §3.4]) says that there is a *spectrum* ("Ω-spectrum", Ex. 1.18) of pointed homotopy types

$$\Big\{ E_n \in \mathrm{Ho}\big(\Delta\mathrm{Sets}^{*/}_{\mathrm{Qu}}\big),\ E_n \xrightarrow[\simeq]{\tilde{\sigma}_n} \Omega E_{n+1} \Big\}_{n\in\mathbb{N}} \tag{2.13}$$

such that the generalized E-cohomology in degree n is equivalently non-abelian cohomology theory in the sense of Def. 2.1 with coefficients in E_n:

$$\underbrace{E^n(-)}_{\text{generalized cohomology}} \simeq H(-;E_n)\,. \tag{2.14}$$

Often one is interested in the special case that the representing spectrum carries the structure of an E_∞-ring (review in [Baker and Richter (2004)][Richter (2022)]), in which case $E^\bullet(-)$ is a *multiplicative cohomology theory* (e.g. [Kono and Tamaki (2006), §2.6]) where, in particular, the generalized cohomology groups (2.14) inherit ordinary ring-structure.

Example 2.11 (Topological K-theory). The classifying space (2.13) representing complex K-cohomology theory KU [Atiyah and Hirzebruch (1959), §2] (review in [Atiyah (1967)]) in degree 0 is [Atiyah and Hirzebruch (1961), §1.3]:

$$\mathrm{KU}_0 \simeq \mathbb{Z}\times B\mathrm{U}\,, \tag{2.15}$$

where

$$B\mathrm{U} := \varinjlim_n B\mathrm{U}(n) \tag{2.16}$$

is the classifying space (2.8) for the infinite unitary group (e.g. [Espinoza and Uribe (2014)]). Hence for the case of complex K-theory, Ex. 2.10 says that:

$$\underbrace{\mathrm{KU}^0(-)}_{\text{topological K-theory}} \simeq H(-;\mathbb{Z}\times B\mathrm{U})\,.$$

Example 2.12 (Iterated K-theory). Given an E_∞-ring spectrum R (Ex. 2.10), one may form its *algebraic K-theory spectrum* $K(R)$ [Elmendorf *et al.* (1997), §VI][Blumberg *et al.* (2013), §9.5][Lurie (2014)] and hence the corresponding generalized cohomology theory (Ex. 2.10). Much like complex topological K-theory (Ex. 2.11) is the K-theory of topological $\mathbb{C}$-module bundles, so $K(R)$-cohomology theory is the K-theory of R-module ∞-bundles [Lind (2016)]. Specifically, for $R = \mathrm{ku}$ the connective spectrum of topological K-theory, its algebraic K-theory $K(\mathrm{ku})$ [Ausoni (2010)][Ausoni and Rognes (2002)][Ausoni and Rognes (2012)] has been argued to be the K-theory of certain categorified complex vector bundles [Baas *et al.* (2004)] [Baas *et al.* (2011)].

Moreover, if R is connective, then $K(R)$ itself carries the structure of a connective E_∞-ring spectrum (by [Schwänzl and Vogt (1994), Thm. 1][Elmendorf *et al.* (1997), Thm. 6.1]), so that the construction may be iterated to yield *iterated algebraic K-theories* [Rognes (2014)] $K^{\circ_2}(R) := K(K(R))$, $K^{\circ_3}(R) := K(K(K(R)))$, et cetera.

For $R = \mathrm{ku}$, this generalizes the above "form of elliptic cohomology" $K(\mathrm{ku})$ to higher degrees [Lind *et al.* (2020)]. By Ex. 2.10, we will regard these (connective) iterated algebraic K-theories $K^{\circ_n}(\mathrm{ku})$ of the complex topological K-theory spectrum as examples of non-abelian cohomology theories (that happen to be abelian):

$$\overset{\text{iterated K-theory}}{K^{\circ_n}(\mathrm{ku})^0(-)} \;\simeq\; H\big(-; K^{\circ_n}(\mathrm{ku})_0\big).$$

Example 2.13 (Stable Cohomotopy). The generalized cohomology theory (Ex. 2.10) represented by the suspension spectra (Ex. 1.19) of n-spheres is called *stable Cohomotopy theory* (e.g. [Stretch (1981)][Nowak (2003)]) or *stable framed Cobordism theory*:

$$\mathbb{S}^n(-) \;=\; H\big(-; (\Sigma^\infty S^n)_0\big). \tag{2.17}$$

Non-abelian cohomology operations.

Definition 2.3 (Non-abelian cohomology operation). For $A_1, A_2 \in \mathrm{Ho}(\Delta\mathrm{Sets}_{\mathrm{Qu}})$ (Ex. 1.14), we say that a natural transformation in non-abelian cohomology (Def. 2.1) from A_1-cohomology theory to A_2-cohomology theory (2.4) is a (non-abelian) *cohomology operation*

$$\phi_* \;:\; H(-;A_1) \longrightarrow H(-;A_2). \tag{2.18}$$

By the Yoneda lemma, these are in bijective correspondence to morphisms of coefficients

$$A_1 \xrightarrow{\;\phi\;} A_2 \quad \in \mathrm{Ho}\big(\Delta\mathrm{Sets}_{\mathrm{Qu}}\big) \tag{2.19}$$

via the covariant functoriality of the hom-sets (2.2):

$$\phi_* \;=\; H(-;\phi) \;:=\; \mathrm{Ho}\big(\Delta\mathrm{Sets}_{\mathrm{Qu}}\big)(-;\phi). \tag{2.20}$$

Example 2.14 (Cohomology of coefficient spaces parametrizes cohomology operations). By the Yoneda lemma (2.20) in $\mathrm{Ho}\big(\Delta\mathrm{Sets}_{\mathrm{Qu}}\big)$ (Ex. 1.14), the set of all cohomology operations (Def. 2.3) from A_1-cohomology theory to A_2-cohomology theory (2.18) coincides with the non-abelian A_2-cohomology (Def. 2.1) *of* the coefficients A_1:

$$\underset{\substack{\text{non-abelian } A_2\text{-cohomology of } A_1 \\ \text{acting as cohomology operations}}}{H(A_1;A_2)} \times H(-;A_1) \xrightarrow{(-)\circ(-)} H(-;A_2) \tag{2.21}$$

acting by composition composition in $\mathrm{Ho}(\Delta\mathrm{Sets}_{\mathrm{Qu}})$.

Example 2.15 (Cohomology operations in ordinary cohomology). In specialization to Ex. 2.1 the non-abelian cohomology operations according to Def. 2.3 reduce to the classical cohomology operations in ordinary cohomology [Steenrod (1972)][Mosher and Tangora (1968)] (review in [May (1999), §22.5]), such as Steenrod operations [Steenrod (1947)][Steenrod (1962)] (review in [Kochman (1996), §2.5]). These operations admit refinements, involving rational/real form data, to differential cohomology operations [Grady and Sati (2018b)].

Example 2.16 (Cohomology operations in generalized cohomology). In specialization to Ex. 2.10, the non-abelian cohomology operations according to Def. 2.3 on a Whitehead-generalized cohomology theory $E^\bullet(-)$ regarded as a system of non-abelian cohomology theories $\{E^n(-)\}$ reduce to the traditional notion of *unstable* cohomology operations on generalized cohomology theories [Boardman *et al.* (1995)], such as the Adams operations in K-theory [Adams (1962)] (review in [Aguilar *et al.* (2002), §10]) or the Quillen operations in stable Cobordism theory (review in [Kochman (1996), §4,5]). For differential refinements see [Grady and Sati (2021b)].

Example 2.17 (Characteristic classes of principal ∞-bundles). For G a topological group, the ordinary group cohomology of G (Ex. 2.3) parametrizes, via Ex. 2.14, the cohomology operations from non-abelian cohomology classifying G-principal bundles (Ex. 2.2, 2.4, 2.6) to ordinary cohomology of the base space (Ex. 2.1):

$$\underset{\substack{\text{group}\\ \text{cohomology}}}{H^n_{\mathrm{Grp}}(G;A)} \times \underset{\substack{G\text{-principal}\\ \text{bundles}}}{H^1(-;G)} \xrightarrow[\text{(2.21)}]{\substack{\text{characteristic}\\ \text{classes}}} \underset{\substack{\text{ordinary}\\ \text{cohomology}}}{H^n(-;A)}\,. \tag{2.22}$$

This is the assignment of *characteristic classes* to principal bundles (principal ∞-bundles). In the case when $A = \mathbb{R}$, this is equivalently the *Chern-Weil homomorphism*, by Chern's fundamental theorem (see Rem. 8.1 and Thm. 8.6 below).

Example 2.18 (Rationalization cohomology operation). For fairly general non-abelian coefficients A (see Def. 5.2, Def. IV.1 for details), their *rationalization*[4] $A \xrightarrow{\eta^{\mathbb{R}}_A} L_{\mathbb{R}}A$ (Def. 5.2, 5.7 below) induces a cohomology operation (Def. 2.3) from non-abelian A-cohomology theory (Def. 2.1) to *non-abelian real cohomology* (Def. 5.14 below):

$$\underset{\substack{\text{non-abelian}\\ \text{cohomology}}}{H(-;A)} \xrightarrow[\text{rationalization}]{(\eta^{\mathbb{R}}_A)_*} \underset{\substack{\text{non-abelian}\\ \text{real cohomology}}}{H(-;L_{\mathbb{R}}A)}\,. \tag{2.23}$$

[4] To make the connection to differential cohomology, we consider rationalization over the *real* numbers; see Rem. 5.2 below.

Remark 2.2 (Rationalization as character map). Up to composition with an equivalence provided by the non-abelian de Rham theorem (Thm. 6.5 below), which serves to bring the right hand side of (2.23) into neat minimal form, this rationalization cohomology operation is the *character map in non-abelian cohomology* (Def. IV.2 below).

Example 2.19 (Stabilization cohomology operation). For $A \in \mathrm{Ho}(\Delta\mathrm{Sets}_{\mathrm{Qu}})$, the non-abelian cohomology operation (Def. 2.3) induced (2.20) by the unit of the derived stabilization adjunction (Ex. 1.19) goes from non-abelian A-cohomology theory (Def. 2.1) to (abelian) generalized cohomology theory (Ex. 2.10) represented by the 0th component space of the suspension spectrum of A:

$$\underset{\text{non-abelian}\atop A\text{-cohomology}}{H(-;A)} \xrightarrow[\text{stabilization}]{} \underset{\text{generalized}\atop \Sigma^\infty A\text{-cohomology}}{H\big(-;(\mathbb{D}\Sigma^\infty A)_0\big)}.$$

Hence a lift through this operation is an enhancement of generalized cohomology to non-abelian cohomology.

Example 2.20 (Non-abelian enhancement of stable Cohomotopy). The canonical non-abelian enhancement (in the sense of Ex. 2.19) of stable Cohomotopy (Ex. 2.13) is actual Cohomotopy theory (Ex. 2.7):

$$\underset{\text{Cohomotopy}}{\pi^n(-)} \xrightarrow[\text{stabilization}]{} \underset{\text{stable}\atop\text{Cohomotopy}}{\mathbb{S}^n(-)}.$$

Example 2.21 (Hurewicz homomorphism and Hopf degree theorem). By definition of Eilenberg-MacLane spaces (2.6) there is, for $n \in \mathbb{N}$, a canonical map

$$S^n \xrightarrow{e^{(n)}} K(\mathbb{Z},n) \qquad \in \mathrm{Ho}(\Delta\mathrm{Sets}_{\mathrm{Qu}}),$$

which represents the element $1 \in \mathbb{Z} \simeq \pi_n\big(K(\mathbb{Z},n)\big)$. The non-abelian cohomology operation (Def. 2.3) induced by this, from degree n Cohomotopy (Ex. 2.7) to degree n ordinary cohomology (Ex. 2.1)

$$\pi^n(-) \xrightarrow{e_*^{(n)}} H^n(-;\mathbb{Z})$$

is the cohomological version of the Hurewicz homomorphism. The *Hopf degree theorem* (e.g. [Kosinski (1993), §IX (5.8)]) is the statement that the non-abelian cohomology operation $e_*^{(n)}$ becomes an isomorphism on connected, orientable closed manifolds of dimension n. These maps, together with their differential refinements, are analyzed in more detail via Postnikov towers in [Grady and Sati (2021a)].

Structured non-abelian cohomology.

Remark 2.3 (Structured non-abelian cohomology). More generally, it makes sense to consider the analog of Def. 2.1 for the homotopy category $\mathrm{Ho}(\mathbf{H})$ of a model category which is a *homotopy topos* [Toën and Vezzosi (2005)][Lurie (2009a)][Rezk (2010)].

(i) This yields *structured* non-abelian cohomology [Simpson (1997b)][Simpson (2002)][Toën (2002)][Sati *et al.* (2012)][Nikolaus *et al.* (2015a)][Nikolaus *et al.* (2015b)][Schreiber (2013)][Fiorenza *et al.* (2020b)][Sati and Schreiber (2020c)]:

$$\underset{\substack{\text{structured}\\ \text{non-abelian cohomology}}}{H(\mathscr{X};\mathbf{A})} := \underset{\text{homotopy topos}}{\mathrm{Ho}(\mathbf{H})}\big(\underset{\infty\text{-stacks}}{\mathscr{X},\mathbf{A}}\big),$$

including the stacky non-abelian cohomology originally considered in [Giraud (1971)][Breen (1990)] ("gerbes", see [Nikolaus *et al.* (2015a), §4.4]), and, more generally, *differential-, étale-, and equivariant-* nonabelian cohomology theories (see [Sati and Schreiber (2020c), p. 6]) based on ∞-stacks.

(ii) In good cases (cohesive homotopy toposes [Schreiber (2013)][Sati and Schreiber (2020c), §3.1]), the homotopy topos $\mathrm{Ho}(\mathbf{H})$ comes equipped with a *shape* operation down to the classical homotopy category (Ex. 1.14):

$$\begin{array}{ccc} \overset{\text{homotopy topos}}{\mathrm{Ho}(\mathbf{H})} & \xrightarrow{\quad\mathrm{Shp}\quad} & \overset{\text{classical homotopy category}}{\mathrm{Ho}\big(\Delta\mathrm{Sets}_{\mathrm{Qu}}\big)} \\ \underset{\substack{\text{structured}\\ \text{non-abelian cohomology}}}{H(\mathscr{X};\mathbf{A})} & \longmapsto & \underset{\substack{\text{plain}\\ \text{non-abelian cohomology}}}{H\big(\mathrm{Shp}(\mathscr{X});\mathrm{Shp}(\mathbf{A})\big)} \end{array} \tag{2.24}$$

which takes, for well-behaved group ∞-stacks G, the *classifying stacks* $\mathbf{B}G$ of G-principal bundles to the traditional classifying spaces $BG \simeq \mathrm{Shp}(\mathbf{B}G)$ of underlying topological groups (2.8). This gives a forgetful functor from structured non-abelian cohomology to plain non-abelian cohomology in the sense of Def. 2.1. A classical example is the map from non-abelian Čech cohomology with coefficients in a well-behaved group G to homotopy classes of maps to the classifying space of G, in which case this comparison map is a bijection (Ex. 2.2).

(iii) All constructions on non-abelian cohomology have their structured analogues, for instance non-abelian cohomology operations (Def. 2.3) in structured cohomology

$$H(\mathscr{X};\mathbf{A}_1) \xrightarrow{\ \phi_*\ } H(\mathscr{X};\mathbf{A}_2) \tag{2.25}$$

are induced by postcomposition with morphisms $\mathbf{A}_1 \xrightarrow{\phi} \mathbf{A}_2$ of coefficient stacks.

Ultimately, one is interested in working with structured non-abelian cohomology on the left of (2.24). However, since this is rich and intricate, it behooves us to study its projection into plain non-abelian cohomology on the right of (2.24). This is what we are mainly concerned with here. But we provide in Chapter 9 a brief discussion of non-abelian differential cohomology on smooth ∞-stacks.

Chapter 3

Twisted non-abelian cohomology

For $\mathscr{C}$ any category and $B \in \mathscr{C}$ any object, there is the *slice category* $\mathscr{C}^{/X}$, whose objects are morphisms in $\mathscr{C}$ to X and whose morphisms are commuting triangles over X in $\mathscr{C}$. Basic as this is, hom-sets in the *homotopy category* $\mathrm{Ho}(\mathbf{C}^{/B})$ (Def. 1.8) of a slice model category $\mathbf{C}^{/B}$ (Ex. 1.5) are of paramount interest:

The slicing imposes *twisting* on the corresponding non-abelian cohomology (Def. 2.1), in that the slicing of the domain space serves as a twist, the slicing of the coefficient space as a local coefficient bundle, and the slice morphisms as twisted cocycles.

Proposition 3.1 (∞-Actions on homotopy types [Dror *et al.* (1980)][Prezma (2012), §5][Nikolaus *et al.* (2015a), §4][Sharma (2019)][Sati and Schreiber (2020c), §2.2]**).** *For any $A \in \mathrm{Ho}\big(\Delta\mathrm{Sets}_{\mathrm{Qu}}\big)$ (Ex. 1.14) and G a topological group, homotopy-coherent actions of G on A are equivalent to fibrations ρ with homotopy fiber A (Def. 1.14) over the classifying space BG* (2.8)

$$\begin{array}{ccc} A & \longrightarrow & A /\!\!/ G \\ & & \big\downarrow{\scriptstyle \rho} \\ & & BG\,. \end{array} \tag{3.1}$$

Here

$$A /\!\!/ G \;\simeq\; \big(A \times EG\big)_{/\mathrm{diag}G}$$

is the homotopy quotient (Borel construction) of the action.

Definition 3.2 (Twisted non-abelian cohomology [Nikolaus *et al.* (2015a), §4][Fiorenza *et al.* (2020b), (10)][Sati and Schreiber (2020c), Rem. 2.94]**).** For $X, A \in \mathrm{Ho}\big(\Delta\mathrm{Sets}_{\mathrm{Qu}}\big)$ (Def. 1.14) we say:

(i) A *local coefficient bundle* for twisted A-cohomology is an A-fibration ρ over a classifying space BG (2.8) as in Prop. 3.1:

$$\begin{array}{ccc} A & \longrightarrow & A /\!\!/ G \\ & {\scriptstyle \text{local coefficient bundle}} & \big\downarrow{\scriptstyle \rho} \\ & & BG\,. \end{array} \tag{3.2}$$

(ii) A *twist* for non-abelian A-cohomology theory on X with local coefficient bundle ρ over BG is a map

$$X \xrightarrow{\ \tau\ } BG \quad \in \mathrm{Ho}\big(\Delta\mathrm{Sets}_{\mathrm{Qu}}\big). \tag{3.3}$$

(iii) The *non-abelian τ-twisted A-cohomology* of X with local coefficients ρ is the hom-set from τ (3.3) to ρ (3.1)

$$\underset{\text{twisted non-abelian cohomology}}{H^{\tau}(X;A)} := \mathrm{Ho}\Big(\Delta\mathrm{Sets}_{\mathrm{Qu}}^{/BG}\Big)(\tau,\rho) = \left\{ \begin{array}{ccc} X & \overset{\text{cocycle } c}{\dashrightarrow} & A /\!/ G \\ & \Leftarrow\!\simeq & \\ \text{twist } \tau \searrow & & \swarrow \rho \text{ local coefficients} \\ & BG & \end{array} \right\}_{/\text{homotopy relative } BG} \tag{3.4}$$

in the homotopy category (Def. 1.8) of the slice model category over BG (Ex. 1.5) of the classical model category on topological spaces (Ex. 1.1).

Definition 3.3 (Associated coefficient bundle [Nikolaus *et al.* (2015a), §4.1][Sati and Schreiber (2020c), Prop. 2.92]**).** Given a local coefficient A-fiber bundle ρ (3.2) and a twist τ (3.3) on a domain space X, the corresponding *associated A-fiber bundle* over X is the homotopy pullback (Def. 1.15) of ρ along τ, sitting in a homotopy pullback square (1.42) of this form:

$$\begin{array}{ccc} \underset{\text{associated }A\text{-fiber bundle}}{E} & \longrightarrow & \underset{\text{local coefficient bundle}}{A /\!/ G} \\ {\scriptstyle \mathbb{R}\tau^*\rho}\downarrow & \text{(hpb) homotopy pullback} & \downarrow{\scriptstyle \rho} \\ X & \underset{\text{twist}}{\xrightarrow{\ \tau\ }} & BG \end{array} \tag{3.5}$$

We write

$$\underset{\text{sections of associated bundle}}{\Gamma_X(E)_{/\sim}} := \mathrm{Ho}\Big(\mathrm{TopSp}_{\mathrm{Qu}}^{/X}\Big)\big(\mathrm{id}_X, \mathbb{R}\tau^*\rho\big) = \left\{ \begin{array}{ccc} & \overset{\text{section } \sigma}{\nearrow} & E \ \text{associated bundle} \\ & & \downarrow \\ X & = & X \end{array} \right\}_{/\text{vertical homotopy}} \tag{3.6}$$

for the set of vertical homotopy classes of section of the associated bundle, hence for the hom-set, from the identity on X to the associated bundle projection, in the homotopy category (Def. 1.8) of the slice model category over X (Ex. 1.5) of the classical model category on topological spaces (Ex. 1.1).

Proposition 3.4 (Twisted non-abelian cohomology is sections of associated coefficient bundle [Nikolaus *et al.* (2015a), Prop. 4.17]**).** *Given a local coefficient bundle* ρ (3.2) *and a twist* τ (3.3), *the τ-twisted non-abelian cohomology (Def. 3.2) with local coefficient in ρ is equivalent to the vertical homotopy classes of sections* (3.6) *of the associated coefficient bundle E (Def. 3.3):*

$$\underset{\text{twisted non-abelian cohomology}}{H^{\tau}(X;A)} \simeq \underset{\text{sections of associated bundle}}{\Gamma_X(E)_{/\sim}}. \tag{3.7}$$

Proof. Consider the following sequence of bijections:

$$\begin{aligned} H^{\tau}(X;A) &= \mathrm{Ho}\Big(\mathrm{TopSp}_{\mathrm{Qu}}^{/BG}\Big)(\tau\,,\rho) \\ &\simeq \mathrm{Ho}\Big(\mathrm{TopSp}_{\mathrm{Qu}}^{/BG}\Big)\big(\mathbb{D}\tau_{*}\mathrm{id}_X\,,\rho\big) \\ &\simeq \mathrm{Ho}\Big(\mathrm{TopSp}_{\mathrm{Qu}}^{/X}\Big)\big(\mathrm{id}_X\,,\mathbb{D}\tau^{*}\rho\big) \\ &= \Gamma_X(E)_{/\sim}\,. \end{aligned}$$

Here the first line is the definition (3.4). Then the first step is the observation that every slice object is the derived left base change (Ex. 1.7, Prop. 1.13) along itself of the identity on its domain, by (1.28). With this, the second step is the hom-isomorphism (1.2) of the derived base change adjunction $\mathbb{D}\tau_! \dashv \mathbb{R}\tau^*$. The last line is (3.6). □

In twisted generalization of Ex. 2.1 we have:

Example 3.1 (Twisted ordinary cohomology with local coefficients). Let $n \in \mathbb{N}$, let $X \in \mathrm{Ho}\big(\Delta\mathrm{Sets}_{\mathrm{Qu}}\big)$ (Ex. 1.14) be connected and consider a traditional *system of local coefficients* [Steenrod (1943), §3] (see also [May (1977a)][Ando *et al.* (2010)][Grady and Sati (2018c)])

$$\Pi_1(X) \xrightarrow{\;\;t\;\;} \mathrm{AbGrps}\,,$$

namely, a functor from the fundamental groupoid of X to the category of abelian groups. Since the construction $A \mapsto K(A,n)$ of Eilenberg-MacLane spaces (2.6) is itself functorial and using the assumption that X is connected, this induces (see [Bullejos *et al.* (2003), Def. 3.1]) a local coefficient bundle (3.2) of the form

$$\begin{array}{ccc} K(A,n) & \longrightarrow & K(A,n)\,/\!\!/\,\pi_1(X)\,. \\ & & \big\downarrow \rho_t \\ & & B\pi_1(X) \end{array} \tag{3.8}$$

Finally, write $X \xrightarrow{\tau} B\pi_1(X)$ for the classifying map (via Ex. 2.2) of the universal connected cover of X (equivalently: for the 1-truncation projection of X). Then the τ-twisted non-abelian cohomology (Def. 3.2) of X with local coefficients in ρ_t (3.8) is equivalently *t-twisted ordinary cohomology*, traditionally known as ordinary cohomology *with local coefficients t*:

$$\overset{\substack{\text{twisted}\\ \text{ordinary cohomology}}}{H^{n+t}(X;A)} \;\simeq\; H^{\tau}\big(X;K(A,n)\big)\,.$$

This is manifest from comparing Def. 3.2 with the characterization of cohomology with local coefficients found in [Hirashima (1979), Cor. 1.3][Goerss and Jardine (1999), p. 332][Bullejos *et al.* (2003), Lem. 4.2].

Example 3.2 (Classification of tangential structure). Let X be a smooth manifold of dimension n. Its *frame bundle* is an $\mathrm{O}(n)$-principal bundle $\mathrm{Fr}(X) \to X$, whose class

(a diffeomorphism invariant of X)

$$\begin{array}{ccc} \mathrm{O}(n)\mathrm{Bundles}(X)_{/\sim} & \xrightarrow{\ \simeq\ } & H\big(X;\, B\mathrm{O}(n)\big) \\ \big[\mathrm{Fr}(X)\big] & \leftrightarrow & \big[\tau_{\mathrm{fr}}\big] \end{array} \tag{3.9}$$

gives, by Ex. 2.2, the class of a twist τ_{fr} (3.3) in the non-abelian $\mathrm{O}(n)$-cohomology of X.

Now for BG any connected homotopy type (Prop. 2.2) and for $BG \xrightarrow{\rho} B\mathrm{O}(n)$ any map (equivalently the delooping of a morphism of ∞-groups $G \longrightarrow \mathrm{O}(n)$), we get a local coefficient bundle (3.2) with (homotopy-)coset space fiber [Fiorenza *et al.* (2020b), Lem. 2.7]:

$$\begin{array}{ccc} \mathrm{O}(n)/\!\!/G & \xrightarrow{\ \mathrm{hofib}(\rho)\ } & BG \\ & & \big\downarrow{\scriptstyle \rho} \\ & & B\mathrm{O}(n)\,. \end{array} \tag{3.10}$$

The relative homotopy class of a homotopy lift of the frame bundle classifier τ_{fr} (3.2) through this map ρ

$$\left[\begin{array}{ccc} X & \overset{\text{tangential structure}}{\dashrightarrow} & BG \\ & \Swarrow g & \\ {\scriptstyle \tau_{\mathrm{fr}}}\searrow & & \swarrow{\scriptstyle \rho} \\ & B\mathrm{O}(n) & \end{array} \right] \in G\mathrm{TangentialStructures}(X) \tag{3.11}$$

is known a *topological G-structure* or *tangential ρ-structure* on X (e.g. [Kochman (1996), §1.4][Galatius *et al.* (2009), §5][Sati and Schreiber (2020c), Def. 4.48]). For instance, for ρ a stage in the Whitehead tower of $\mathrm{O}(n)$, this is, in turn, *Orientation*, *Spin structure*, *String structure*, *Fivebrane structure* [Sati *et al.* (2012)] and higher structures (see [Sati (2015)][Sati and Wheeler (2018)]):

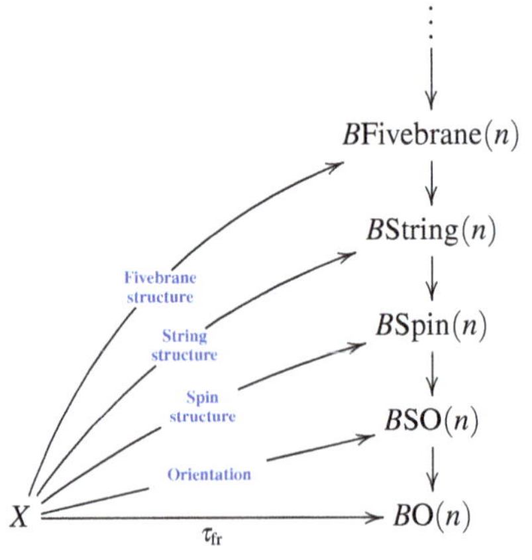

By comparison of (3.11) with (3.4) we see that tangential G-structures on X are classified by twisted non-abelian cohomology (Def. 3.2) with coefficients in (homotopy-)coset spaces $\mathrm{O}(n) /\!\!/ G$ (3.10) and twisted by the class τ_{fr} of the frame bundle (3.9):

$$G\mathrm{TangentialStructures}(X) \;\simeq\; H^{\tau_{\mathrm{fr}}}\big(X; \mathrm{O}(n) /\!\!/ G\big)\,. \tag{3.12}$$

According to Prop. 2.2, this example is actually universal for τ_{fr}-twisted non-abelian cohomology.

As a special case of Ex. 3.1 and in twisted generalization of Ex. 2.8, 2.9 we have:

Example 3.3 (Orientifold gerbes). Consider the action $\sigma_{\mathrm{U}(1)}$ of $\mathbb{Z}_2$ on the circle group $\mathrm{U}(1) \subset \mathbb{C}^\times$ given by complex conjugation. This deloops (see [Fiorenza *et al.* (2015a), §4.4]) to an action $\sigma_{B^n\mathrm{U}(1)}$ of $\mathbb{Z}_2$ on the classifying spaces $B^n\mathrm{U}(1)$ (2.8). By Prop. 3.1 there is a corresponding local coefficient bundle

$$\begin{array}{ccc} B^n\mathrm{U}(1) & \longrightarrow & B^n\mathrm{U}(1) /\!\!/ \mathbb{Z}_2 \\ & & \big\downarrow{\scriptstyle \sigma_{B^n\mathrm{U}(1)}} \\ & & B\mathbb{Z}_2 \end{array} \tag{3.13}$$

Moreover, consider a smooth manifold X, with orientation bundle classified by $X \xrightarrow{\mathrm{or}} B\mathbb{Z}_2$. Then the or-twisted cohomology (Def. 3.2) of X...

(i) ...with local coefficients in $\sigma_{B^2\mathrm{U}(1)}$ classifies what is equivalently known as Jandl gerbes [Schreiber *et al.* (2007)][Gawędzki *et al.* (2011)] or real gerbes [Hekmati *et al.* (2019)] or orientifold B-fields;
(ii) ...with local coefficients in $\sigma_{B^3\mathrm{U}(1)}$ classifies what is equivalently known as topological sectors of orientifold C-fields [Fiorenza *et al.* (2015a), §4.4].
More generally, one can consider twisted Deligne cohomology [Grady and Sati (2018c)] as well as higher-twisted periodic integral- and Deligne-cohomology [Grady and Sati (2019a)] (see also Chapter 9).

Remark 3.1 (The Whitehead principle of non-abelian cohomology). Let $A \in \mathrm{Ho}\big(\Delta\mathrm{Sets}_{\mathrm{Qu}}\big)$ be connected, so that $A \simeq BG$ (Prop. 2.2).

(i) If A is also n-truncated (1.56), then its Postnikov tower (Prop. 1.21) says that A is the total space of a local coefficient bundle (3.2) of the form

$$\begin{array}{ccc} K(\pi_n(A), n) & \xrightarrow{\;\mathrm{hfib}(p_n)\;} & A \\ & & \big\downarrow{\scriptstyle p_n^A} \\ & & A(n-1) \simeq B\big(G(n-2)\big) \end{array}$$

with homotopy fiber an Eilenberg-MacLane space (2.6).
(ii) Accordingly, non-abelian cohomology with coefficients in A (Def. 2.1) is equivalently the disjoint union, over the space of twists τ_n (3.3) in non-abelian cohohomology with

coefficients in $A(n-1)$, of τ-twisted non-abelian cohomology (Def. 3.2) with coefficients in $K(\pi_n(A),n)$:

$$\overset{\substack{\text{non-abelian cohomology}\\\text{in higher degree}}}{H(X;A)} \;\simeq\; \underset{\substack{\tau_n \in H(X;A(n-1))\\ \text{twist in}\\ \text{non-abelian cohomology}\\ \text{of lower degree}}}{\coprod} \overset{\substack{\text{higher twisted}\\\text{ordinary cohomology}}}{H^{\tau_n}\big(X;\, K(\pi_n(A),n)\big)}\,. \tag{3.14}$$

(iii) But notice that this is just the first step, and that iterating this unravelling yields unwieldy formulas:

$$\begin{aligned}
&\text{first } H(X;A) \simeq \coprod_{\tau_n \in \coprod\limits_{\tau_{n-1}\in H(X;A(n-2))} H^{\tau_{n-1}}\left(X;K(\pi_{n-1}(A),n-1)\right)} H^{\tau_n}\big(X;\,K(\pi_n(A),n)\big),\\
&\text{then } H(X;A) \simeq \coprod_{\tau_n \in \coprod\limits_{\tau_{n-1}\in \coprod\limits_{\tau_{n-2}\in H\left(X;A(n-3)\right)} H^{\tau_{n-2}}\left(X;K(\pi_{n-2}(A),n-2)\right)} H^{\tau_{n-1}}\left(X;K(\pi_{n-1}(A),n-1)\right)} H^{\tau_n}\big(X;\,K(\pi_n(A),n)\big)\\
&\text{etc.}
\end{aligned} \tag{3.15}$$

(iv) Thus, non-abelian cohomology in higher degrees (Ex. 2.6) decomposes as a tower of consecutively higher twisted but otherwise ordinary cohomology theories, starting with a twist in non-abelian cohomology in degree 1. This phenomenon has been called the *Whitehead principle of non-abelian cohomology* [Toën (2002), p. 8] and has been interpreted as saying that "nonabelian cohomology occurs essentially only in degree 1" [Simpson (1997a), p. 1].

(v) But the above formulas (3.14), (3.15), make manifest that there are two perspectives on this phenomenon. On the one hand: non-abelian cohomology in higher degrees may be *computed* by brute force as a sequence of consecutively higher twisted abelian cohomologies, with lowest twist starting in degree-1 non-abelian cohomology. On the other hand, conversely: intricate such systems of consecutively twisted abelian cohomology theories are neatly *understood* as unified by non-abelian cohomology.

(vi) Similarly, even though Postnikov towers do exist (Prop. 1.21) in the classical homotopy category (Ex. 1.14), the latter is far from being equivalent to the stable homotopy category (1.64) "up to twists in degree 1".

In twisted generalization of Ex. 2.11, we have:

Example 3.4 (Twisted topological K-theory). The classifying space $\mathrm{KU}_0 \simeq \mathbb{Z} \times B\mathrm{U}$ (2.15) for complex topological K-theory (Ex. 2.11) is the fiber of a local coefficient bundle (3.2) over $K(\mathbb{Z},3) \simeq B^3\mathrm{U}(1)$ (2.12):

$$\begin{array}{ccc}
\mathrm{KU}_0 & \longrightarrow & \mathrm{KU}_0 /\!\!/ B\mathrm{U}(1) \\
 & & \downarrow \\
 & & B^2\mathrm{U}(1)
\end{array} \tag{3.16}$$

For $X \xrightarrow{\tau} B^2\mathrm{U}(1)$ a corresponding twist (3.3) (hence equivalently a bundle gerbe, by Ex. 2.8), the corresponding twisted non-abelian cohomology (Def. 3.2) is twisted complex topological K-theory [Karoubi (1968)][Donovan and Karoubi (1970)]:

$$\underset{\text{twisted topological K-theory}}{\mathrm{KU}^{\tau}(-)} \simeq H^{\tau}\big(-;\mathbb{Z}\times B\mathrm{U}\big). \tag{3.17}$$

This is manifest from comparing (3.4) with [Freed *et al.* (2008), (2.6)]. Alternatively, under Prop. 3.4, this is manifest from comparing the equivalent right hand side of (3.7) with [Rosenberg (1989), Prop. 2.1] (using [Nikolaus *et al.* (2015a), Cor. 4.18]) or, more directly, with [Atiyah and Segal (2004), §3][Ando *et al.* (2010), §2.1].

Generally, in twisted generalization of Ex. 2.10, we have:

Example 3.5 (Local coefficient bundle for twisted Whitehead-generalized cohomology). Let R be an E_∞-ring spectrum (Ex. 2.10) and write $\mathrm{GL}(1,R)$ for its *∞-group of units* [Schlichtkrull (2004), §2.3][May and Sigurdsson (2006), §22.2][Ando *et al.* (2008), §3][Ando *et al.* (2014b), §2], defined as the homotopy pullback (Def. 1.15) of the component space $R_0 = \mathbb{D}\Omega^\infty R$ (1.65) fibered over its 0-truncation (i.e. its 1-coskeleton (1.54)) to the ordinary group of units of this ordinary ring of connected components:

$$\begin{array}{ccc} \underset{\infty\text{-group of units}}{\mathrm{GL}(1,R)} & \longrightarrow & \underset{E_\infty\text{-ring space}}{R_0} \\ \downarrow & \text{(hpb)} & \downarrow{\scriptstyle p_0} \\ \underset{\text{ordinary group of units}}{\mathrm{GL}(1,\pi_0(R_0))} & \hookrightarrow & \underset{\text{ordinary ring of connected components}}{\pi_0(R_0)} \end{array} \tag{3.18}$$

This makes $\mathrm{GL}(1,R)$ an ∞-group (as in Ex. 2.6) with group operation induced from the *multiplicative* structure on R_0. The canonical action of $\mathrm{GL}(1,R)$ on R_0 is given, via Prop. 3.1, by a local coefficient bundle (3.2) of this form:

$$\begin{array}{ccc} R_0 & \longrightarrow & (R_0)/\!\!/\mathrm{GL}(1,R) \\ & & \downarrow{\scriptstyle \rho_R} \\ & & B\mathrm{GL}(1,R). \end{array} \tag{3.19}$$

Proposition 3.5 (Twisted non-abelian cohomology subsumes twisted generalized cohomology). *For R an E_∞-ring spectrum (Ex. 2.10), the twisted non-abelian cohomology (Def. 3.2) with local coefficient bundle ρ_R from Ex. 3.5 is, equivalently, twisted generalized R-cohomology in the traditional sense (e.g. [May and Sigurdsson (2006), §22.1]):*

$$\underset{\text{twisted Whitehead-generalized cohomology}}{R^{\tau}(-)} \simeq \underset{\text{twisted non-abelian cohomology with local } \rho_R\text{-coefficients}}{H^{\tau}(-;\rho_R)}. \tag{3.20}$$

Proof. Given any twist $X \xrightarrow{\tau} B\mathrm{GL}(1,R)$ (3.2), write $P \to X$ for the homotopy pullback (Def. 1.15) along τ of the essentially unique point inclusion:

$$\begin{array}{ccc} P & \longrightarrow & * \\ \downarrow & \text{(hpb)} & \downarrow \\ X & \xrightarrow[\tau]{} & B\mathrm{GL}(1,R) \end{array} \qquad \begin{array}{ccc} (P \times R_0) /\!\!/_{\mathrm{diag}} \mathrm{GL}(1,R) \simeq E & \longrightarrow & R_0 /\!\!/ \mathrm{GL}(1,R) \\ {\scriptstyle \mathbb{R}\tau^* \rho_R}\downarrow & \text{(hpb)} & \downarrow {\scriptstyle \rho_R} \\ X & \xrightarrow[\tau]{} & B\mathrm{GL}(1,R) \end{array} \tag{3.21}$$

This P is the $\mathrm{GL}(1,R)$-principal ∞-bundle which is classified by τ, [Nikolaus *et al.* (2015a), Thm. 3.17], to which the coefficient bundle E (3.5) is $\mathrm{GL}(1,R)$-associated [Nikolaus *et al.* (2015a), Prop. 4.6], as shown on the right of (3.21). Consider then the following sequence of natural bijections:

$$\begin{aligned} H^\tau(X; R_0) &\simeq \Gamma_X(E) \\ &\simeq \mathrm{Ho}\big(\mathrm{GL}(1,R)\,\mathrm{Actions}\big)(P; R_0) \\ &\simeq \mathrm{Ho}\big(R\mathrm{Modules}\big)(M\tau; R) \\ &\simeq R^\tau(X). \end{aligned} \tag{3.22}$$

Here the first step is Prop. 3.4, while the second step is [Nikolaus *et al.* (2015a), Cor. 4.18]. The third step is [Ando *et al.* (2008), (2.15)][Ando *et al.* (2014b), (3.15)], with $M\tau$ denoting the R-Thom spectrum of τ [Ando *et al.* (2008), Def. 2.6][Ando *et al.* (2014b), Def. 3.13]. The last step is [Ando *et al.* (2008), §2.5] [Ando *et al.* (2014b), §1.4][Ando *et al.* (2014a), §2.7]. The composite of these natural bijections is the desired (3.20). □

Example 3.6 (Higher Cohomotopy-twisted K-theory). For complex topological K-theory $R = \mathrm{KU}$ (Ex. 2.11) with $\mathrm{KU}_0 = \mathbb{Z} \times B\mathrm{U}$ (2.15) – where the $\mathbb{Z}$-factor encodes the virtual rank of vector bundles and the multiplicative operation in the ring structure corresponds to tensor product of vector bundles – the ∞-group of units (3.18) classifies the virtual vector bundles of invertible rank in $\{\pm 1\} = \mathrm{GL}(1,\mathbb{Z}) \subset \mathbb{Z}$:

$$\mathrm{GL}(1,\mathrm{KU}) \;\simeq\; \big(\{\pm 1\} \times B\mathrm{U}\big)_{\otimes}. \tag{3.23}$$

(Here the subscript just indicates the ∞-group structure, now with respect to the multiplicative operation corresponding to tensor product of virtual vector bundles.) Since delooping $B(-)$ shifts up homotopy groups by one, it follows that the homotopy groups of $B\mathrm{GL}(1,\mathrm{KU})$, appearing in (3.19), are freely generated by the powers of the Bott generator

$\beta \in \pi_2(\mathrm{KU})$, shifted up in degree by one:

$$\pi_\bullet(\mathrm{KU}_0) \simeq \begin{cases} \mathbb{Z} \simeq \langle \beta^k \rangle & | \; n \text{ even} \\ 0 & | \; n \text{ odd} \end{cases}$$

$$\Rightarrow \pi_n(\mathrm{GL}(1,\mathrm{KU})) \simeq \begin{cases} \mathbb{Z}_2 & | \quad n=1 \\ \mathbb{Z} \simeq \langle \beta^{2k} \rangle & | \quad n = 2k+3 \\ 0 & \quad n \text{ even} \end{cases} \tag{3.24}$$

It follows that, parameterized by any odd-dimensional sphere S^{2k+1} for positive $k \in \mathbb{N}_+$, there are exactly $\mathbb{Z}$ worth of higher twists of complex K-theory, up to equivalence, embodied by the local coefficient bundles which are the homotopy pullback of (3.19) along the classifying maps of the elements (3.24). The universal one among these is the pullback along the classifying map for the suspended power of the Bott generator itself:

$$\begin{array}{ccccc} \mathrm{KU}_0 & \longrightarrow & \overset{\text{local coefficient bundle for Cohomotopy-twisted K-theory}}{(\mathrm{KU}_0)/\!\!/\Omega S^{2k+1}} & \longrightarrow & \overset{\text{universal local coefficient bundle for twisted complex K-theory}}{(\mathrm{KU}_0)/\!\!/\mathrm{GL}(1,\mathrm{KU})} \\ & & \Big\downarrow \rho_{\mathrm{KU}}^{S^{2k+1}} & \text{(hpb)} & \Big\downarrow \rho_{\mathrm{KU}} \\ & & S^{2k+1} & \xrightarrow[\Sigma(\beta^{2k})]{\text{shifted power of Bott generator}} & B\mathrm{GL}(1,\mathrm{KU}) \end{array} \tag{3.25}$$

By Def. 3.2, these local coefficient bundles encode higher twists of complex K-theory (Ex. 2.11) by classes in unstable/non-abelian Cohomotopy (Ex. 2.7) in degree $2k+1$:

$$\underset{\text{twist in Cohomotopy}}{[\lambda] \in \pi^{2k+1}(X)} \quad \vdash \quad \underset{\text{cohomotopically-twisted topological K-theory}}{\mathrm{KU}^\lambda(X) = H^\lambda(X; \mathrm{KU}_0)}. \tag{3.26}$$

This cohomotopically higher twisted K-theory has been considered in [Macdonald *et al.* (2021), Def. 2.5].

In twisted generalization of Ex. 2.12, we have:

Example 3.7 (Twisted iterated K-theory). Let $r \in \mathbb{N}$, $r \geq 1$. By [Lind *et al.* (2020), Prop. 1.5, Def. 1.7] and using Prop. 3.5, there is a local coefficient bundle (3.2) of the form

$$\begin{array}{ccc} \big(K^{2r-2}(\mathrm{ku})\big)_0 & \longrightarrow & \Big(\big(K^{2r-2}(\mathrm{ku})\big)_0\Big)/\!\!/B^{2r-1}\mathrm{U}(1) \\ & & \Big\downarrow \rho_{\mathrm{lsw}_{2r-1}} \\ & & B^{2r}\mathrm{U}(1), \end{array} \tag{3.27}$$

where $K^{2r-2}(\mathrm{ku})_0$ is the 0th space in the spectrum (2.13) representing iterated K-theory (Ex. 2.12) and $B^{2r}\mathrm{U}(1) \simeq K(\mathbb{Z}, 2r+1)$ is the classifying space for bundle $(2r-1)$-gerbes (Ex. 2.9). This means that for $X \xrightarrow{\tau} B^{2r}\mathrm{U}(1)$ a classifying map for such a higher gerbe, the τ-twisted non-abelian cohomology (Def. 3.2) with local coefficients in (3.27) is equivalently (still by Prop. 3.5) integrally twisted iterated K-theory according to [Lind *et al.* (2020)]:

$$\overset{\text{twisted iterated K-theory}}{\big(K^{\circ_{2r-1}}(\mathrm{ku})\big)^{\tau}(-)} \;\simeq\; H^{\tau}\Big(-;K^{\circ_{2r-2}}(\mathrm{ku})_0\Big)\,.$$

In twisted generalization of Ex. 2.7, we have:

Example 3.8 (Twisted Cohomotopy theory [Fiorenza *et al.* (2020b), §2.1]**).** For $n \in \mathbb{N}$, consider the canonical action of the orthogonal group $\mathrm{O}(n+1)$ on the homotopy type of the n-sphere, via the defining action on the unit sphere in $\mathbb{R}^{n+1}$, which restricts along the canonical inclusion $\mathrm{O}(n) \hookrightarrow \mathrm{O}(n+1)$ to the defining action of $\mathrm{O}(n)$ on the one-point compactification $(\mathbb{R}^n)^{\mathrm{cpt}} = S^n$. By Prop. 3.1, this corresponds to local coefficient bundles (3.2) for twisting Cohomotopy theory (Ex. 2.7):

$$\begin{array}{ccccc} S^n & \longrightarrow & S^n /\!\!/ \mathrm{O}(n) & \longrightarrow & S^n /\!\!/ \mathrm{O}(n+1) \\ & & {\scriptstyle \rho_J}\big\downarrow & \text{(hpb)} & \big\downarrow \\ & & B\mathrm{O}(n) & \longrightarrow & B\mathrm{O}(n+1) \end{array} \tag{3.28}$$

The classifying map $B\mathrm{O}(n) \xrightarrow{J} B\mathrm{Aut}(S^n)$ of ρ_J is the unstable *J-homomorphism* (e.g. [Kono and Tamaki (2006), §4.4]). For X a smooth manifold of dimension $d \geq n+1$, and equipped with tangential $\mathrm{O}(n+1)$-structure (e.g. [Sati and Schreiber (2020c), Def. 4.48])

$$\begin{array}{ccc} X & \xrightarrow{\quad\tau\quad} & B\mathrm{O}(n+1) \\ & {\scriptstyle TX}\searrow \;\; \Leftarrow_{\simeq} \;\; \swarrow{\scriptstyle Bi} & \\ & B\mathrm{O}(d) & \end{array}$$

the τ-twisted non-abelian Cohomology (Def. 3.2) with local coefficients in (3.28) is the *tangentially twisted Cohomotopy theory* of [Fiorenza *et al.* (2020b)][Fiorenza *et al.* (2021b)][Sati and Schreiber (2021b)]:

$$\overset{\text{tangentially twisted Cohomotopy}}{\pi^{\tau}(-)} \;:=\; H^{\tau}\big(-;S^n\big)\,.$$

This twisted Cohomotopy theory in degree $n = 4$ encodes, in particular, the shifted flux quantization condition of the C-field [Fiorenza *et al.* (2020b), Prop. 3.13] and the vanishing of the residual M5-brane anomaly [Sati and Schreiber (2021b)]; while tangentially twisted Cohomotopy in degree $n = 7$ encodes, in particular, level quantization of the Hopf-Wess-Zumino term on the M5-brane [Fiorenza *et al.* (2021b)].

Twisted non-abelian cohomology operations. In generalization of Def. 2.3, we set:

Definition 3.6 (Twisted non-abelian cohomology operation). Given a transformation of local coefficient bundles (3.2) presented (under localization (1.24) to homotopy types (1.51)) as a strictly commuting diagram

$$\begin{array}{ccc} A_1 /\!\!/ G_1 & \xrightarrow{\phi_t} & A_2 /\!\!/ G_2 \\ {\scriptstyle \rho_1}\big\downarrow & & \big\downarrow{\scriptstyle \rho_2} \\ BG_1 & \xrightarrow{\phi_b} & BG_2 \end{array} \quad \in \mathrm{TopSp}_{\mathrm{Qu}}, \tag{3.29}$$

pasting composition induces,[1] for each twist $X \xrightarrow{\tau} BG_1$ (3.3), a map

$$\phi_* \; : \; H^{\tau}(X; A_1) \xrightarrow{(\phi_t \circ (-)) \circ (\rho_1)_*} H^{\phi_b \circ \tau}(X; A_2) \tag{3.30}$$

of twisted non-abelian cohomology sets (Def. 3.2). We call these *twisted non-abelian cohomology operations.*

Example 3.9 (Total non-abelian class of twisted cocycles). For any coefficient bundle ρ (3.2) there is the tautological transformation (3.29) to its total space regarded as fibered over the point:

$$\begin{array}{ccc} A /\!\!/ G & = & A /\!\!/ G \\ {\scriptstyle \rho}\big\downarrow & & \big\downarrow \\ BG & \longrightarrow & * . \end{array}$$

The induced twisted non-abelian cohomology operation (3.30) goes from twisted cohomology to non-twisted cohomology with coefficient in the total space:

$$H^{\tau}(X; A) \xrightarrow{\rho_*} H\big(X; A /\!\!/ G\big). \tag{3.31}$$

Example 3.10 (Hopf cohomology operation in twisted Cohomotopy [Fiorenza *et al.* (2020b), §2.3]**).** The quaternionic Hopf fibration $S^7 \xrightarrow{h_{\mathbb{H}}} S^4$ is equivariant under the symplectic unitary group $\mathrm{Sp}(2) \simeq \mathrm{Spin}(5)$, so that after passage to classifying spaces it induces a morphism of local coefficient bundles (3.29) for twisted Cohomotopy (3.28) in degrees 4 and 7:

$$\begin{array}{ccc} S^7 /\!\!/ \mathrm{Sp}(2) & \xrightarrow[h_{\mathbb{H}} /\!\!/ \mathrm{Sp}(2)]{\text{Borel-equivariantized quaternionic Hopf fibration}} & S^4 /\!\!/ \mathrm{Sp}(2) \\ {\scriptstyle J_7}\big\downarrow & & \big\downarrow{\scriptstyle J_4} \\ B\mathrm{Sp}(2) & = & B\mathrm{Sp}(2) . \end{array} \tag{3.32}$$

[1] We postpone discussing the details of forming pasting composites to Part V, where they are provided by Def. V.2.

Via (3.30), this induces for each Spin 8-manifold X equipped with tangential Sp(2)-structure (Ex. 3.2)

$$\begin{array}{ccc} X & \xrightarrow{\tau} & B\mathrm{Sp}(2) \\ & {\scriptstyle TX}\searrow \;\Leftarrow_{\simeq}\; \swarrow{\scriptstyle Bi} & \\ & BO(8) & \end{array} \tag{3.33}$$

a twisted non-abelian cohomology operation (Def. 3.6)

$$\pi^{\tau_7}(X) \xrightarrow{(h_{\mathbb{H}} /\!\!/ \mathrm{Sp}(2))_*} \pi^{\tau^4}(X) \tag{3.34}$$

in twisted non-abelian Cohomotopy theory (Ex. 3.8). Lifting through the twisted non-abelian cohomology transformation (3.34) encodes vanishing of C-field flux up to C-field background charge [Fiorenza *et al.* (2020b), Prop. 3.14].

Example 3.11 (Twistorial Cohomotopy [Fiorenza *et al.* (2022), §3.2]**).** The equivariantized Hopf morphism (3.32) of coefficient bundles factors through Borel-equivariantizations of the complex Hopf fibration $h_{\mathbb{C}}$ followed by that of the *twistor fibration $t_{\mathbb{H}}$*

$$\begin{array}{ccccc} S^7 /\!\!/ \mathrm{Sp}(2) & \xrightarrow[h_{\mathbb{C}} /\!\!/ \mathrm{Sp}(2)]{\text{Borel-equivariantized complex Hopf fibration}} & \mathbb{C}P^3 /\!\!/ \mathrm{Sp}(2) & \xrightarrow[t_{\mathbb{H}} /\!\!/ \mathrm{Sp}(2)]{\text{Borel-equivariantized twistor fibration}} & S^4 /\!\!/ \mathrm{Sp}(2) \\ \downarrow{\scriptstyle J_{S^7}} & & \downarrow{\scriptstyle J_{\mathbb{C}P^3}} & & \downarrow{\scriptstyle J_{S^4}} \\ B\mathrm{Sp}(2) & = & B\mathrm{Sp}(2) & = & B\mathrm{Sp}(2) \end{array} \tag{3.35}$$

The twisted non-abelian cohomology theory (Def. 3.2) with local coefficients in the bundle appearing in this factorization is the *Twistorial Cohomotopy* of [Fiorenza *et al.* (2022)]

$$\overset{\text{Twistorial Cohomotopy}}{\mathcal{T}^{\tau}(-)} := H^{\tau}(-;\mathbb{C}P^3).$$

Via (3.30), the morphisms (3.35) induce, for each spin 8-manifold X equipped with tangential Sp(2)-structure (3.33), twisted non-abelian cohomology operations (Def. 3.6)

$$\overset{\text{tang. twisted 7-Cohomotopy}}{\pi^{\tau^7}(X)} \xrightarrow{(h_{\mathbb{C}} /\!\!/ \mathrm{Sp}(2))_*} \overset{\text{Twistorial Cohomotopy}}{\mathcal{T}^{\tau}(X)s} \xrightarrow{(t_{\mathbb{H}} /\!\!/ \mathrm{Sp}(2))_*} \overset{\text{tang. twisted 4-Cohomotopy}}{\pi^{\tau^4}(X)} \tag{3.36}$$

between tangentially twisted non-abelian Cohomotopy theory (Ex. 3.8) and Twistorial Cohomotopy.

We turn to the differential refinement of this statement in Chapter 12 below.

PART III

Non-abelian de Rham cohomology

Chapter 4

Dgc-algebras and L_∞-algebras

We formulate (twisted) non-abelian de Rham cohomology (Def. 6.3, Def. 6.9) of differential forms with values in L_∞-algebras (Ex. 4.13) and prove the (twisted) non-abelian de Rham theorem (Thm. 6.5, Thm. 6.15), as a consequence of the fundamental theorem of dg-algebraic rational homotopy theory, which we recall (Prop. 5.6).

Here we fix notation and conventions for the following system of categories and functors:

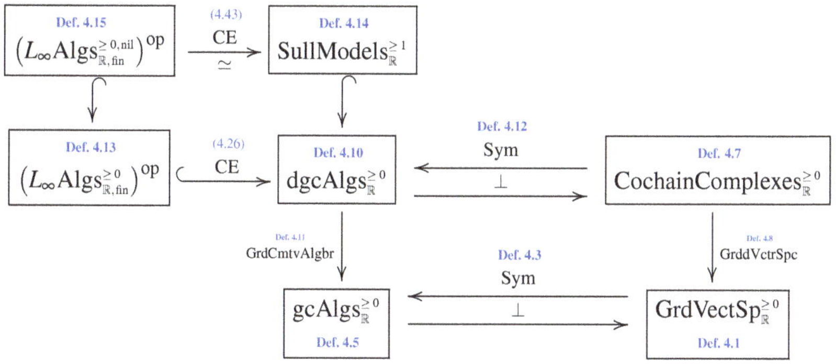

$$
\begin{array}{ccc}
\boxed{\begin{array}{c}\text{Def. 4.15}\\ \left(L_\infty\mathrm{Algs}^{\geq 0,\mathrm{nil}}_{\mathbb{R},\mathrm{fin}}\right)^{\mathrm{op}}\end{array}} & \xrightarrow[\simeq]{\overset{(4.43)}{\mathrm{CE}}} & \boxed{\begin{array}{c}\text{Def. 4.14}\\ \mathrm{SullModels}^{\geq 1}_{\mathbb{R}}\end{array}}
\end{array}
$$

$$
\boxed{\begin{array}{c}\text{Def. 4.13}\\ \left(L_\infty\mathrm{Algs}^{\geq 0}_{\mathbb{R},\mathrm{fin}}\right)^{\mathrm{op}}\end{array}} \overset{\overset{(4.26)}{\mathrm{CE}}}{\hookrightarrow} \boxed{\begin{array}{c}\text{Def. 4.10}\\ \mathrm{dgcAlgs}^{\geq 0}_{\mathbb{R}}\end{array}} \underset{\perp}{\overset{\overset{\text{Def. 4.12}}{\mathrm{Sym}}}{\rightleftarrows}} \boxed{\begin{array}{c}\text{Def. 4.7}\\ \mathrm{CochainComplexes}^{\geq 0}_{\mathbb{R}}\end{array}}
$$

$$
\overset{\text{Def. 4.11}}{\underset{\mathrm{GrdCmtvAlgbr}}{\downarrow}} \qquad \overset{\text{Def. 4.8}}{\underset{\mathrm{GrddVctrSpc}}{\downarrow}}
$$

$$
\boxed{\begin{array}{c}\mathrm{gcAlgs}^{\geq 0}_{\mathbb{R}}\\ \text{Def. 4.5}\end{array}} \underset{\perp}{\overset{\overset{\text{Def. 4.3}}{\mathrm{Sym}}}{\rightleftarrows}} \boxed{\begin{array}{c}\mathrm{GrdVectSp}^{\geq 0}_{\mathbb{R}}\\ \text{Def. 4.1}\end{array}}
$$

$$\tag{4.1}$$

Remark 4.1 (Homotopical grading). Our grading conventions, to be detailed in the following, are strictly *homotopy theoretic*, in that all algebraic data in degree n always corresponds to homotopy groups in that same degree:

(i) Every graded-algebraic object discussed here corresponds, under the equivalences of rational homotopy theory laid out in Chapter 5 below, to a rational space, such that algebraic generators in degree n correspond to homotopy groups in the same degree n. Since homotopy groups of spaces are in non-negative degree $n \in \mathbb{N}$, all dg-algebraic objects discussed both homological as well as cohomological, we take to be concentrated in non-negative degree. This implies that we take linear duality of (co)chain complexes (Def. 4.3) to preserve the degree as opposed to changing it by a sign.

(ii) In particular, our L_∞-algebras are in non-negative degree, hence are *connective*, naturally accommodating (as in [Lada and Markl (1995)] [Buijs *et al.* (2011), §2.9]) the rationalized Whitehead homotopy Lie algebras $\pi_\bullet(\Omega X)\otimes_{\mathbb{Z}}\mathbb{R}$ of connected spaces X, with their natural non-negative grading induced from that of the homotopy groups of ΩX. See Prop. 5.11 and Prop. 5.13 below.

(iii) Accordingly, all Chevalley-Eilenberg (CE) dgc-algebras (Ex. 4.10) are taken to be in non-negative degree, as usual, so that their generators in degree n correspond to dual homotopy groups in degree n. For example, the CE-algebra model for an Eilenberg-MacLane space $K(\mathbb{Z},n)$ has a single generator which is in degree $+n$ (Ex. 5.4).

Graded vector spaces.

Definition 4.1 (Connective graded vector spaces). **(i)** We write

$$\mathrm{GrdVectSp}_{\mathbb{R}}^{\geq 0} \in \mathrm{Cats} \tag{4.2}$$

for the category whose objects are $\mathbb{N}$-graded (i.e. non-negatively $\mathbb{Z}$-graded) vector spaces over the real numbers; and we write

$$\mathrm{GrdVectSp}_{\mathbb{R}}^{\geq 0,\mathrm{fin}} \hookrightarrow \mathrm{GrdVectSp}_{\mathbb{R}}^{\geq 0} \in \mathrm{Cats} \tag{4.3}$$

for its full subcategory on those objects which are of *finite type*, namely degree-wise finite-dimensional.

(ii) For $V \in \mathrm{GrdVectSp}_{\mathbb{R}}^{\geq 0}$ and $k \in \mathbb{N}$, we write

$$V^k \in \mathrm{VectSp}_{\mathbb{R}}$$

for the component vector space in degree k.

Example 4.1 (The zero-object in graded vector spaces). We write

$$0 \in \mathrm{GrdVectSp}_{\mathbb{R}}^{\geq 0} \tag{4.4}$$

for the graded vector space which is the zero vector space in each degree. This is both the initial as well as the terminal object (hence the zero object) in $\mathrm{GrdVectSp}_{\mathbb{R}}^{\geq 0}$.

Example 4.2 (Graded linear basis). For $n_1, n_2, \ldots, n_k \in \mathbb{N}$ a finite sequence of non-negative integers, we write

$$\big\langle \alpha_{n_1}, \alpha_{n_2}, \ldots, \alpha_{n_k} \big\rangle \in \mathrm{GrdVectSp}_{\mathbb{R}}^{\geq 0,\mathrm{fin}}$$

for the graded vector space (Def. 4.1) spanned by elements α_{n_i} in degree n_i, respectively.

Definition 4.2 (Tensor product of graded vector spaces). The category of $\mathrm{GrdVectSp}_{\mathbb{R}}^{\geq 0}$ (Def. 4.1) becomes a symmetric monoidal category under the graded tensor product given by

$$(V \otimes W)^k := \bigoplus_{n_1+n_2=k} V^{n_1} \otimes W^{n_2}\,.$$

and the symmetric braiding isomorphism given by

$$
\begin{array}{ccc}
V \otimes W & \xrightarrow[\simeq]{\sigma^{V,W}} & W \otimes V \\
\uparrow & & \uparrow \\
V^{n_1} \otimes W^{n_2} & \xrightarrow[\simeq]{\sigma^{V,W}_{n_1,n_2}} & W^{n_2} \otimes V^{n_1} \\
\Cup & & \Cup \\
(v,w) & \longmapsto & (-1)^{n_1 n_2} \cdot (w,v)
\end{array}
\tag{4.5}
$$

We denote this by

$$\left(\mathrm{GrdVectSp}_{\mathbb{R}}^{\geq 0}, \otimes, \sigma\right) \in \mathrm{SymmetricMonoidalCategories}\,. \tag{4.6}$$

Definition 4.3 (Degreewise linear dual). For $V \in \mathrm{GrdVectSp}_{\mathbb{R}}^{\geq 0,\mathrm{fin}}$ (Def. 4.1) we write

$$V^\vee \in \mathrm{GrdVectSp}_{\mathbb{R}}^{\geq 0,\mathrm{fin}}$$

for its degree-wise linear dual:[1]

$$(V^\vee)^k := (V^k)^*\,. \tag{4.7}$$

Definition 4.4 (Degree shift). For $V \in \mathrm{GrdVectSp}_{\mathbb{R}}^{\geq 0}$ (Def. 4.1) we write

$$\mathfrak{b}V \in \mathrm{GrdVectSp}_{\mathbb{R}}^{\geq 0} \tag{4.8}$$

for the result of shifting degrees up by 1:

$$(\mathfrak{b}V)^k := \begin{cases} V^{k-1} & \mid k \geq 1, \\ 0 & \mid k = 0. \end{cases}$$

Graded-commutative algebras.

Definition 4.5 (Graded-commutative algebras). We write

$$\mathrm{gcAlgs}_{\mathbb{R}}^{\geq 0} := \mathrm{CommMonoids}\left(\mathrm{GrdVectSp}_{\mathbb{R}}^{\geq 0}, \otimes, \sigma\right) \in \mathrm{Cats} \tag{4.9}$$

for the category whose objects are non-negatively $\mathbb{Z}$-graded, graded-commutative unital algebras over the real numbers (hence commutative unital monoids with respect to the braided tensor product of Def. 4.2); and we write

$$\mathrm{gcAlgs}_{\mathbb{R}}^{\geq 0,\mathrm{fin}} \hookrightarrow \mathrm{gcAlgs}_{\mathbb{R}}^{\geq 0} \in \mathrm{Cats} \tag{4.10}$$

for its full sub-category in those objects which are of *finite type*, namely degree-wise finite dimensional.

[1]This is in contrast to the intrinsic duality $(-)^*$ in the monoidal category of graded vector spaces in *unbounded* degree (not considered here), which instead goes along with inversion of the degree: $(V^*)^k = (V^{-k})^*$.

Definition 4.6 (Underlying graded vector space). We write

$$\mathrm{gcAlgs}_{\mathbb{R}}^{\geq 0} \xrightarrow{\mathrm{GrddVctrSpc}} \mathrm{GrdVectSp}_{\mathbb{R}}^{\geq 0} \tag{4.11}$$

for the functor on graded algebras (Def. 4.5) that forgets the algebra structure and remembers only the underlying graded vector space (Def. 4.1).

Example 4.3 (Free graded-commutative algebras). For $V \in \mathrm{GrdVectSp}_{\mathbb{R}}^{\geq 0}$ (Def. 4.1), we write

$$\mathrm{Sym}(V) \in \mathrm{gcAlgs}_{\mathbb{R}}^{\geq 0} \tag{4.12}$$

for the graded-commutative algebra (Def. 4.5) freely generated by V, hence that whose underlying graded vector space (4.11) is

$$\mathrm{GrddVctrSpc}\big(\mathrm{Sym}(V)\big) = \mathbb{R} \oplus V \oplus \big(V \otimes V\big)_{/\mathrm{Sym}(2)} \oplus \big(V \otimes V \otimes V\big)_{/\mathrm{Sym}(3)} \oplus \cdots,$$

where the symmetric groups $\mathrm{Sym}(n)$ act via the braiding (4.5).

Example 4.4 (Graded Grassmann algebra). For $V \in \mathrm{GrdVectSp}_{\mathbb{R}}^{\geq 0}$ (Def. 4.1), we write

$$\wedge^{\bullet} V := \mathrm{Sym}\big(\mathfrak{b}V\big) \in \mathrm{gcAlgs}_{\mathbb{R}}^{\geq 0}$$

for the free graded-commutative algebra (Def. 4.3) on V shifted up in degree (Def. 4.4); and we call this the *graded Grassmann-algebra* on V.

Example 4.5 (Graded polynomial algebra). For $n_1, n_2, \ldots, n_k \in \mathbb{N}$ a finite sequence of non-negative integers, we write

$$\mathbb{R}\big[\alpha_{n_1}, \alpha_{n_2}, \ldots, \alpha_{n_k}\big] := \mathrm{Sym}\Big(\big\langle \alpha_{n_1}, \alpha_{n_2}, \ldots, \alpha_{n_k} \big\rangle\Big) \in \mathrm{gcAlgs}_{\mathbb{R}}^{\geq 0, \mathrm{fin}}$$

for the free graded-commutative algebras (Def. 4.3) the graded vector space spanned by the α_{n_i} (Def. 4.2).

Remark 4.2 (Incarnations of Grassmann algebras). With these notation conventions from Ex. 4.3, 4.4, 4.5, an ordinary Grassmann algebra on k generators is equivalently:

$$\wedge^{\bullet}\big(\mathbb{R}^k\big) = \mathrm{Sym}\big(\mathfrak{b}\mathbb{R}^k\big) = \mathbb{R}\big[\theta_1^{(1)}, \theta_1^{(2)}, \ldots, \theta_1^{(k)}\big].$$

Cochain complexes.

Definition 4.7 (Connective cochain complexes). We write

$$\mathrm{CochainComplexes}_{\mathbb{R}}^{\geq 0} \in \mathrm{Cats}$$

for the category of cochain complexes (i.e. with differential of degree $+1$) of real vector spaces in non-negative degree.

Definition 4.8 (Underlying graded vector space). We write

$$\mathrm{CochainComplexes}_{\mathbb{R}}^{\geq 0} \xrightarrow{\mathrm{GrddVctrSpc}} \mathrm{GrdVectSp}_{\mathbb{R}}^{\geq 0} \tag{4.13}$$

for the forgetful functor on connective cochain complexes (Def. 4.7) which forgets the differential and remembers only the underlying connective graded vector space (Def. 4.1).

Definition 4.9 (Tensor product on cochain complexes). The tensor product and braiding of graded vector spaces from Def. 4.2 lifts, through (4.13), to a tensor product and braiding on $\mathrm{CochainComplexes}_{\mathbb{R}}^{\geq 0}$ (Def. 4.7), making it a symmetric monoidal category:

$$\left(\mathrm{CochainComplexes}_{\mathbb{R}}^{\geq 0}, \otimes, \sigma\right) \in \mathrm{SymmetricMonoidalCategories}\,. \tag{4.14}$$

Differential graded commutative algebras.

Definition 4.10 (Connective differential graded commutative algebras [Gelfand and Manin (1996), V.3.1]**).** We write

$$\mathrm{dgcAlgs}_{\mathbb{R}}^{\geq 0} := \mathrm{CommMonoids}\left(\mathrm{CochainComplexes}_{\mathbb{R}}^{\geq 0}, \otimes, \sigma\right) \in \mathrm{Cats}$$

for the category whose objects are differential-graded, graded-commutative, unital algebras over the real numbers concentrated in non-negative degrees (hence commutative unital monoids in the symmetric monoidal category of Def. 4.9).

Definition 4.11 (Underlying graded-commutative algebra). We write

$$\mathrm{dgcAlgs}_{\mathbb{R}}^{\geq 0} \xrightarrow{\mathrm{GrddCmmttvAlgbr}} \mathrm{gcAlgs}_{\mathbb{R}}^{\geq 0} \tag{4.15}$$

for the functor on dgc-algebras (Def. 4.10) that forgets the differential and remembers only the underlying graded-commutative algebra (Def. 4.5).

Definition 4.12 (Free differential graded algebras). For $V^\bullet$ in $\mathrm{CochainComplexes}_{\mathbb{R}}^{\geq 0}$ (Def. 4.7) we write

$$\mathrm{Sym}(V^\bullet) \in \mathrm{dgcAlgs}_{\mathbb{R}}^{\geq 0}$$

for the free differential graded-commutative algebra on $V^\bullet$, (Def. 4.10), hence whose underlying graded-commutative algebra algebra (4.15) is as in Ex. 4.3.

Example 4.6 (Initial algebra). The real algebra of real numbers, regarded as concentrated in degree-0

$$\mathbb{R} \in \mathrm{gcAlgs}_{\mathbb{R}}^{\geq 0} \hookrightarrow \mathrm{dgcAlgs}_{\mathbb{R}}^{\geq 0}$$

is the *initial* object: For any other $A \in \mathrm{gcAlgs}_{\mathbb{R}}^{\geq 0}$ (Def. 4.9) or $\in \mathrm{dgcAlgs}_{\mathbb{R}}^{\geq 0}$ (Def. 4.10) there is a unique morphism $\mathbb{R} \overset{i_{\mathbb{R}}}{\hookrightarrow} A$ (because our algebras are unital and homomorphims need to preserve the unit element).

Example 4.7 (The terminal algebra). We write

$$0 \in \mathrm{gcAlgs}_{\mathbb{R}}^{\geq 0} \hookrightarrow \mathrm{dgcAlgs}_{\mathbb{R}}^{\geq 0} \tag{4.16}$$

for the unique graded-commutative algebra (Def. 4.5) or dgc-algebra (Def. 4.10) whose underlying graded vector space (Def. 4.6) is the zero-vector space[2] (4.4). This is the terminal object[3] in $\mathrm{gcAlgs}_{\mathbb{R}}^{\geq 0}$: For every $A \in \mathrm{gcAlgs}_{\mathbb{R}}^{\geq 0}$, there is a unique morphism $A \xrightarrow{\exists !} 0$.

Example 4.8 (Product and co-product algebras). In the categories $\mathrm{gcAlgs}_{\mathbb{R}}^{\geq 0}$ (Def. 4.5) and $\mathrm{dgcAlgs}_{\mathbb{R}}^{\geq 0}$ (Def. 4.10):

(i) the coproduct is given by the tensor product (Def. 4.2),
(ii) the product is given by the direct sum
on underlying graded vector spaces (Def. 4.6).
(The first follows by [Johnstone (2002), p. 478, Cor. 1.1.9], while the second holds since (4.11) is a right adjoint.)

Example 4.9 (Smooth de Rham complex (e.g. [Bott and Tu (1982)]**)).** For X be a smooth manifold, its de Rham algebra of smooth differential forms is a dgc-algebra in the sense of Def. 4.10, to be denoted here:

$$\Omega^{\bullet}_{\mathrm{dR}}(X) \in \mathrm{dgcAlgs}_{\mathbb{R}}^{\geq 0}.$$

Example 4.10 (Chevalley-Eilenberg algebras of Lie algebras). For $(\mathfrak{g}, [-,-])$ a finite-dimensional real Lie algebra, its Chevalley-Eilenberg algebra is a dgc-algebra (Def. 4.10):

$$\mathrm{CE}(\mathfrak{g}) := \left(\wedge^{\bullet}\mathfrak{g}^*, \, d|_{\wedge^1\mathfrak{g}^*} = [-,-]^*\right) \in \mathrm{dgcAlgs}_{\mathbb{R}}^{\geq 0}$$

with underlying graded-commutative algebra (Def. 4.5) the Grassmann algebra on the linear dual space $\mathfrak{g}^*$ (Def. 4.4, Rem. 4.2), and with differential given on $\wedge^1\mathfrak{g}^*$ by the linear dual of the Lie bracket and necessarily extended from there to all of $\wedge^{\bullet}\mathfrak{g}^*$ by the graded Leibniz rule.

More explicitly, for $\{v_a\}_{a=1}^{\dim_{\mathbb{R}}(\mathfrak{g})}$ a linear basis for the underlying vector space of the Lie algebra

$$\mathfrak{g} \simeq \langle v_1, v_2, \dots, v_{\dim(\mathfrak{g})} \rangle, \tag{4.17}$$

with Lie brackets

$$[v_a, v_b] = f^c_{ab} v_c, \quad \text{for structure constants } f^c_{ab} \in s\mathbb{R} \tag{4.18}$$

we have

$$\mathrm{CE}(\mathfrak{g}) \simeq \mathbb{R}\big[\theta_1^{(1)}, \theta_1^{(2)}, \dots, \theta_1^{(\dim(\mathfrak{g}))}\big] / \big(d\,\theta_1^{(c)} = f_{ab}{}^c\, \theta_1^{(b)} \wedge \theta_1^{(a)}\big), \tag{4.19}$$

[2]Notice that the algebra 0 (4.16) is indeed a unital algebra (4.9).
[3]Beware that the corresponding statement in [Gelfand and Manin (1996), p. 335] is incorrect.

which means, for instance, that in this dg-algebra we have relations such as

$$\begin{aligned} d\big(\theta_1^{(c_1)} \wedge \theta_2^{(c_2)}\big) &= \big(d\theta_1^{(c_1)}\big) \wedge \theta_1^{(c_2)} - \theta_1^{(c_1)} \wedge \big(d\theta_1^{(c_2)}\big) \\ &= f_{a_1 b_1}{}^{c_1}\, \theta_1^{(a_1)} \wedge \theta_1^{(b_1)} \wedge \theta_1^{(c_2)} - f_{a_2 b_2}{}^{c_2}\, \theta_1^{(c_1)} \wedge \theta_1^{(a_2)} \wedge \theta_1^{(b_2)} \end{aligned}$$

etc.

Using such manuipulations, one readily observes that the Jacobi identity condition on the Lie bracket $[-,-]$ is equivalent to the condition that the differential $d := [-,-]^*$ squares to zero. This means that (4.19) being a dgc-algebra is actually equivalent to $(\mathfrak{g}, [-,-])$ being a Lie algebra.

This construction is evidently contravariantly functorial and constitutes a full subcategory inclusion

$$\mathrm{LieAlgebras}_{\mathbb{R},\mathrm{fin}} \xhookrightarrow{\;\mathrm{CE}\;} \big(\mathrm{dgcAlgs}_{\mathbb{R}}^{\geq 0}\big)^{\mathrm{op}}\,, \tag{4.20}$$

meaning that, in addition, homomorphisms of Lie algebras are in natural bijection to dgc-algebra morphisms between their CE-algebras.

This observation is the golden route to approaching L_∞-algebras:

L_∞-algebras.

Definition 4.13 (Chevalley-Eilenberg algebras of L_∞-algebras [Lada and Markl (1995), Thm. 2.3][Sati *et al.* (2009), Def. 13][Buijs *et al.* (2011), §2]**).** In direct generalization of (4.20), consider those $A \in \mathrm{dgcAlgs}_{\mathbb{R}}^{\geq 0}$ (Def. 4.10) whose underlying graded-commutative algebra (4.15) is free (Ex. 4.3, Rem. 4.2) on the degreewise dual $\mathfrak{b}\mathfrak{g}^\vee$ (Def. 4.3) of the degree shift $\mathfrak{b}\mathfrak{g}$ (Def. 4.4) of some connective finite-type graded vector space (Def. 4.1)

$$\mathfrak{g} \in \mathrm{GrdVectSp}_{\mathbb{R}}^{\geq 0,\mathrm{fin}} \tag{4.21}$$

in that

$$A := \big(\wedge^\bullet \mathfrak{g}^\vee, d\big) := \big(\mathrm{Sym}(\mathfrak{b}\mathfrak{g}^\vee), d\big) \in \mathrm{dgcAlgs}_{\mathbb{R}}^{\geq 0}. \tag{4.22}$$

In this case the differential d restricted to $\wedge^1 \mathfrak{g}^\vee$ defines, under linear dualization, a sequence of n-ary graded-symmetric multilinear maps $\{-,\dots,-\}$ on $\mathfrak{g}$:

$$\begin{array}{ccl} d|_{\wedge^1\mathfrak{g}^\vee}(-) & = & \{-\}^* + \{-,-\}^* + \{-,-,-\}^* + \cdots \\ \wedge^1\mathfrak{g}^\vee \xrightarrow{\;d\;} & & \wedge^1\mathfrak{g}^\vee \oplus \wedge^2\mathfrak{g}^\vee \oplus \wedge^3\mathfrak{g}^\vee \oplus \cdots = \wedge^\bullet\mathfrak{g}^\vee = \mathrm{Sym}(\mathfrak{b}\mathfrak{g}^\vee)\,, \end{array} \tag{4.23}$$

and the condition $d \circ d = 0$ imposes a sequence of compatibility conditions on these brackets, generalizing the Jacobi identity in Ex. 4.10. The corresponding graded skew-symmetric n-ary brackets ([Lada and Stasheff (1993), (3)])

$$[a_1,\dots,a_n] := (-1)^{n+\sum_{i\leq n/2}\deg(a_i)}\{a_1,\dots,a_n\} \tag{4.24}$$

subject to these conditions give $\mathfrak{g}$ the structure of an L_∞*-algebra* (or *strong homotopy Lie algebra*):

$$\Big(\mathfrak{g}, [-], [-,-], [-,-,-], \cdots\Big) \in L_\infty\mathrm{Algs}^{\geq 0}_{\mathbb{R},\mathrm{fin}}, \tag{4.25}$$

which makes A in (4.22) its Chevalley-Eilenberg algebra:

$$\begin{aligned} \mathrm{CE}(\mathfrak{g}) &:= \big(\wedge^\bullet \mathfrak{g}^\vee, d = \{-\}^* + \{-,-\}^* + \{-,-,-\}^* + \cdots\big) \\ &= \big(\mathrm{Sym}(\mathfrak{b}\mathfrak{g}^\vee), d_{\mathrm{CE}}\big). \end{aligned} \tag{4.26}$$

This construction constitutes a full subcategory inclusion

$$L_\infty\mathrm{Algs}^{\geq 0}_{\mathbb{R},\mathrm{fin}} \overset{\mathrm{CE}}{\hookrightarrow} \big(\mathrm{dgcAlgs}^{\geq 0}_{\mathbb{R}}\big)^{\mathrm{op}}. \tag{4.27}$$

into the category of dgc-algebras of the category of connective finite-type L_∞-algebras, with the homotopy-correct morphisms between them (known as "weak maps" [Lada and Markl (1995), Rem. 5.4], "sh maps" [Merkulov (2004), §2.11] or "L_∞-morphisms" [Kontsevich (2003), p. 12]).

Example 4.11 (Differential graded Lie algebras). A differential graded Lie algebra is an L_∞-algebra (4.25) whose only possibly non-vanishing brackets are the unary bracket $\partial := [-]$ (its differential) and the binary bracket $[-,-]$ (its graded Lie bracket). In further specialization, a plain Lie algebra (Ex. 4.10) is an L_∞-algebra/dg-Lie algebra concentrated in degree 0:

$$\mathrm{LieAlgebras}_{\mathbb{R},\mathrm{fin}} \hookrightarrow \mathrm{DiffGradedLieAlgebras}^{\geq 0}_{\mathbb{R},\mathrm{fin}} \hookrightarrow L_\infty\mathrm{Algs}^{\geq 0}_{\mathbb{R},\mathrm{fin}}. \tag{4.28}$$

Example 4.12 (Line Lie n-algebra). For $n \in \mathbb{N}$, the *line Lie* $(n+1)$*-algebra* is the L_∞-algebra (Def. 4.13)

$$\mathfrak{b}^n\mathbb{R} \in L_\infty\mathrm{Algs}^{\geq 0}_{\mathbb{R},\mathrm{fin}} \tag{4.29}$$

whose Chevalley-Eilenberg algebra (4.26) is the polynomial dgc-algebra (Ex. 4.14) on a single closed generator in degree $n+1$:

$$\mathrm{CE}\big(\mathfrak{b}^n\mathbb{R}\big) := \mathbb{R}[c_{n+1}]/(d\,c_{n+1} = 0). \tag{4.30}$$

More generally, for $V \in \mathrm{VectSp}^{\mathrm{fin}}_{\mathbb{R}}$, we have

$$\begin{aligned} &\mathfrak{b}^n V \simeq \bigoplus_{\dim(V)} \mathfrak{b}^n\mathbb{R} \in L_\infty\mathrm{Algs}^{\geq 0}_{\mathbb{R},\mathrm{fin}}, \\ &\text{with } \mathrm{CE}\big(\mathfrak{b}^n V\big) \simeq \mathbb{R}\big[c^{(1)}_{n+1}, c^{(2)}_{n+1}, \ldots, c^{(\dim V)}_{n+1}\big] \Big/ \begin{pmatrix} d\,c^{(1)}_{n+1} = 0, \\ \vdots \\ d\,c^{(\dim V)}_{n+1} = 0 \end{pmatrix}. \end{aligned} \tag{4.31}$$

Example 4.13 (String Lie 2-algebra [Baez *et al.* (2007), §5][Henriques (2008), §1.2][Fiorenza *et al.* (2014b), App.]**).** Let $\mathfrak{g} \in \mathrm{LieAlgebras}_{\mathbb{R},\mathrm{fin}}$ be semisimple (such as

$\mathfrak{g} = \mathfrak{su}(n+1), \mathfrak{so}(n+3)$, for $n \in \mathbb{N}$), hence equipped with a non-degenerate, symmetric, $\mathfrak{g}$-invariant bilinear form ("Killing form")

$$\mathfrak{g} \otimes \mathfrak{g} \xrightarrow{\langle -,- \rangle} \mathbb{R} . \tag{4.32}$$

Then the element

$$\mu := \langle -, [-,-] \rangle \in \mathrm{CE}(\mathfrak{g})$$

in the Chevalley-Eilenberg (4.10) is closed (is a Lie algebra cocycle)

$$d\mu = 0 .$$

In terms of a linear basis $\{v_a\}$ (4.17) with structure constants $\{f^c_{ab}\}$ (4.18) and inner product $k_{ab} := \langle v_a, v_b \rangle$ we have, in terms of (4.19):

$$\mu := f_{ab}{}^{c'} k_{c'c}\, \theta_1^{(c)} \wedge \theta_1^{(b)} \wedge \theta_1^{(a)} .$$

Hence we get an L_∞-algebra (Def. 4.13)

$$\mathfrak{string}_{\mathfrak{g}} \in L_\infty \mathrm{Algs}^{\geq 0}_{\mathbb{R},\mathrm{fin}} \tag{4.33}$$

with the following Chevalley-Eilenberg algebra (4.26):

$$\mathrm{CE}(\mathfrak{string}_{\mathfrak{g}}) := \mathbb{R}\begin{bmatrix} \{\theta_1^a\} \\ b_2 \end{bmatrix} \Big/ \left(\begin{aligned} d\,\theta_1^{(c)} &= f^c_{ab}\, \theta_1^{(b)} \wedge \theta_1^{(a)} \\ d b_2 &= \underbrace{f^{c'}_{ab} k_{c'c}\, \theta_1^{(c)} \wedge \theta_1^{(b)} \wedge \theta_1^{(a)}}_{=\mu} \end{aligned} \right) . \tag{4.34}$$

This is known as the *string Lie 2-algebra*, since it is [Baez *et al.* (2007)][Henriques (2008)] the L_∞-algebra of the String 2-group of Ex. 2.4.

Sullivan models and nilpotent L_∞-algebras.

Example 4.14 (Polynomial dgc-algebras). For $A \in \mathrm{dgcAlgs}^{\geq 0}_{\mathbb{R}}$ (Def. 4.10), and

$$\mu \in A^{n+1} \subset A, \qquad d\mu = 0 \tag{4.35}$$

a closed element of homogeneous degree $n+1$, we write

$$A[\alpha_n] / (d\,\alpha_n = \mu\,) \in \mathrm{dgcAlgs}^{\geq 0}_{\mathbb{R}} \tag{4.36}$$

for the dgc-algebra obtained by adjoining a generator α_n of degree n to the underlying graded-commutative algebra (4.15) of A and extending the differential from A to $A[\alpha_n]$ by taking its value on the new generator to be μ. The polynomial dgc-algebra (4.36) receives a canonical algebra inclusion of A:

$$A \overset{i_A}{\hookrightarrow} A[\alpha_n]/(d\,\alpha_n = \mu) . \tag{4.37}$$

Example 4.15 (Multivariate polynomial dgc-algebras). Let $A \in \mathrm{dgcAlgs}_{\mathbb{R}}^{\geq 0}$ (Def. 4.10), $\mu^{(1)} \in A^{n_1+1}$, $d\mu^{(1)} = 0$, with corresponding polynomial dgc-algebra (4.36) as in Ex. 4.14. Then, for

$$\mu^{(2)} \in A\big[\alpha^{(1)}_{n_1}\big]/\big(d\,\alpha^{(1)}_{n_1} = \mu^{(1)}\big), \quad d\,\mu^{(2)} = 0$$

another closed element of some homogeneous degree $n_2 + 1$, in the new algebra (4.36) we may iterate the construction of Ex. 4.14 to obtain the bivariate polynomial dgc-algebra over A, to be denoted:

$$A\begin{bmatrix}\alpha^{(2)}_{n_2}\\ \alpha^{(1)}_{n_1}\end{bmatrix}\Big/\begin{pmatrix}d\,\alpha^{(2)}_{n_1} = \mu^{(2)}_{n_2}\\ d\,\alpha^{(1)}_{n_1} = \mu^{(1)}_{n_1}\end{pmatrix} := \Big(A\big[\mu^{(1)}_{n_1}\big]/\big(d\,\alpha^{(1)}_{n_1} = \mu^{(1)}\big)\Big)\big[\alpha^{(2)}\big]/\big(d\,\alpha^{(2)}_{n_2} = \mu^{(2)}\big).$$

Iterating further, we have multivariate polynomial dgc-algebras over A, to be denoted as follows:

$$A\begin{bmatrix}\alpha^{(k)}_{n_k},\\ \vdots\\ \alpha^{(2)}_{n_2}\\ \alpha^{(1)}_{n_1}\end{bmatrix}\Big/\begin{pmatrix}d\,\alpha^{(k)}_{n_k} = \mu^{(k)}\\ \vdots\\ d\,\alpha^{(2)}_{n_1} = \mu^{(2)},\\ d\,\alpha^{(1)}_{n_1} = \mu^{(1)}\end{pmatrix} \in \mathrm{dgcAlgs}_{\mathbb{R}}^{\geq 0} \tag{4.38}$$

with

$$\mu^{r} \in A\begin{bmatrix}\alpha^{(r-1)}_{n_{r-1}},\\ \vdots\\ \alpha^{(1)}_{n_1}\end{bmatrix}, \qquad \text{for } 1 \leq r \leq k.$$

These multivariate polynomial algebras (4.38) receive the canonical inclusion (4.37) of A:

$$A \overset{i_A}{\hookrightarrow} A\begin{bmatrix}\alpha^{(k)}_{n_k},\\ \vdots\\ \alpha^{(2)}_{n_2}\\ \alpha^{(1)}_{n_1}\end{bmatrix}\Big/\begin{pmatrix}d\,\alpha^{(k)}_{n_k} = \mu^{(k)},\\ \vdots\\ d\,\alpha^{(2)}_{n_1} = \mu^{(2)},\\ d\,\alpha^{(1)}_{n_1} = \mu^{(1)}\end{pmatrix}, \tag{4.39}$$

these being the composites of the stage-wise inclusions (4.37).

Definition 4.14 (Semifree dgc-Algebras/Sullivan models/FDAs). The multivariate polynomial dgc-algebras of Ex. 4.15 are sometimes called **(i)** *semi-free dgc-algebras* over A (since their underlying graded-commutative algebra (4.15) is free, as in Ex. 4.3), but they are traditionally known **(ii)** in rational homotopy theory as *relative Sullivan models* (due to [Sullivan (1977)], review in [Félix *et al.* (2001), II][Menichi (2015)][Félix and Halperin (2017)]), or, **(iii)** in supergravity theory (following [van Nieuwenhuizen (1983)][D'Auria and Fré (1982)]), as *FDAs*[4] [Castellani *et al.* (1991)], (for translation see

[4]Beware that "FDA" in the supergravity literature is meant to be short-hand for "free differential algebra", which is misleading, because what is really meant are not free dgc-algebras as in Ex. 4.12 (in general) but just "semi-free" dcg-algebras, only whose underlying graded-commutative algebras (4.15) is required to be free (Ex. 4.3).

[Fiorenza *et al.* (2015c)][Fiorenza *et al.* (2017)][Fiorenza *et al.* (2018)][Huerta *et al.* (2019)][Braunack-Mayer *et al.* (2019)][Fiorenza *et al.* (2019)]). Here we write:

$$\mathrm{SullModels}^{\geq 1}_{\mathbb{R}} \hookrightarrow \mathrm{SullModels}_{\mathbb{R}} \hookrightarrow \mathrm{dgcAlgs}^{\geq 0}_{\mathbb{R}} \tag{4.40}$$

for, from right to left, **(a)** the full subcategory of connective dgc-algebras (Def. 4.10) on those which are isomorphic to a multivariate polynomial dgc-algebra over $\mathbb{R}$, as in Ex. 4.15 (i.e., the ordering of the generators in (4.38) is not part of the data of a Sullivan model, only the resulting dgc-algebra); and **(b)** for the further full subcategory on those Sullivan model that are generated in positive degree ≥ 1.

Example 4.16 (Polynomial dgc-algebras as pushouts). For $A \in \mathrm{dgcAlgs}^{\geq 0}_{\mathbb{R}}$ (Def. 4.10) the polynomial dgc-algebras over A (Def. 4.14) are pushouts in $\mathrm{dgcAlgs}^{\geq 0}_{\mathbb{R}}$ of the following form:

$$\begin{array}{ccc}
A\big[\alpha_n\big]\big/\big(d\,\alpha_n = \mu\big) & \xleftarrow[\mu \leftarrow c_{n+1}]{\alpha_n \leftarrow \alpha_n} & \mathbb{R}\begin{bmatrix}\alpha_n, \\ c_{n+1}\end{bmatrix}\Big/\begin{pmatrix} d\,\alpha_n & = c_{n+1} \\ d\,c_{n+1} & = 0\end{pmatrix} \\
{\scriptstyle i_A}\Big\uparrow & \text{(po)} & \Big\uparrow {\scriptstyle c_{n+1} \,\mapsfrom\, c_{n+1}} \\
A & \xleftarrow[\mu \leftarrow c_{n+1}]{} & \mathbb{R}\big[c_{n+1}\big]\big/\big(d\,c_{n+1} = 0\big)
\end{array} \tag{4.41}$$

Here on the right we have multivariate polynomial dgc-algebras (Ex. 4.15) over $\mathbb{R}$ (Ex. 4.6) as shown. The horizontal morphisms encode the choice of $\mu \in A$ (4.35) and the left vertical morphism is the canonical inclusion (4.37).

Example 4.17 (Chevalley-Eilenberg algebras of nilpotent Lie algebras). Beware that not every Lie algebra $\mathfrak{g}$ has Chevalley-Eilenberg algebra (Ex. 4.10) which satisfies the stratification in the Def. 4.15 of multivariate polynomial dgc-algebras.

- **(i)** For instance, the Lie algebra $\mathfrak{su}(2)$ has

$$\mathrm{CE}\big(\mathfrak{su}(2)\big) = \mathbb{R}\big[\theta_1, \theta_2, \theta_3\big]\Big/\Big(d\,\theta_i = \sum_{j,k} \varepsilon_{ijk}\theta_j \wedge \theta_k\Big)$$

  and no ordering of $\{1,2,3\}$ brings this into the iterative form required in (4.38).

- **(ii)** Instead, those Lie algebras whose CE-algebra is of the form (4.38) are precisely the nilpotent Lie algebras.

In generalization of Ex. 4.17, we may say (by [Berglund (2015), Thm. 2.3] this matches [Getzler (2009), Def. 4.2]):

Definition 4.15 (Nilpotent L_∞-algebras). An L_∞-algebra (4.25) is *nilpotent* if its CE-algebra (Def. 4.13) is a multivariate polynomial dgc-algebra (Ex. 4.15), hence is in the sub-category of $\mathrm{SullModels}_{\mathbb{R}}$ (4.40):

$$
\begin{array}{ccc}
L_\infty \mathrm{Algs}^{\geq 0, \mathrm{nil}}_{\mathbb{R}, \mathrm{fin}} & \xhookrightarrow{\;\;\mathrm{CE}\;\;} & \big(\mathrm{SullModels}_{\mathbb{R}}\big)^{\mathrm{op}} \\
\big\downarrow & \scriptstyle (\mathrm{pb}) & \big\downarrow \\
L_\infty \mathrm{Algs}^{\geq 0}_{\mathbb{R}, \mathrm{fin}} & \xhookrightarrow{\;\;\mathrm{CE}\;\;} & \big(\mathrm{dgcAlgs}^{\geq 0}_{\mathbb{R}}\big)^{\mathrm{op}}
\end{array}
\tag{4.42}
$$

In fact, from (4.22) it is clear that every connected Sullivan model, hence with generators in degrees ≥ 1, is the Chevalley-Eilenberg algebra of a unique nilpotent L_∞-algebra, so that the defining inclusion at the top of (4.42) further restricts to an equivalence of homotopy categories:

$$
L_\infty \mathrm{Algs}^{\geq 0, \mathrm{nil}}_{\mathbb{R}, \mathrm{fin}} \xrightarrow[\simeq]{\;\;\mathrm{CE}\;\;} \big(\mathrm{SullModels}^{\geq 1}_{\mathbb{R}}\big)^{\mathrm{op}} . \tag{4.43}
$$

Homotopy theory of connective dgc-Algebras. We recall the homotopy theory of connective differential graded-commutative algebras, making free use of model category theory [Quillen (1967)]; for a review see [Hovey (1999)][Lurie (2009a), A.2] and Chapter 1.

Definition 4.16 (Homotopical structure on connective cochain complexes). Consider the following sub-classes of morphisms in the category $\mathrm{CochainComplexes}^{\geq 0}_{\mathbb{R}}$ (Def. 4.7):

(i) W – *weak equivalences* are the quasi-isomorphisms;
(ii) Fib – *fibrations* are the degreewise surjections;
(iii) Cof – *cofibrations* are the injections in positive degrees.
We call this the *injective* homotopical structure on $\mathrm{CochainComplexes}^{\geq 0}_{\mathbb{R}}$.

Proposition 4.17 (Injective model structure on connective cochain complexes [Hess (2007), p. 6]**).** *Equipped with the injective homotopical structure of Def. 4.16 the category of* $\mathrm{CochainComplexes}^{\geq 0}_{\mathbb{R}}$ *(Def. 4.7) becomes a model category (Def. 1.3) which is right proper (Def. 1.4). We denote this by:*

$$
\big(\mathrm{CochainComplexes}^{\geq 0}_{\mathbb{R}}\big)_{\mathrm{inj}} \in \mathrm{ModelCategories}.
$$

Proof. The proof of the model structure itself is formally dual to the proof of the projective model structure on connective chain complexes [Quillen (1967), II.4][Goerss and Schemmerhorn (2007), Thm. 1.5]; it is spelled out in [Dungan (2010), Thm. 2.4.5]. (Here we are using that for modules over a field of characteristic zero, as in our case, the condition that kernels of epimorphisms be injective is automatic.) A proof of right properness with respect to degreewise surjections is spelled out in [Strickland (2020), Prop. 24]. □

Definition 4.18 (Homotopical structure on connective dgc-algebras [Bousfield and Gugenheim (1976), §4.2][Gelfand and Manin (1996), §V.3.4]**).** Consider the following sub-classes of morphisms in the category of $\mathrm{dgcAlgs}^{\geq 0}_{\mathbb{R}}$ (Def. 4.10):

(i) W – *weak equivalences* are the quasi-isomorphisms;
(ii) Fib – *fibrations* are the degreewise surjections;
We call this the *projective* homotopical structure on $\mathrm{dgcAlgebras}^{\geq 0}_{\mathbb{R}}$.

Proposition 4.19 (Model structure on connective dgc-algebras). *Equipped with the homotopical structure from Def. 4.18, the category of* $\mathrm{dgcAlgs}_{\mathbb{R}}^{\geq 0}$ *(Def. 4.10), becomes a model category (Def. 1.3) which is right proper (Def. 1.4), in fact this is the case over any ground field k of characteristic 0.*

$$\left(\mathrm{dgcAlgs}_{k}^{\geq 0}\right)_{\mathrm{trinj}} \in \mathrm{ModelCategories}. \tag{4.44}$$

Proof. The model structure itself is due to [Bousfield and Gugenheim (1976), §4.3], the proof is spelled out in [Gelfand and Manin (1996), V.3.4]. Right properness follows from the right properness of the injective model structure on cochain complexes (Prop. 4.17) since the free/forgetful adjunction (4.45) implies that underlying pullbacks of dgc-algebras are pullbacks of the underlying cochain complexes. □

Proposition 4.20 (Quillen adjunction between dgc-algebras and cochain complexes). *The adjunction* (4.1) *between* $\mathrm{dgcAlgs}_{\mathbb{R}}^{\geq 0}$ *(Def. 4.10) and* $\mathrm{CochainComplexes}_{\mathbb{R}}^{\geq 0}$ *(Def. 4.7) is a Quillen adjunction (Def. 1.12) with respect to the model category structures from Prop. 4.17 and Prop. 4.19:*

$$\left(\mathrm{dgcAlgs}_{\mathbb{R}}^{\geq 0}\right)_{\mathrm{trinj}} \underset{\text{underlying}}{\overset{\mathrm{Sym}}{\underset{\longrightarrow}{\overset{\longleftarrow}{\bot_{\mathrm{Qu}}}}}} \left(\mathrm{CochainComplexes}_{\mathbb{R}}^{\geq 0}\right)_{\mathrm{inj}}. \tag{4.45}$$

Proof. It is immediate from Def. 4.16 and Def. 4.18 that the forgetful right adjoint preserves the classes W and Fib. □

Remark 4.3 (All dgc-algebras are projectively fibrant). Every object $A \in \left(\mathrm{dgcAlgs}_{\mathbb{R}}^{\geq 0}\right)_{\mathrm{trinj}}$ (4.44) is fibrant: By Ex. 4.7 the terminal morphism is to the 0-algebra, and this is clearly surjective, hence is a fibration, by Def. 4.18: $A \xrightarrow[\in \mathrm{Fib}]{} 0$.

Cofibrant dgc-algebras. With all dgc-algebras being fibrant (Rem. 4.3), the crucial property is cofibrancy.

Lemma 4.1 (Generating cofibrations). *The following inclusions of multivariate polynomial dgc-algebras (Ex. 4.15) are cofibrations in* $\left(\mathrm{dgcAlgs}_{\mathbb{R}}^{\geq 0}\right)_{\mathrm{trinj}}$ *(Def. 4.19)*

$$\mathbb{R}[c_{n+1}]/(d\,c_{n+1} = 0) \xhookrightarrow[\in \mathrm{Cof}]{c_{n+1} \mapsto c_{n-1}} \mathbb{R}\begin{bmatrix} \alpha_n, \\ c_{n+1} \end{bmatrix} \Big/ \begin{pmatrix} d\,\alpha_n = c_{n+1}, \\ d\,c_{n+1} = 0 \end{pmatrix} \quad \textit{for } n \in \mathbb{N}. \tag{4.46}$$

(In fact, these are the *generating* cofibrations of dgc-algebras, in the sense of cofibrant generation of model categories, but we do not further need this notion here.)

Proof. Consider the following morphisms of cochain complexes, for $n \in \mathbb{N}$:

$$
\left[\begin{array}{c} \vdots \\ 0 \\ \uparrow d \\ 0 \\ \uparrow d \\ \langle c_{n+1} \rangle \\ \uparrow d \\ 0 \\ \uparrow d \\ 0 \\ \uparrow d \\ \vdots \\ \uparrow d \\ 0 \end{array}\right]
\overset{i_n}{\hookrightarrow}
\left[\begin{array}{c} \vdots \\ 0 \\ \uparrow d \\ 0 \\ \uparrow d \\ \langle c_{n+1} \rangle \\ \uparrow d \\ \langle \alpha_n \rangle \\ \uparrow d \\ 0 \\ \uparrow d \\ \vdots \\ \uparrow d \\ 0 \end{array}\right]
\quad \text{with } d\alpha_n = c_{n+1}. \tag{4.47}
$$

Since these are injections, they are cofibrations in $\big(\mathrm{CochainComplexes}_{\mathbb{R}}^{\geq 0}\big)_{\mathrm{inj}}$ (Prop. 4.17), by Def. 4.16. Thus also their images under Sym (Def. 4.12) are cofibrations in $\big(\mathrm{dgcAlgs}_{\mathbb{R}}^{\geq 0}\big)_{\mathrm{trinj}}$ (Prop. 4.19) because Sym is a left Quillen functor, by Prop. 4.20. But $\mathrm{Sym}(i_n)$ manifestly equals (4.46), and so the claim follows. □

Proposition 4.21 (Relative Sullivan algebras are cofibrations). *For a multivariate polynomial dgc-algebra from Ex. 4.15, the canonical inclusion* (4.39) *of the base algebra is a cofibration in* $\big(\mathrm{dgcAlgs}_{\mathbb{R}}^{\geq 0}\big)_{\mathrm{trinj}}$ *(Prop. 4.19):*

$$
A \underset{\in \mathrm{Cof}}{\overset{i_A}{\hookrightarrow}} A\left[\begin{array}{c} \alpha_{n_k}^{(k)}, \\ \vdots \\ \alpha_{n_1}^{(1)} \end{array}\right] \Big/ \left(\begin{array}{c} d\,\alpha_{n_k}^{(k)} = \mu^{(k)}, \\ \vdots \\ d\,\alpha_{n_1}^{(1)} = \mu^{(1)} \end{array}\right). \tag{4.48}
$$

In particular, since $\mathbb{R} \in \mathrm{dgcAlgs}_{\mathbb{R}}^{\geq 0}$ *is the initial object (Ex. 4.6), all multivariate polynomial dgc-algebras over* $\mathbb{R}$ *(the Sullivan models, Def. 4.14) are cofibrant objects in* $\big(\mathrm{dgcAlgs}_{\mathbb{R}}^{\geq 0}\big)_{\mathrm{trinj}}$.

Proof. By Lem. 4.1, the right vertical morphisms in the pushout diagram (4.41) are cofibrations. Since the class of cofibrations is preserved under pushout, so are hence the left vertical morphisms in (4.41), which are the base algebra inclusions (4.37) of polynomial dgc-algebras. The base algebra inclusions into general multivariate polynomial dgc-algebras are composites of these, and since the class of cofibrations is preserved under composition, the claim follows. □

Lemma 4.2 (Pushout along relative Sullivan algebras preserves quasi-isomorphisms [Félix *et al.* (2001), Prop. 6.7(ii), Lem. 14.2]). *The operation of pushout* (1.6) *along* *the canonical inclusion* (4.39) *of a base dgc-algebra into a multivariate polynomial dgc-algebra (Ex. 4.15) preserves quasi-isomorphisms. In fact, it sends quasi-isomorphism between base algebras to quasi-isomorphisms of multivariate polynomial dgc-algebras:*

$$
\begin{array}{ccc}
A & \xrightarrow[i_A]{\in \mathrm{Cof}} & A\left[\begin{matrix}\alpha^{(k)}_{n_k}, \\ \vdots \\ \alpha^{(1)}_{n_1}\end{matrix}\right] \Big/ \left(\begin{matrix} d\,\alpha^{(k)}_{n_k} = \mu^{(k)}, \\ \vdots \\ d\,\alpha^{(1)}_{n_1} = \mu^{(1)} \end{matrix}\right) \\
{\scriptstyle f}\downarrow{\scriptstyle \in \mathrm{W}} & \text{(po)} & \downarrow{\scriptstyle (i_A)_* f} \\
A' & \xrightarrow[i_A]{\in \mathrm{Cof}} & A'\left[\begin{matrix}\alpha^{(k)}_{n_k}, \\ \vdots \\ \alpha^{(1)}_{n_1}\end{matrix}\right] \Big/ \left(\begin{matrix} d\,\alpha^{(k)}_{n_k} = \mu^{(k)}, \\ \vdots \\ d\,\alpha^{(1)}_{n_1} = \mu^{(1)} \end{matrix}\right)
\end{array}
\qquad \Rightarrow \qquad (i_a)_* f \,\in\, \mathrm{W}\,. \tag{4.49}
$$

Lemma 4.3 (Weak equivalences of nilpotent L_∞-algebras [Félix *et al.* (2001), Prop. 14.13]**).** *A morphism between Chevalley-Eilenberg algebras (Def. 4.13) of nilpotent L_∞-algebras (Def. 4.15), is a quasi-isomorphism of dgc-algebras (hence a weak equivalence according to Def. 4.18) precisely if the corresponding morphism* (4.20) *of L_∞-algebras is a quasi-isomorphism between the chain complexes given by the unary bracket operation $\partial := [-]$* (4.24)*:*

$$
\mathrm{CE}(\mathfrak{g}) \xleftarrow[\in \mathrm{W}]{\mathrm{CE}(\phi)} \mathrm{CE}(\mathfrak{h}) \quad \Leftrightarrow \quad \big(\mathfrak{g}, [-]_{\mathfrak{g}}\big) \xrightarrow[\in \mathrm{W}]{\phi} \big(\mathfrak{h}, [-]_{\mathfrak{h}}\big).
$$

Remark 4.4 (Homotopy theory of nilpotent L_∞-algebras inside all L_∞-algebras).
(i) Prop. 4.21, with Rem. 4.3 and Def. 4.13, allows to identify the homotopy category of finite-type nilpotent connective L_∞-algebras (Def. 4.15), with a full subcategory of the homotopy category (Def. 1.8) of the opposite (Ex. 1.4) of dgc-algebras (Prop. 4.19):

$$
\mathrm{Ho}\big(L_\infty\mathrm{Algs}^{\geq 0,\mathrm{nil}}_{\mathbb{R},\mathrm{fin}}\big) \overset{\mathrm{CE}}{\hookrightarrow} \mathrm{Ho}\Big(\big(\mathrm{dgcAlgs}^{\geq 0}_{\mathbb{R}}\big)^{\mathrm{op}}_{\mathrm{trinj}}\Big)\,. \tag{4.50}
$$

(ii) There is also the homotopy theory of more general L_∞-algebras [Hinich (2001)][Pridham (2010)][Vallette (2020)][Rogers (2020)], whose weak equivalences are the quasi-isomorphisms on chain complexes formed by the unary bracket $[-]$ (4.24). Lem. 4.3 says that the homotopy theory (4.50) of finite-type, nilpotent connective L_∞-algebras that we are concerned with here is fully faithfully embedded into this more general L_∞ homotopy theory:

$$
\mathrm{Ho}\big(L_\infty\mathrm{Algs}^{\geq 0,\mathrm{nil}}_{\mathbb{R},\mathrm{fin}}\big) \hookrightarrow \mathrm{Ho}\big(L_\infty\mathrm{Algs}_{\mathbb{R}}\big)\,.
$$

Minimal Sullivan models

Definition 4.22 (Minimal Sullivan models [Bousfield and Gugenheim (1976), Def. 7.2][Hess (2007), Def. 1.10]**).** A connected (relative) Sullivan model dgc-algebra $A \in \mathrm{SullModels}^{\geq 1}_{\mathbb{R}}$ (Def. 4.14) is called *minimal* if it is given by a multivariate polynomial dgc-algebra as in (4.38) the degrees n_i of whose generators $\alpha^{(i)}_{n_i}$ are of monotonically increasing degrees n_i

$$i < j \;\; \Rightarrow \;\; n_j \leq n_j .$$

Remark 4.5 (Meaning of minimality). The condition in Def. 4.22 means that a dgc-algebra is minimal precisely if it may be obtained by a sequence of pushouts as in Ex. 4.16 such that the generators are adjoined in order of increasing degree, hence such that generator $\alpha^{(j)}_{n_j}$ adjoined in the jth step has degree no lower than the degrees of the generators already adjoined in the previous steps.

Often there are no generators in degree 1, in which case the minimality condition has the following simpler and more popular form:

Example 4.18 (Minimal models of simply connected dgc-algebras [Bousfield and Gugenheim (1976), Prop. 7.4]**).** If $A \in \mathrm{SullModels}^{\geq 1}_{\mathbb{R}}$ (Def. 4.14) is trivial in degree 1, then it is minimal (Def. 4.22) precisely if the unary bracket $[-]$ (4.23) of the corresponding L_∞-algebra (4.43) vanishes:

$$A^1 = 0 \;\; \Rightarrow \;\; \big(A \text{ is minimal} \;\; \Leftrightarrow \;\; [-] = 0\big) .$$

Finally, the condition $[-] = 0$ is often written in its dual form as

$$dA \subset A \wedge A ,$$

expressing that the CE-differential of a single generator is (either zero or) the wedge product of at least two generators.

Proposition 4.23 (Existence of minimal Sullivan models [Bousfield and Gugenheim (1976), Prop. 7.7, 7.8][Félix *et al.* (2001), Thm. 14.12]**).** *If* $A \in \mathrm{dgcAlgs}^{\geq 0}_{\mathbb{R}}$ *is cohomologically connected, in that* $H^0(A) = \mathbb{R}$*, then:*

(i) *There exists a minimal Sullivan model* A_{min} *(Def. 4.22) with weak equivalence in* $\big(\mathrm{dgcAlgs}^{\geq 0}_{\mathbb{R}}\big)_{\mathrm{trinj}}$ (4.44) *to* A

$$A_{\mathrm{min}} \xrightarrow{\;\; p^{\mathrm{min}}_A \in \mathrm{W} \;\;} A . \tag{4.51}$$

(ii) *This* A_{min} *is unique up to isomorphisms of* $\mathrm{dgcAlgs}^{\geq 0}_{\mathbb{R}}$ *compatible with the weak equivalences in* (4.51)*: Any two* $p^{\mathrm{min}}_A, p^{\mathrm{min}'}_A$ *in* (4.51)*, make a commuting diagram of this form ([Félix et al. (2001), Thm. 14.12]):*

$$\begin{array}{ccc} A_{\mathrm{min}} & \xrightarrow{p^{\mathrm{min}}_A} & \\ \simeq\downarrow & & A . \\ A_{\mathrm{min}'} & \xrightarrow{p^{\mathrm{min}'}_A} & \end{array} \tag{4.52}$$

More generally:

Proposition 4.24 (Existence of minimal relative Sullivan models [Félix *et al.* (2001), Thm. 14.12]**).** *Let* $B \xrightarrow{\phi} A$ *be a morphism in* $\mathrm{dgcAlgs}^{\geq 0}_{\mathbb{R}}$ *(Def. 4.10) such that*

(a) *A and B are cohomologically connected, in that $H^0(A) = \mathbb{R}$ and $H^0(B) = \mathbb{R}$,*
(b) *$H^1(\phi) : H^1(B) \longrightarrow H^1(A)$ is an injection.*

Then:

(i) *There exists a minimal relative Sullivan model $B \hookrightarrow A_{\mathrm{min}_B}$ (Def. 4.22) equipped with a weak equivalence to ϕ in $\left(\mathrm{dgcAlgs}^{\geq 0}_{\mathbb{R}}\right)_{\mathrm{trinj}}$ (Def. 4.44):*

$$\begin{array}{ccc} A_{\mathrm{min}_B} & \xrightarrow{\;\in \mathrm{W}\;} & A \\ & \nwarrow \qquad \nearrow_{\phi} & \\ & B & \end{array} \tag{4.53}$$

(ii) *This A_{min_B} is unique up to isomorphism in the coslice category $\left(\mathrm{dgcAlgs}^{\geq 0}_{\mathbb{R}}\right)^{B/}$ compatible with the weak equivalence in* (4.53), *in cosliced generalization of* (4.52).

Chapter 5

$\mathbb{R}$-rational homotopy theory

We recall fundamental facts of dg-algebraic rational homotopy theory [Sullivan (1977)][Bousfield and Gugenheim (1976)][Griffiths and Morgan (2013)] (review in [Félix *et al.* (2001)][Hess (2007)][Félix *et al.* (2008)] [Félix and Halperin (2017)]),[1] with emphasis on its incarnation over the *real* numbers (Rem. 5.2) and streamlined towards the application to non-abelian de Rham theory below in Chapter 6 and thus to the non-abelian character map in Part IV. For the usual technical reasons (Rem. 5.1), we focus on the following class of homotopy types (with little to no restriction in practice):

Definition 5.1 (Connected nilpotent spaces of finite rational type [Bousfield and Gugenheim (1976), 9.2]**).** Write

$$\mathrm{Ho}\big(\Delta\mathrm{Sets}_{\mathrm{Qu}}\big)^{\mathrm{fin}_{\mathbb{Q}}}_{\geq 1,\mathrm{nil}} \hookrightarrow \mathrm{Ho}\big(\Delta\mathrm{Sets}_{\mathrm{Qu}}\big)$$

for the full subcategory of homotopy types of topological spaces X (1.51) on those which are:

(i) *connected*: $\pi_0(X) \simeq *$;

(ii) *nilpotent*: $\pi_1(X) \in \mathrm{NilpotentGroups}$, and $\pi_{n\geq 2}(X)$ are nilpotent $\pi_1(X)$-modules (e.g. [Hilton (1982)]);

(iii) *finite rational type*: $\dim_{\mathbb{Q}}\big(H^n(X;\mathbb{Q})\big) < \infty$, for all $n \in \mathbb{N}$.

Remark 5.1 (Technical assumptions). The connectedness assumption in Def. 5.1 is a pure convenience; for non-connected spaces all of the following applies just by iterating over connected components. On the other hand, the nilpotency and $\mathbb{R}$-finiteness condition in Def. 5.1 are strictly necessary for the plain dg-algebraic formulation of rational homotopy theory (due to [Bousfield and Gugenheim (1976)][Sullivan (1977)]) to satisfy the fundamental theorem (Thm. 5.6 below). The generalizations required to drop these assumptions are known, but considerably more unwieldy.

[1] One may naturally understand rational homotopy theory also within the theory of derived algebraic ∞-stacks [Toën (2006)][Lurie (2011), §1]. The real character map in Part IV instead expresses rational homotopy theory within smooth (differential-geometric) ∞-stacks in non-abelian generalization of the way it appears in differential cohomology theory, see Chapter 9.

(i) To drop the nilpotency assumption, all dgc-algebra models need to be equipped with the action of the fundamental group (see [Félix *et al.* (2015)]).
(ii) To drop the finite-type assumption one needs dgc-coalgebras in place of dgc-algebras, as in the original [Quillen (1969)].

Therefore, the construction of the (twisted) non-abelian character map, below in sections Part IV and Part V, works also without imposing these technical assumptions, but a discussion in that generality is beyond the scope of the present article, we discuss this elsewhere.

Example 5.1 (Examples of nilpotent spaces [Hilton (1982), §3][May and Ponto (2012), §3.1]**).** Such examples (Def. 5.1) include:

- **(i)** every simply connected space X, $\pi_1(X) = 1$;
- **(ii)** every simple space X, i.e. with abelian fundamental group acting trivially, such as tori;
- **(iii)** hence every connected H-space;
- **(iv)** hence every loop space $X \simeq \Omega Y$, and hence every ∞-group (Prop. 2.2);
- **(v)** hence every infinite-loop space, i.e., every component space E_n of a spectrum E (2.13);
- **(vi)** the classifying spaces BG (2.8) of nilpotent Lie groups G;
- **(vii)** the mapping spaces $\mathrm{Maps}(X,A)$ out of manifolds X into nilpotent spaces A.

Rational homotopy theory is concerned with understanding the following notion:

Definition 5.2 (Rationalization [Bousfield and Kan (1972b), p. 133][Bousfield and Gugenheim (1976), §11.1][Hess (2007), §1.4, §1.7]**).**
(i) A connected nilpotent homotopy type $X \in \mathrm{Ho}\big(\mathrm{TopSp}_{\mathrm{Qu}}\big)_{\geq 1,\mathrm{nil}}$ (Def. 5.1) is called *rational* if the following equivalent conditions hold [Bousfield and Kan (1972b), §V 3.3][Bousfield and Gugenheim (1976), §9.2]:

- the higher homotopy groups $\pi_{\bullet \geq 2}(X)$ have the structure of $\mathbb{Q}$-vector spaces, and the fundamental group $\pi_1(X)$ is *uniquely divisible* in that each element g has a unique nth root x, i.e. with $x^n = g$, for all $n \in \mathbb{N}_+$;
- the integral homology groups $H_{\bullet \geq 1}(X;\mathbb{Z})$ all carry the structure of $\mathbb{Q}$-vector spaces;

(ii) A *rationalization* of X is a map

$$X \xrightarrow{\eta_X^{\mathbb{Q}}} L_{\mathbb{Q}}(X) \quad \in \mathrm{Ho}\big(\mathrm{TopSp}_{\mathrm{Qu}}\big)_{\geq 1,\mathrm{nil}} \tag{5.1}$$

such that:

(a) $L_{\mathbb{Q}}$ is rational in the above sense;
(b) the map $\eta_X^{\mathbb{Q}}$ induces an isomorphism on rational cohomology groups:

$$H^\bullet\big(L_{\mathbb{Q}}X;\mathbb{Q}\big) \xrightarrow[\simeq]{H^\bullet(\eta_X^{\mathbb{Q}};\mathbb{Q})} H^\bullet(X;\mathbb{Q}).$$

Rationalization exists essentially uniquely, and defines a reflective subcategory inclusion

$$\underset{\text{connected, nilpotent, rational homotopy types}}{\mathrm{Ho}\big(\mathrm{TopSp}_{\mathrm{Qu}}\big)^{\mathbb{Q}}_{\geq 1,\mathrm{nil}}} \underset{\hookrightarrow}{\overset{L_{\mathbb{Q}}}{\underset{\bot}{\longleftarrow}}} \underset{\text{connected, nilpotent homotopy types}}{\mathrm{Ho}\big(\mathrm{TopSp}_{\mathrm{Qu}}\big)_{\geq 1,\mathrm{nil}}} \tag{5.2}$$

whose adjunction unit (1.3) is (5.1).

PL de Rham theory. At the heart of dg-algebraic rational homotopy theory is the observation that a variant of the de Rham dg-algebra of a smooth manifold (Ex. 4.9) applies to general topological spaces: the *PL de Rham complex*[2] (Def. 5.3). This satisfies an appropriate *PL de Rham theorem* (Prop. 5.4) and makes dg-algebras of PL differential forms detect rational homotopy type (Prop. 5.6). At the same time, over a smooth manifold the PL de Rham complex is suitably equivalent to the smooth de Rham complex (Lem. 6.4).

Definition 5.3 (PL de Rham complex and PL de Rham cohomology [Bousfield and Gugenheim (1976), pp. 1-7][Griffiths and Morgan (2013), §9.1]**).** Let k be a field of characteristic zero.

(i) The *simplicial dgc-algebra of k-polynomial differential forms on the standard simplices* ([Sullivan (1977), p. 297][Bousfield and Gugenheim (1976), p. 1][Griffiths and Morgan (2013), p. 83]) is:

$$\begin{array}{rcl} \Omega^\bullet_{k\mathrm{pdR}}\big(\Delta^{(-)}\big) \;:\; \Delta^{\mathrm{op}} & \longrightarrow & \mathrm{dgcAlgs}_k^{\geq 0} \\[6pt] [n] & \longmapsto & k\Big[t_0^{(0)},\ldots,t_0^{(n)},\,\theta_1^{(0)},\ldots,\theta_1^{(n)}\Big]\Big/\left(\begin{smallmatrix}\sum_i t_0^{(i)} = 1,\\ \forall_i\, d t_0^{(i)} = \theta_1^{(i)}\end{smallmatrix}\right) \\[6pt] {\scriptstyle f}\Big\downarrow & & \Big\uparrow \;\; \sum\limits_{i\in f^{-1}(\{j\})} t_0^{(i)} \;\leftarrow\!\shortmid\; t_0^{(j)} \\[6pt] [m] & \longmapsto & k\Big[t_0^{(0)},\ldots,t_0^{(m)},\,\theta_1^{(0)},\ldots,\theta_1^{(m)}\Big]\Big/\left(\begin{smallmatrix}\sum_j t_0^{(j)} = 1,\\ \forall_j\, d t_0^{(j)} = \theta_1^{(j)}\end{smallmatrix}\right) \end{array} \tag{5.3}$$

(ii) For $S \in \Delta\mathrm{Sets}$, its *piecewise k-linear* de Rhm complex, or *PL de Rham complex* for short, is the hom-object of simplicial objects from S to $\Omega^\bullet_{k\mathrm{pdR}}\big(\Delta^{(-)}\big)$ (5.3), hence is the following *end* (e.g. [Borceux (1994), Def. 6.6.8]) in $\mathrm{dgcAlgs}_{\mathbb{R}}^{\geq 0}$:

$$\underset{\substack{\text{piecewise } k\text{-linear}\\ \text{de Rham complex}\\ \text{of simplicial set}}}{\Omega^\bullet_{\mathrm{P}k\mathrm{LdR}}(S)} \;:=\; \int_{[n]\in\Delta} \prod_{S_n} \underset{\substack{k\text{-polynomial}\\ \text{de Rham complex}\\ \text{of the } n\text{-simplex}}}{\Omega^\bullet_{k\mathrm{pdR}}(\Delta^n)}\,. \tag{5.4}$$

[2] The terminology "PL" or "P.L." for this construction seems to have been silently introduced in [Bousfield and Gugenheim (1976)], as shorthand for "piecewise linear", and has become widely adopted (e.g. [Griffiths and Morgan (2013), §9]). Our subscript "PkL" is for "piecewise k-linear", in this sense. But beware that this refers to the piecewise-linear structure that a choice of triangulation (Ex. 1.16) induces on a topological space; while the actual differential forms in the PL de Rham complex are piecewise *polynomial* with respect to this piecewise linear structure.

This means that an element $\omega \in \Omega^\bullet_{\mathrm{P}k\mathrm{LdR}}(S)$ is a k-polynomial differential form $\omega^{(n)}_\sigma \in \Omega^\bullet_{k\mathrm{pdR}}(\Delta^n)$ (5.3) on each n-simplex $\sigma \in S_n$ for all $n \in \mathbb{N}$, such that these are compatible under pullback along all simplex face inclusions δ_i and along all degenerate simplex projections σ_i:

$$\Omega^\bullet_{\mathrm{P}k\mathrm{LdR}}(S) \;=\; \left\{ \begin{array}{ccc} \vdots & & \vdots \\ S_2 & \overset{\omega^{(2)}_{(-)}}{\dashrightarrow} & \Omega^\bullet_{k\mathrm{pdR}}(\Delta^2) \\ {\scriptstyle\delta_0^*}\!\downarrow\;\; {\scriptstyle\sigma_0^*}\!\uparrow\;\; {\scriptstyle\delta_1^*}\!\downarrow\;\; {\scriptstyle\sigma_1^*}\!\uparrow\;\; {\scriptstyle\delta_2^*}\!\downarrow & & {\scriptstyle\delta_0^*}\!\downarrow\;\; {\scriptstyle\sigma_0^*}\!\uparrow\;\; {\scriptstyle\delta_1^*}\!\downarrow\;\; {\scriptstyle\sigma_1^*}\!\uparrow\;\; {\scriptstyle\delta_2^*}\!\downarrow \\ S_1 & \overset{\omega^{(1)}_{(-)}}{\dashrightarrow} & \Omega^\bullet_{k\mathrm{pdR}}(\Delta^1) \\ {\scriptstyle\delta_0^*}\!\downarrow\;\; {\scriptstyle\sigma_0^*}\!\uparrow\;\; {\scriptstyle\delta_1^*}\!\downarrow & & {\scriptstyle\delta_0^*}\!\downarrow\;\; {\scriptstyle\sigma_0^*}\!\uparrow\;\; {\scriptstyle\delta_1^*}\!\downarrow \\ S_0 & \overset{\omega^{(0)}_{(-)}}{\dashrightarrow} & \Omega^\bullet_{k\mathrm{pdR}}(\Delta^0) \end{array} \right\}.$$

(iii) For $X \in \mathrm{TopSp}$, its *PL de Rham complex* is that of its singular simplicial set, according to (5.4):

$$\Omega^\bullet_{\mathrm{P}k\mathrm{LdR}}(X) := \Omega^\bullet_{\mathrm{P}k\mathrm{LdR}}\big(\mathrm{Sing}(X)\big). \tag{5.5}$$

By pullback of differential forms, this extends to a functor of the form

$$\Omega^\bullet_{\mathrm{P}k\mathrm{LdR}} \;:\; \Delta\mathrm{Sets} \longrightarrow \big(\mathrm{dgcAlgs}^{\geq 0}_{\mathbb{R}}\big)^{\mathrm{op}}. \tag{5.6}$$

(iv) We write

$$H^\bullet_{\mathrm{P}k\mathrm{LdR}}(-) := H\Omega^\bullet_{\mathrm{P}k\mathrm{LdR}}(-) \tag{5.7}$$

for *PL de Rham cohomology*, the cochain cohomology of the PL de Rham complex.

Proposition 5.4 (PL de Rham theorem [Bousfield and Gugenheim (1976), Thm. 2.2] [Griffiths and Morgan (2013), Thm. 9.1]**).** *The evident operation of integrating differential forms over simplices induces a quasi-isomorphism*

$$\Omega^\bullet_{\mathrm{P}k\mathrm{LdR}}(-) \xrightarrow{\;\in \mathrm{qIso}\;} C^\bullet(-;k)$$

from the PL de Rham complex (Def. 5.3) to the cochain complex of ordinary singular cohomology with coefficients in k. Hence on cochain cohomology this induces an isomorphism

$$H^\bullet_{\mathrm{P}k\mathrm{LdR}}(-) \xrightarrow{\;\simeq\;} H^\bullet(-;k)$$

between PL de Rham cohomology (5.7) *and ordinary cohomology with coefficients in k.*

Example 5.2 (PL de Rham complex of the interval). The PL de Rham complex (Def. 5.3) of the 1-simplex, hence the polyonial differential forms (5.3) on Δ^1, is isomorphic to the multivariate polynomial dgc-algebra (Ex. 4.15) of the form

$$\Omega^\bullet_{\mathrm{PL}k\mathrm{dR}}(\Delta^1) \simeq \Omega^\bullet_{k\mathrm{pdR}}(\Delta^1) \simeq k\big[t_0,\theta_1\big]\big/\big(d\,t_0 = \theta_1\big)\,. \tag{5.8}$$

For $A \in \big(\mathrm{dgcAlgs}_k^{\geq 0}\big)_{\mathrm{trinj}}$ (Prop. 1.27), its tensor product with (5.8)

$$A \otimes_k \Omega^\bullet_{\mathrm{PL}k\mathrm{dR}}(\Delta^1) \simeq A\big[t_0,\,\theta_1\big]\big/\big(d\,t_0 = \theta_1\big)$$

is a path space object for A (Def. 1.5), in that it fits into the following diagram (we are notationally suppressing here the differentials just for readability):

$$A \xrightarrow[\in \mathrm{W}]{a \mapsto a} A\big[t_0,\,\theta_1\big] \xrightarrow[\in \mathrm{Fib}]{a \mapsto \big(a(t_0=0,\theta_1=0)+a(t_0=1,\theta_1=0)\big)} A \oplus A \simeq A \times A, \qquad \underbrace{\qquad}_{\Delta_A} \tag{5.9}$$

where the morphism on the right is given by evaluation of polynomials as shown, and where the equivalence on the right is by Ex. 4.8.

Here the morphism on the right is a degreewise surjection (a pre-image for $(a_0,a_1) \in A \oplus A$ is $(1-t_0)\wedge a_0 + t_0 \wedge a_1 \in A[t_0,\theta_1]$), hence a fibration according to Def. 4.18; while the morphism on the left is a quasi-isomorphism, in fact a chain homotopy equivalence (with homotopy operator $t_0 \cdot \frac{\partial}{\partial \theta_1}$), hence a weak equivalence according to Def. 4.18.

The following type of argument will be greatly expanded on in Chapter 6:

Lemma 5.1 (Homotopical formulation of ordinary cohomology).
(i) *The cochain cohomology of any* $A \in \mathrm{dgcAlgs}_{\mathbb{R}}^{\geq 0}$ *(Def. 4.10) in positive degree is naturally and* ℝ*-linearly identified,*

$$H^{\bullet+1}(A) \simeq \mathrm{Ho}\Big(\big(\mathrm{dgcAlgs}_{\mathbb{R}}^{\geq 0}\big)_{\mathrm{trinj}}\Big)\big(\mathrm{CE}(\mathfrak{b}^\bullet\mathbb{R}), A\big)\,, \tag{5.10}$$

with the hom-sets out of the CE-algebras (4.30) *of the line Lie* $(\bullet+1)$*-algebras (Ex. 4.12) in the homotopy category (Def. 1.8) of the dgc model category (Prop. 4.19).*
(ii) *For* $X \in \mathrm{TopSp} \xrightarrow{\mathrm{Sing}} \Delta\mathrm{Sets}$*, its real cohomology in positive degree is naturally identified with these hom-sets into its PL de Rham complex (Def. 5.3)*

$$H^{\bullet+1}(X;\mathbb{R}) \simeq \mathrm{Ho}\Big(\big(\mathrm{dgcAlgs}_k^{\geq 0}\big)_{\mathrm{trinj}}\Big)\big(\mathrm{CE}(\mathfrak{b}^\bullet\mathbb{R}), \Omega^\bullet_{\mathrm{PL}\mathbb{R}\mathrm{dR}}(X)\big) \tag{5.11}$$

(iii) *If the above* X *is equipped with a base-point* $* \xrightarrow{x} X$*, then the real cohomology of* X *in positive degrees is equivalently computed by the homotopy classes of morphisms of augmented dgc-algebras, hence with respect to the slice model structure (Ex. 1.5) over the* *initial dgc-algebra* $\mathbb{R}$ *(Ex. 4.6), as follows:*

$$H^{\bullet+1}(X;\mathbb{R}) \simeq \mathrm{Ho}\Big(\big(\mathrm{dgcAlgs}_{\mathbb{R}}^{\geq 0}\big)^{/\mathbb{R}}_{\mathrm{trinj}}\Big)\left(\mathrm{CE}(\mathfrak{b}^{\bullet}\mathbb{R}),\, \Omega^{\bullet}_{\mathrm{PRLdR}}(X) \underset{\Omega^{\bullet}_{\mathrm{PRLdR}}(*)}{\times} \mathbb{R}\right). \tag{5.12}$$

Proof. Observing that

(a) $\mathrm{CE}(\mathfrak{b}^n\mathbb{R})$ is cofibrant, by Prop. 4.21;

(b) A is fibrant, by Rem. 4.3;

(c) $A[t_0,\theta_1]/(d\,t_0 = \theta_1)$ is a path space object for A, by Ex. 5.2;

we may identify, by Prop. 1.10, the morphisms in the homotopy category with equivalence classes of dgc-algebra morphisms $\mathrm{CE}(\mathfrak{b}^n\mathbb{R}) \xrightarrow{c} A$ under the corresponding equivalence relation of right homotopy (Def. 1.6):

$$c \sim_r c' \quad \Leftrightarrow \quad \begin{array}{ccc} & & A \\ & \overset{c}{\nearrow} & \uparrow (-)_{|t_0=0,\theta_1=0} \\ \mathrm{CE}(\mathfrak{b}^n\mathbb{R}) & \overset{\exists}{\dashrightarrow} & \dfrac{A[t_0,\theta_1]}{(d\,t_0=\theta_1)} \\ & \underset{c'}{\searrow} & \downarrow (-)_{|t_0=1,\theta_1=0} \\ & & A\,. \end{array}$$

Since $\mathrm{CE}(\mathfrak{b}^n\mathbb{R})$ is free on a single generator θ_n in degree n, subject only to the differential relation $d\,\theta_n = 0$, dgc-algebra homomorphisms $\mathrm{CE}(\mathfrak{b}^n\mathbb{R}) \to A$ are in bijection to closed degree-n elements of A (see also Ex. 6.2). Hence, under this identification it remains to see that existence of coboundaries is equivalent to existence of right homotopies:

$$\underset{h\in A}{\exists}\; c' = c + dh \quad \Leftrightarrow \quad \underset{\eta \in \frac{A[t_0,\theta_1]}{(d\,t_0=\theta_1)}}{\exists} \begin{cases} d\eta = 0, \\ \eta(t_0 = 0, \theta_1 = 0) = c, \\ \eta(t_0 = 1, \theta_1 = 0) = c'. \end{cases}$$

Indeed: If h is given as on the left, then $\eta := t_0 \wedge c' + (1-t_0)\wedge c + \theta_1 \wedge (c - c')$ is as required on the right; while if any η is given as on the right, then $h := \int_0^1 \eta\, dt_0$ is as required on the left (by Stokes, as in Lem. 6.1). This proves the first statement, whence the second follows via Prop. 5.4.

To deduce from this the third statement, observe that:

(a) for $A \xrightarrow{\varepsilon_A} \mathbb{R}$ an augmented dgc-algebra, the canonical path object (5.9) for A yields a path space object in the slice over $\mathbb{R}$ by equipping it with the induced augmentation $\frac{A[t_0,\theta_1]}{(d\,t_0=\theta_1)} \xrightarrow{a \mapsto a(t_0=0,\theta_1=0)} A \xrightarrow{\varepsilon_A} \mathbb{R}$,

(b) the projection

$$\Omega^{\bullet}_{\mathrm{PRLdR}}(X) \underset{\Omega^{\bullet}_{\mathrm{PRLdR}}(*)}{\times} \mathbb{R} \xrightarrow{\in \mathrm{W}} \Omega^{\bullet}_{\mathrm{PRLdR}}(X) \tag{5.13}$$

is a quasi-isomorphism, by right-properness (Def. 1.4) of the model structure (Prop. 4.19), since this is the pullback of the quasi-isomorphism $\mathbb{R} \xrightarrow{\in \mathrm{W}} \Omega^{\bullet}_{\mathrm{PRLdR}}$ (by Prop. 5.4) along the morphism $\Omega^{\bullet}_{\mathrm{PRLdR}}(* \to X)$, which is a projective fibration by the fact that $* \to X$ is an injection and hence a cofibration (Ex. 1.2) and that $\Omega^\bullet_{\mathrm{P}\mathbb{R}\mathrm{LdR}}$ is a left Quillen functor (Prop. 5.5) to the opposite model structure (Ex. 1.4).

Therefore, since the generators $\mathrm{CE}(\mathfrak{b}^n\mathbb{R})$ are in positive degree and hence unaffected by the augmentation slicing, the right homotopy classes in the slice (5.12) are computed as in case (2) above and hence yield the cochain cohomology of $\Omega^\bullet_{\mathrm{P}\mathbb{R}\mathrm{LdR}}(X) \times_{\Omega^\bullet_{\mathrm{P}\mathbb{R}\mathrm{LdR}}(*)} \mathbb{R}$, which by (5.13) equals the real cohomology of X. □

In fact, before passing to cochain cohomology, the PL de Rham complex captures the full rational homotopy type. This is the Fundamental Theorem which we recall as Prop. 5.6:

Lemma 5.2 (Extension lemma for polynomial differential forms [Griffiths and Morgan (2013), Lem. 9.4]**).** *For $n \in \mathbb{N}$, the operation of pullback of piecewise polynomial differential forms (Def. 6.4) along the boundary inclusion of the n-simplex $\partial\Delta^n \xrightarrow{i_n} \Delta^n$ is an epimorphism:*

$$\Omega^\bullet_{\mathrm{P}k\mathrm{LdR}}(\Delta^n) \xrightarrow{\;\; i_n^* \;\;} \Omega^\bullet_{\mathrm{P}k\mathrm{LdR}}(\partial\Delta^n)\,.$$

Proposition 5.5 (PL de Rham Quillen adjunction [Bousfield and Gugenheim (1976), §8]**).** *For all ground fields k of characteristic zero, the PL de Rham complex functor (Def. 5.3) is the left adjoint in a Quillen adjunction (Def. 1.12)*

$$\big(\mathrm{dgcAlgs}_k^{\geq 0}\big)^{\mathrm{op}}_{\mathrm{trinj}} \underset{B\exp_{\mathrm{P}k\mathrm{L}}}{\overset{\Omega^\bullet_{\mathrm{P}k\mathrm{LdR}}}{\underset{\longrightarrow}{\overset{\longleftarrow}{\bot_{\mathrm{Qu}}}}}} \Delta\mathrm{Sets}_{\mathrm{Qu}} \tag{5.14}$$

between the opposite (Def. 1.4) of the model category of dgc-algebras (Prop. 4.19) and the classical model structure on simplicial sets (Prop. 1.2); where the right adjoint sends a dgc-algebra A to

$$B\exp_{\mathrm{P}k\mathrm{L}}(A) = \Big(\Delta[n] \longmapsto \mathrm{dgcAlgs}_k^{\geq 0}\big(\Omega^\bullet_{\mathrm{P}k\mathrm{LdR}}(\Delta^n), A\big)\Big) \quad \in \Delta\mathrm{Sets}\,. \tag{5.15}$$

Proof. That the right adjoint exists and is given as in (5.15) follows by general nerve/realization theory [Kan (1958)], or else by direct inspection.

For the left adjoint to preserve cofibrations means to take injections of simplicial sets to degreewise surjections of dgc-algebras. This follows from the extension lemma (Lem. 5.2). Moreover, the left adjoint preserves even all weak equivalences, by the PL de Rham theorem (Prop. 5.4). □

Proposition 5.6 (Fundamental theorem of dgc-algebraic rational homotopy theory). *For $k = \mathbb{Q}$, the derived adjunction (Prop. 1.13)*

$$\mathrm{Ho}\Big(\big(\mathrm{dgcAlgs}_{\mathbb{Q}}^{\geq 0}\big)^{\mathrm{op}}_{\mathrm{trinj}}\Big) \underset{\mathbb{D}B\exp_{\mathrm{P}\mathbb{Q}\mathrm{L}}}{\overset{\mathbb{D}\Omega^\bullet_{\mathrm{P}\mathbb{Q}\mathrm{LdR}}}{\underset{\longrightarrow}{\overset{\longleftarrow}{\bot}}}} \mathrm{Ho}\big(\Delta\mathrm{Sets}_{\mathrm{Qu}}\big) \tag{5.16}$$

of the Quillen adjunction (5.14) *from Prop. 5.5 is such that:*

(i) *on connected, nilpotent, $\mathbb{Q}$-finite homotopy types (Def. 5.1) the derived PLdR-adjunction unit* (1.38) *is equivalently the unit* (5.1) *of rationalization (Def. 5.2):*

$$
\begin{array}{ccc}
X & \xrightarrow[\mathbb{D}\eta_X^{\mathrm{PQLdR}}]{\text{derived unit of rational PL de Rham adjunction}} & \mathbb{D}B\exp_{\mathrm{PQL}} \circ \Omega^\bullet_{\mathrm{PQLdR}}(X) \\
\Big\| & & \wr\!\mid \\
X & \xrightarrow[\text{rationalization unit}]{\eta_X^{\mathbb{Q}}} & L_{\mathbb{Q}}X
\end{array}
\qquad \in \mathrm{Ho}\big(\mathrm{TopSp}_{\mathrm{Qu}}\big)^{\mathrm{fin}_{\mathbb{Q}}}_{\geq 1,\mathrm{nil}}\,. \tag{5.17}
$$

(ii) *For X, A nilpotent, connected, $\mathbb{Q}$-finite homotopy types (Def. 5.1), the PL de Rham complex functor* (5.6) *from Def. 5.3 induces natural bijections*

$$
\mathrm{Ho}\big(\mathrm{TopSp}_{\mathrm{Qu}}\big)(X, L_{\mathbb{Q}}A) \xrightarrow[\Omega^\bullet_{\mathrm{PQLdR}}]{\simeq} \mathrm{Ho}\Big(\big(\mathrm{dgcAlgs}^{\geq 0}_{\mathbb{Q}}\big)_{\mathrm{trinj}}\Big)\Big(\Omega^\bullet_{\mathrm{PQLdR}}(A)\,, \Omega^\bullet_{\mathrm{PQLdR}}(X)\Big). \tag{5.18}
$$

Proof. (i) This is [Bousfield and Gugenheim (1976), Thm. 11.2].
(ii) This follows via [Bousfield and Gugenheim (1976), Thm. 9.4(i)], which says that the derived adjunction (5.16) restricts on connected, nilpotent, $\mathbb{Q}$-finite (Def. 5.1) rational homotopy types (Def. 5.2) to an equivalence of homotopy categories:

$$
\mathrm{Ho}\Big(\big(\mathrm{dgcAlgs}^{\geq 0}_{\mathbb{Q}}\big)^{\mathrm{op}}_{\mathrm{trinj}}\Big)^{\geq 1}_{\mathrm{fin}} \underset{\mathbb{D}B\exp_{\mathrm{PQL}}}{\overset{\mathbb{D}\Omega^\bullet_{\mathrm{PQLdR}}}{\underset{\simeq}{\rightleftarrows}}} \mathrm{Ho}\big(\Delta\mathrm{Sets}_{\mathrm{Qu}}\big)^{\mathbb{Q},\mathrm{fin}_{\mathbb{Q}}}_{\geq 1,\mathrm{nil}}\,. \tag{5.19}
$$

In detail, this is witnessed by the following sequence of natural bijections of hom-sets:

$$
\begin{aligned}
& \mathrm{Ho}\left(\mathrm{TopSp}_{\mathrm{Qu}}\right)\left(X\,, L_{\mathbb{Q}}A\right) \\
& \simeq \mathrm{Ho}\left(\Delta\mathrm{Sets}_{\mathrm{Qu}}\right)\left(\mathrm{Sing}(X)\,, L_{\mathbb{Q}}\mathrm{Sing}(A)\right) \\
& \simeq \mathrm{Ho}\left(\Delta\mathrm{Sets}_{\mathrm{Qu}}\right)\left(L_{\mathbb{Q}}\mathrm{Sing}(X)\,, L_{\mathbb{Q}}\mathrm{Sing}(A)\right) \\
& \simeq \mathrm{Ho}\left(\Delta\mathrm{Sets}_{\mathrm{Qu}}\right)\left(\mathbb{D}B\exp_{\mathrm{PQL}} \circ \Omega^\bullet_{\mathrm{PQLdR}}(X)\,, \mathbb{D}B\exp_{\mathrm{PL}} \circ \Omega^\bullet_{\mathrm{PQLdR}}(A)\right) \\
& \simeq \mathrm{Ho}\left(\left(\mathrm{dgcAlgs}^{\geq 0}_{\mathbb{Q}}\right)^{\mathrm{op}}_{\mathrm{trinj}}\right)\left(\Omega^\bullet_{\mathrm{PQLdR}} \circ \mathbb{D}B\exp_{\mathrm{PQL}} \circ \Omega^\bullet_{\mathrm{PQLdR}}(X)\,,\right. \\
& \qquad\qquad\qquad \left. \Omega^\bullet_{\mathrm{PQLdR}} \circ \mathbb{D}B\exp_{\mathrm{PQL}} \circ \Omega^\bullet_{\mathrm{PQLdR}}(A)\right) \\
& \simeq \mathrm{Ho}\left(\left(\mathrm{dgcAlgs}^{\geq 0}_{\mathbb{Q}}\right)^{\mathrm{op}}_{\mathrm{trinj}}\right)\left(\Omega^\bullet_{\mathrm{PQLdR}}(X)\,, \Omega^\bullet_{\mathrm{PQLdR}}(A)\right) \\
& \simeq \mathrm{Ho}\left(\left(\mathrm{dgcAlgs}^{\geq 0}_{\mathbb{Q}}\right)_{\mathrm{trinj}}\right)\left(\Omega^\bullet_{\mathrm{PQLdR}}(A)\,, \Omega^\bullet_{\mathrm{PQLdR}}(X)\right).
\end{aligned} \tag{5.20}
$$

Here the first step is (1.51); the second step uses that rationalization is a reflection (5.2); the third step uses (5.17); the fourth is the equivalence (5.19) along $\mathbb{D}\Omega^\bullet_{\mathrm{PQLdR}}$ (using, with Ex. 1.10, that every simplicial set is already cofibrant (1.15), Ex. 1.2); the fifth step is the statement from (5.19) that $\mathbb{D}B\exp_{\mathrm{PQL}}$ is the inverse equivalence. The last step is just the definition of the opposite of a category. The composite of the bijections (5.20) is the desired bijection (5.18). □

In view of (5.17) the following notation is convenient, keeping in mind that L_k is a localization in the sense of localization of spaces only for $k = \mathbb{Q}$:

Definition 5.7 (Rationalization over $\mathbb{R}$). For k a field of characteristic zero, we write $L_k := \mathbb{D}B\exp_{\mathrm{P}k\mathrm{L}} \circ \mathbb{D}\Omega^\bullet_{\mathrm{P}k\mathrm{LdR}}$ for the monad given by the derived functors (Prop. 1.13) of the k-PL de Rham Quillen adjunction (Prop. 5.5). Our focus here is on the case over the real numbers:

$$(-) \xrightarrow{\eta^{\mathbb{R}} := \mathbb{D}\eta^{\mathrm{PL}\mathbb{R}dR}} L_{\mathbb{R}}(-) := \mathbb{D}B\exp_{\mathrm{P}\mathbb{R}\mathrm{L}} \circ \mathbb{D}\Omega^\bullet_{\mathrm{P}\mathbb{R}\mathrm{LdR}}(0)\,. \tag{5.21}$$

We may refer to $L_{\mathbb{R}}$ as *rationalization over* $\mathbb{R}$. Because, while the derived PLdR-adjunction (Prop. 5.5) is a localization of homotopy types only over $k = \mathbb{Q}$, (Prop. 5.6, Rem. 5.5), for general k it is the suitable *change of scalars* of $\mathbb{Q}$-localization:

Lemma 5.3 (Derived change of scalars [Bousfield and Gugenheim (1976), Lem. 11.6]**).** *For k a field of characteristic zero, the extension/restriction of scalars-adjunction along $\mathbb{Q} \hookrightarrow k$ is a Quillen adjunction (Def. 1.12) between the corresponding model categories of dgc-algebras (from Prop. 4.19):*

$$\big(\mathrm{dgcAlgs}_k^{\geq 0}\big)_{\mathrm{trinj}} \underset{\mathrm{res}_{\mathbb{Q}}}{\overset{(-)\otimes_{\mathbb{Q}} k}{\underset{\longrightarrow}{\overset{\longleftarrow}{\bot_{\mathrm{Qu}}}}}} \big(\mathrm{dgcAlgs}_{\mathbb{Q}}^{\geq 0}\big)_{\mathrm{trinj}}\,.$$

Proof. Since restriction of scalars $\mathrm{res}_{\mathbb{Q}}$ is the identity on the underlying sets of a dgc-algebra, it manifestly preserves all fibrations (since these are the surjections of underlying sets) and all weak equivalences (since these are the bijections on underlying cochain cohomology groups). □

Proposition 5.8 (PkLdR-Adjunction factors through rationalization). *The following diagram of derived functors (Prop. 1.13 – with the left derived functors from Prop. 5.5 and the right derived functor from Lem. 5.3) commutes up to natural isomorphism:*

$$\begin{array}{ccc}
\mathrm{Ho}\Big(\big(\mathrm{dgcAlg}_{\mathbb{Q}}^{\geq 0}\big)^{\mathrm{op}}_{\mathrm{trinj}}\Big) & \xleftarrow{\mathbb{D}\Omega^\bullet_{\mathrm{P}\mathbb{Q}\mathrm{LdR}}} & \mathrm{Ho}\big(\Delta\mathrm{Sets}_{\mathrm{Qu}}\big)^{\mathrm{fin}_{\mathbb{Q}}} \\
{\scriptstyle \mathbb{D}((-)\otimes_{\mathbb{Q}}\mathbb{R})}\big\downarrow & & \big\| \\
\mathrm{Ho}\Big(\big(\mathrm{dgcAlg}_{\mathbb{R}}^{\geq 0}\big)^{\mathrm{op}}_{\mathrm{trinj}}\Big) & \underset{\mathbb{D}\Omega^\bullet_{\mathrm{P}\mathbb{R}\mathrm{LdR}}}{\longleftarrow} & \mathrm{Ho}\big(\Delta\mathrm{Sets}_{\mathrm{Qu}}\big)^{\mathrm{fin}_{\mathbb{Q}}}\,.
\end{array} \tag{5.22}$$

Proof. Via the formula for derived functors in terms of (co)fibrant replacement (Ex. 1.10) and using that Sullivan models are cofibrant in $\big(\mathrm{dgcAlgs}_k^{\geq 0}\big)_{\mathrm{trinj}}$ (Prop. 4.21), hence fibrant in $\big(\mathrm{dgcAlgs}_k^{\geq 0}\big)^{\mathrm{op}}_{\mathrm{trinj}}$, this follows by [Bousfield and Gugenheim (1976), Lem. 11.7]. □

Remark 5.2 (Rational homotopy theory over the real numbers). Below in Prop. IV.1 we recast Prop. 5.8 as the statement that the real character map on non-abelian cohomology factors through the rational character map via extension of scalars.

This fact motivates and justifies the focus on *rational homotopy theory over the real numbers* (as in [Deligne *et al.* (1975)] [Griffiths and Morgan (2013)], see also [Wierstra (2017)]) in all of the following. Rational homotopy theory over the real numbers is the version that connects to differential geometry (e.g. [Félix *et al.* (2008)]), since the *smooth* de Rham complex is not defined over $\mathbb{Q}$ but over $\mathbb{R}$ (see Lem. 6.4). The original account [Bousfield and Gugenheim (1976)] of rational homotopy theory is, for the most part, formulated over an arbitrary field k of characteristic zero; and [Bousfield and Gugenheim (1976), Lem. 11.7] (Prop. 5.8) makes explicit that the choice of this base field does not change the form of the classical theorems. For example, the "real-ified" homotopy groups of a space X

$$\pi_\bullet(X) \otimes_{\mathbb{Z}} \mathbb{R} \simeq \big(\pi_\bullet(X) \otimes_{\mathbb{Z}} \mathbb{Q}\big) \otimes_{\mathbb{Q}} \mathbb{R}$$

form a real vector space with real dimension equal to the rational dimension of the corresponding rationalized homotopy groups

$$\dim_{\mathbb{Q}}\big(\pi_\bullet(X) \otimes_{\mathbb{Z}} \mathbb{Q}\big) = \dim_{\mathbb{R}}\big(\pi_\bullet(X) \otimes_{\mathbb{Z}} \mathbb{R}\big),$$

and hence the rational Whitehead L_∞-algebras (Prop. 5.11 below) have the same set of generators and their Chevalley-Eilenberg algebras (Def. 4.13) have the same structure constants, irrespective of whether they come as algebras over $\mathbb{Q}$ or over $\mathbb{R}$.[3] Therefore, we regard the case $k = \mathbb{R}$ as our default and abbreviate the PL de Rham Quillen adjunction (Prop. 5.5) in this case by:

$$\Omega^\bullet_{\mathrm{PLdR}} \dashv_{\mathrm{Qu}} B\exp_{\mathrm{PL}} \quad := \quad \Omega^\bullet_{\mathrm{P}\mathbb{R}\mathrm{LdR}} \dashv_{\mathrm{Qu}} B\exp_{\mathrm{P}\mathbb{R}\mathrm{L}}. \tag{5.23}$$

Remark 5.3 (Real homotopy theory and schematic homotopy type). In contrast to the *rational homotopy theory over* $\mathbb{R}$ with which we are concerned here (Rem. 5.2) is *real homotopy theory* in the sense of [Brown and Szczarba (1995)], given by localization of the category of simplicial *spaces* at *real cohomology*-equivalences seen in *continuous cohomology* [Brown and Szczarba (1989)], making use of the Euclidean topology on the real number coefficients. The "most convincing motivation" [Brown and Szczarba (1995), pp. 882-883] for this construction was formal analogy, and the main application in

---

[3] While rational homotopy theory has the same form over all ground fields of characteristic zero, there is, of course, a difference between rational homotopy equivalences over different ground fields: Two minimal Sullivan models (Prop. 4.23) over the real numbers may be isomorphic as real dgc-algebras but not as rational dgc-algebras. This happens when the isomorphism is given by an irrational linear transformation between the generators. For example, for any $b \in \mathbb{R}$, $b \geq 0$ there is a dgc-algebra isomorphism over the real numbers

$$\mathbb{R}\begin{bmatrix} \omega_3 \\ \alpha_2, \beta_2 \end{bmatrix} \Big/ \begin{pmatrix} d\,\omega_3 = \alpha_2 \wedge \alpha_2 + \beta_2 \wedge \beta_2 \\ d\,\alpha_2 = 0,\ d\beta_2 = 0 \end{pmatrix} \xrightarrow[\simeq]{\substack{\omega_3 \mapsto \omega_3 \\ \alpha_2 \mapsto \alpha_2 \\ \beta_2 \mapsto \sqrt{b}\beta_2}} \mathbb{R}\begin{bmatrix} \omega_3 \\ \alpha_2, \beta_2 \end{bmatrix} \Big/ \begin{pmatrix} d\,\omega_3 = \alpha_2 \wedge \alpha_2 + b\,\beta_2 \wedge \beta_2 \\ d\,\alpha_2 = 0,\ d\beta_2 = 0 \end{pmatrix}.$$

But over the rational numbers this exists only when the square root of b is rational.

Notice that the comparison between the homomotopy types over, in this order, the integers, the rational numbers and then the real numbers is provided by the character map (Def. 9.2 below); and the theory which embodies the distinction between these coefficients is that of homotopy fiber products of the character map, which is the theory of non-abelian differential cohomology (Def. 9.3 below), where for instance the homotopy fiber $\mathbb{R}/\mathbb{Q} \to B\mathbb{Q} \to B\mathbb{R}$ is being detected (e.g. [Grady and Sati (2019c)][Grady and Sati (2019b)]).

[Brown and Szczarba (1995), Thm. 8.2] is a re-derivation of the traditional result [Deligne *et al.* (1975), §6], originally derived in the $\mathbb{R}$-rational homotopy theory of concern here.

With hindsight, one may observe that continuous $\mathbb{R}$-cohomology of simplicial topological spaces is an approximation to cohomology of *topological stacks* with coefficients in the higher topological stack $\mathbf{B}^n\mathbb{R}$ (as in footnote 2 to Ex. 2.3 above), to which it reduces when the domain is fine enough (i.e., cofibrant as a simplicial presheaf on topological spaces). Understood in this stacky refinement, real homotopy type is a topological version of *schematic homotopy type* in algebraic geometry [Toën (2006)]; the general and smooth version of which is discussed in [Stel (2010)].

These localizations of geometric stacks at the stacky ground field are interesting and closely related to the $\mathbb{R}$-rational homotopy theory of concern here, but further discussion of the relation is beyond the scope of this text.

PS de Rham theory. The point of using piecewise *polynomial* differential forms in the PL de Rham complex (Def. 5.3) is that these, but not the piecewise smooth differential forms, can be defined over the field $\mathbb{Q}$ of rational numbers. But since we may and do use the real numbers as the rational ground field (Rem. 5.2), it is expedient to also consider piecewise smooth de Rham complexes:

Definition 5.9 (PS de Rham complex [Griffiths and Morgan (2013), p. 91]**).** For $n \in \mathbb{N}$, we write, in variation of (5.3),

$$\Omega^\bullet_{\mathrm{dR}}\big(\mathbb{R}^n \times \Delta^{(-)}\big) \;:\; \Delta^{\mathrm{op}} \longrightarrow \mathrm{dgcAlgs}^{\geq 0}_{\mathbb{R}}$$

for the simplicial dgc-algebra of smooth differential forms on the product manifold of n-dimensional Cartesian space with the standard simplices (i.e., of smooth differential forms on an ambient Cartesian space (Ex. 4.9), restricted to the simplex and identified there if they agree on some open neighbourhood). As in Def. 5.3, this induces for each $S \in \Delta\mathrm{Sets}$ the corresponding *piecewise smooth de Rham complexes*

$$\Omega^\bullet_{\mathrm{PSdR}}(\mathbb{R}^n \times S) \;:=\; \int_{[k]\in\Delta} \prod_{S_n} \Omega^\bullet_{\mathrm{dR}}(\mathbb{R}^n \times \Delta^n) \tag{5.24}$$

and by pullback of differential forms these extend to functors

$$\Delta\mathrm{Sets} \xrightarrow{\;\Omega^\bullet_{\mathrm{PSdR}}(\mathbb{R}^n \times (-))\;} \big(\mathrm{dgcAlgs}^{\geq 0}_{\mathbb{R}}\big)^{\mathrm{op}}. \tag{5.25}$$

Proposition 5.10 (Fundamental theorem for piecewise smooth de Rham complexes). *For all $n \in \mathbb{N}$ the piecewise smooth de Rham complex functors (Def. 5.9) participate in a Quillen adjunction analogous to the PL de Rham adjunction (Prop. 5.5) over the real numbers*

$$\big(\mathrm{dgcAlgs}^{\geq 0}_{\mathbb{R}}\big)^{\mathrm{op}}_{\mathrm{trinj}} \underset{B\exp_{\mathrm{PS},n}}{\overset{\Omega^\bullet_{\mathrm{PSdR}}(\mathbb{R}^n \times (-))}{\rightleftarrows}} \Delta\mathrm{Sets}_{\mathrm{Qu}} \qquad (\perp_{\mathrm{Qu}}) \tag{5.26}$$

with right adjoint given as in (5.15)*:*

$$B\exp_{\mathrm{PS},n}(A) \;=\; \Big(\Delta[k] \longmapsto \mathrm{dgcAlgs}^{\geq 0}_{\mathbb{R}}\big(\Omega^\bullet_{\mathrm{PLdR}}(\mathbb{R}^n \times \Delta^k), A\big)\Big) \quad \in \Delta\mathrm{Sets}, \tag{5.27}$$

whose derived functors (Prop. 1.13) are naturally equivalent to those of the PL de Rham adjunction (5.16) *over the real numbers:*

$$\mathbb{D}\Omega^\bullet_{\mathrm{PSdR}}(\mathbb{R}^n \times (-)) \simeq \mathbb{D}\Omega^\bullet_{\mathrm{PSdR}}(-) \simeq \mathbb{D}\Omega^\bullet_{\mathrm{P}\mathbb{R}\mathrm{LdR}}(-), \tag{5.28}$$

$$\mathbb{D}B\exp_{\mathrm{PS},n} \simeq \mathbb{D}B\exp_{\mathrm{PS}} \simeq \mathbb{D}B\exp_{\mathrm{P}\mathbb{R}\mathrm{L}}. \tag{5.29}$$

Proof. **(i)** The proofs of the PL de Rham theorem (Prop. 5.4) as well as of the extension Lemma (Lem. 5.2) apply essentially verbatim also to piecewise-smooth differential forms ([Griffiths and Morgan (2013), Prop. 9.8]) and hence so does the proof of the PL de Rham Quillen adjunction in the form given in Prop. 5.5.
(ii) We have evident natural transformations

$$\Omega^\bullet_{\mathrm{P}\mathbb{R}\mathrm{LdR}}(S) \xrightarrow[\in \mathrm{W}]{} \Omega^\bullet_{\mathrm{PSdR}}(S) \xrightarrow[\in \mathrm{W}]{} \Omega^\bullet_{\mathrm{PSdR}}(\mathbb{R}^n \times S), \qquad \text{for } S \in \Delta\mathrm{Sets},$$

given by inclusion of polynomial differential forms into smooth differential forms, and then by pullback of differential forms along the projections $\mathbb{R}^n \times \Delta^k \longrightarrow \Delta^k$. The corresponding component morphisms are quasi-isomorphisms ([Griffiths and Morgan (2013), Cor. 9.9]), hence are weak equivalences in $(\mathrm{dgcAlgs}^{\geq 0}_{\mathbb{R}})_{\mathrm{trinj}}$ (Def. 4.18). Under passage to homotopy categories (Def. 1.8) and derived functors (Ex. 1.10), these natural weak equivalences become the natural isomorphisms (5.28) (by Prop. 1.9). By essential uniqueness of adjoint functors, this implies the natural isomorphisms (5.29). □

Whitehead L_∞-algebras.

Proposition 5.11 (Real Whitehead L_∞-algebras). *For $X \in \mathrm{Ho}(\Delta\mathrm{Sets}_{\mathrm{Qu}})^{\mathrm{fin}_{\mathbb{Q}}}_{\geq 1,\mathrm{nil}}$ (Def. 5.1), there exists a nilpotent L_∞-algebra (Def. 4.15)*

$$\mathfrak{l}X \in L_\infty\mathrm{Algs}^{\geq 0,\mathrm{nil}}_{\mathbb{R},\mathrm{fin}}, \tag{5.30}$$

unique up to isomorphism, whose Chevalley-Eilenberg algebra (Def. 4.13) is the minimal model (Def. 4.22) of the PL de Rham complex of X (Def. 5.3):

$$\mathrm{CE}(\mathfrak{l}X) := \left(\Omega^\bullet_{\mathrm{PLdR}}(X)\right)_{\min} \xrightarrow[\in \mathrm{W}]{p_X^{\min}} \Omega^\bullet_{\mathrm{PLdR}}(X). \tag{5.31}$$

Proof. By the PL de Rham theorem (Prop. 5.4) and the assumption that X is connected, it follows that we have $H\Omega^0_{\mathrm{PLdR}}(X) = \mathbb{R}$. Therefore Prop. 4.23 applies and says that $\left(\Omega^\bullet_{\mathrm{PLdR}}(X)\right)_{\min} \in \mathrm{SullModels}^{\geq 1}_{\mathbb{R}}$ exists, and is unique up to isomorphism. With this, the equivalence (4.43) says that $\mathfrak{l}X$ exists and is unique up to isomorphism. □

Proposition 5.12 ($\mathbb{R}$-Rationalization as integration of Whitehead L_∞-algebras). *For $X \in \mathrm{Ho}(\Delta\mathrm{Sets}_{\mathrm{Qu}})^{\mathrm{fin}_{\mathbb{Q}}}_{\geq 1,\mathrm{nil}}$ (Def. 5.1) its rationalization over the real numbers (Def. 5.7) is equivalently the image under $B\exp_{\mathrm{PL}}$* (5.23) *of the CE-algebra* (5.31) *of its Whitehead L_∞-algebra* (5.30):

$$L_{\mathbb{R}}(X) \simeq B\exp_{\mathrm{PL}}\left(\mathrm{CE}(\mathfrak{l}X)\right) \in \mathrm{Ho}(\Delta\mathrm{Sets}_{\mathrm{Qu}}). \tag{5.32}$$

Proof. By Def. 5.7 and the characterization of derived functors (Ex. 1.10), $L_{\mathbb{R}}$ is equivalently the image under $B\exp_{\mathrm{PL}}$ of any cofibrant replacement of $\Omega^\bullet_{\mathrm{PLdR}}(X) \in (\mathrm{dgcAlgs}^{\geq 0}_{\mathbb{R}})_{\mathrm{trinj}}$ (using that every $X \in \Delta\mathrm{Sets}_{\mathrm{Qu}}$ is already cofibrant (1.15)). This is provided by $\mathrm{CE}(\mathfrak{l}X)$, according to (5.31) and by Prop. 4.21. □

Proposition 5.13 (Rational homotopy groups in the rational Whitehead L_∞-algebra). *Let $X \in \mathrm{Ho}(\Delta\mathrm{Sets}_{\mathrm{Qu}})^{\mathrm{fin}_{\mathbb{Q}}}_{\geq 1,\mathrm{nil}}$ (Def. 5.1).*

(i) *If X is simply connected, $\pi_1(X) = 1$ (Ex. 5.1), then there is an isomorphism of graded vector spaces (Def. 4.1) between the graded vector space underlying* (4.21) *the Whitehead L_∞-algebra $\mathfrak{l}X$ (Prop. 5.11) and the rationalized homotopy groups of the based loop space ΩX:*

$$\underbrace{\mathfrak{l}X}_{\substack{\text{Whitehead}\\ L_\infty\text{-algebra}}} \simeq \underbrace{\pi_\bullet(\Omega X) \otimes_{\mathbb{Z}} \mathbb{R}}_{\substack{\text{rationalized}\\ \text{homotopy groups}}} \in \mathrm{GrdVectSp}^{\geq 0}_{\mathbb{R}}.$$

(ii) *If $\pi_1(X)$ is not necessarily trivial but abelian, then this statement still holds with $\mathfrak{l}X$ replaced by its homology with respect to the unary differential $[-]$* (4.23).
(iii) *If $\pi_1(X)$ is not abelian, then (ii) still holds in degrees ≥ 2.*

Proof. Under translation through Prop. 5.11 and Def. 4.13, and using $\pi_\bullet(\Omega X) \simeq \pi_{\bullet+1}(X)$, claim **(i)** is equivalent to the existence of a dual isomorphism:

$$\underbrace{\mathrm{CE}(\mathfrak{l}X)/_{\mathrm{CE}(\mathfrak{l}X)^2}}_{\wedge^1(\mathfrak{l}X)^\vee} \simeq \mathrm{Hom}_{\mathbb{Z}}(\pi_\bullet(X), \mathbb{R}) \quad \in \mathrm{GrdVectSp}^{\geq 0}_{\mathbb{R}}, \tag{5.33}$$

where the left hand side denotes the graded vector space of indecomposable elements in the Chevalley-Eilenberg algebra (the $\alpha^{(i)}_{n_i}$ in (4.38)). In this form, this is the statement of [Bousfield and Gugenheim (1976), Thm. 11.3 with Def. 6.12], in the special case when, with $\pi_1(X) = 1$, the unary differential $[-]$ in $\mathfrak{l}X$ vanishes (Ex. 4.18). The generalizations follow analogously. □

Remark 5.4 (Equivalent L_∞-structures on Whitehead products). The original discussion of the Whitehead algebra structure on the homotopy groups of a space is in terms of differential-graded Lie algebras ([Hilton (1955), Thm. B]), as are the Quillen models of rational homotopy theory [Quillen (1969)].

(i) Notice that dg-Lie algebras (Ex. 4.11) and L_∞-algebras with minimal CE-algebra (Def. 4.22) are two opposite classes of L_∞-algebras: The former has k-ary brackets (4.23) only for $k \leq 2$, the latter only for $k \geq 2$ (in the simply connected case, by Ex. 4.18). Yet, quasi-isomorphisms connect algebras in one class to those in the other ([Kříž and May (1995), p. 28]), such as to make their homotopy theories equivalent ([Pridham (2010)], see also Rem. 4.4). The transmutation of dg-Lie- into minimal L_∞-algebras is described in [Belchí *et al.* (2017), Thm. 2.1]; that from L_∞- to dg-Lie-algebras in [Fiorenza *et al.* (2014a), §1.0.2].
(ii) The minimal L_∞-algebra structure on $\mathfrak{l}X$ that we obtained in Prop. 5.11, 5.13, has the property that its k-ary brackets are, up to possibly a sign, equal to the order-k higher Whitehead products on X [Belchí *et al.* (2017), Prop. 3.1].

Examples of rationalizations over the real numbers. The following fundamental examples of rationalizations serve to illustrate the above notation and terminology and to highlight that rationalization *over the real numbers* (Def. 5.7), even though it is not a localization (Rem. 5.5 below), still acts as real-ification on the homotopy groups of Eilenberg-MacLane spaces (Ex. 5.4 below) and, more generally, of loop spaces (Ex. 5.6 below). This is the crucial fact that makes the real character map on non-abelian cohomology in Part IV reduce to the traditional Chern-Dold characters on abelian generalized cohomology in Chapter 7.

Example 5.3 ($\mathbb{R}$-Rationalization of n-spheres (e.g. [Menichi (2015), §1.2])**).** The Serre finiteness theorem (see [Ravenel (1986), Thm. 1.1.8]) says that the homotopy groups of n-spheres for $n \geq 1$ are of the form

$$\pi_{n+k}\big(S^n\big) \;\simeq\; \begin{cases} \mathbb{Z} & \mid k=0 \\ \mathbb{Z}\oplus \mathrm{fin} & \mid n=2m \;\text{ and }\; k=2m-1 \\ \mathrm{fin} & \mid \text{otherwise} \end{cases} \tag{5.34}$$

where "fin" stands for some finite group. Since finite groups are pure torsion, hence have trivial rationalization, this means that the rational homotopy groups of spheres are:

$$\pi_{n+k}\big(S^n\big)\otimes_{\mathbb{Z}}\mathbb{R} \;\simeq\; \begin{cases} \mathbb{R} \mid k=0 \\ \mathbb{R} \mid n=2m \text{ and } k=2m-1 \\ 0 \mid \text{otherwise}. \end{cases} \tag{5.35}$$

Moreover, the fact that ordinary cohomology is represented by Eilenberg-MacLane spaces (Ex. 2.1) means that

$$H^k\big(S^n;\mathbb{R}\big) \;\simeq\; \begin{cases} \mathbb{R} \mid k\in\{0,n\} \\ 0 \mid \text{otherwise.} \end{cases} \tag{5.36}$$

With this, Prop. 5.13 together with Prop. 5.4 implies that the Whitehead L_∞-algebras of spheres (Prop. 5.11) are dual to the following Sullivan models:

$$\mathrm{CE}\big(\mathfrak{l}S^n\big) \;\simeq\; \mathbb{R}\big[\omega_n\big]/\big(d\,\omega_n = 0\big) \qquad \text{if } n \text{ is odd} \tag{5.37}$$

and

$$\mathrm{CE}\big(\mathfrak{l}S^n\big) \;\simeq\; \mathbb{R}\begin{bmatrix}\omega_{2n-1}\\ \omega_n\end{bmatrix}\Big/\begin{pmatrix} d\,\omega_{2n-1} &= -\omega_n\wedge\omega_n \\ d\,\omega_n &= 0\end{pmatrix} \qquad \text{if } n>1 \text{ is even} \tag{5.38}$$

Example 5.4 ($\mathbb{R}$-Rationalization of Eilenberg-MacLane spaces). Since the homotopy types of Eilenberg-MacLane-spaces $K(A,n+1)=B^{n+1}A$ (see (2.6)) are fully characterized by their homotopy groups (for discrete abelian groups A, e.g. [Aguilar *et al.* (2002), §6]))

$$\pi_k\big(B^{n+1}A\big) \;\simeq\; \begin{cases} A \mid k=n+1 \\ 0 \mid k\neq n+1 \end{cases}$$

we have, for $n \in \mathbb{N}$:

(i) Their Whitehead L_∞-algebra (Prop. 5.11) is, by Prop. 5.13, the direct sum of $\dim(A \otimes_{\mathbb{Z}} \mathbb{R})$ copies of the line Lie n-algebra (Def. 4.12):

$$\mathfrak{l}(B^{n+1}A) \simeq \mathfrak{b}^n(A \otimes_{\mathbb{Z}} \mathbb{R}) \simeq \bigoplus_{\dim(A \otimes_{\mathbb{Z}} \mathbb{R})} \mathfrak{b}^n \mathbb{R}\,. \tag{5.39}$$

(ii) Their rationalization over $\mathbb{R}$ (Def. 5.7) is the Eilenberg-MacLane space on the realification of A:

$$L_{\mathbb{R}}(B^{n+1}A) \simeq B^{n+1}(A \otimes_{\mathbb{Z}} \mathbb{R}) \in \mathrm{Ho}(\mathrm{TopSp}_{\mathrm{Qu}}). \tag{5.40}$$

Observe how this is implied via the machinery that we have set up above: For all $k \in \mathbb{N}_+$ we have:

$$
\begin{aligned}
&\pi_k\Big(L_{\mathbb{R}}(B^{n+1}A)\Big) \\
&= \mathrm{Ho}\big(\mathrm{TopSp}_{\mathrm{Qu}}^{*/}\big)\Big(S^k, L_{\mathbb{R}}(B^{n+1}A)\Big) && \text{by def. of } \pi_k(-) \\
&= \mathrm{Ho}\big(\mathrm{TopSp}_{\mathrm{Qu}}^{*/}\big)\Big(S^k, B\exp_{\mathrm{PL}}\big(\mathrm{CE}(\mathfrak{b}^n(A \otimes_{\mathbb{Z}} \mathbb{R}))\big)\Big) && \text{by Prop. 5.12 with (5.39)} \\
&\simeq \mathrm{Ho}\Big(\big(\mathrm{dgcAlgs}_{\mathbb{R}}^{\geq 0}\big)^{/\mathbb{R}}_{\mathrm{trinj}}\Big)\Big(\mathrm{CE}\big(\mathfrak{b}^n(A \otimes_{\mathbb{Z}} \mathbb{R})\big), \Omega^\bullet_{\mathrm{PLdR}}(S^k) \underset{\Omega^\bullet_{\mathrm{PLdR}}(*)}{\times} \mathbb{R}\Big) && \text{by Props. 5.5, 1.9, 1.13} \\
&\simeq \prod_{\dim(A \otimes_{\mathbb{Z}} \mathbb{R})} \mathrm{Ho}\Big(\big(\mathrm{dgcAlgs}_{\mathbb{R}}^{\geq 0}\big)^{/\mathbb{R}}_{\mathrm{trinj}}\Big)\Big(\mathrm{CE}\big(\mathfrak{b}^n \mathbb{R}\big), \Omega^\bullet_{\mathrm{PLdR}}(S^k) \underset{\Omega^\bullet_{\mathrm{PLdR}}(*)}{\times} \mathbb{R}\Big) && \text{by (4.31)} \\
&\simeq \prod_{\dim(A \otimes_{\mathbb{Z}} \mathbb{R})} H^{n+1}(S^k; \mathbb{R}) && \text{by Lem. 5.1} \\
&\simeq \begin{cases} A \otimes_{\mathbb{Z}} \mathbb{R} & | \; k = n+1 \\ 0 & | \; k \neq n+1 \end{cases} && \text{by (5.36)}.
\end{aligned}
$$

The same computation, but with S^k replaced by the point $*$ and without the slicing, shows that $\pi_0\Big(L_{\mathbb{R}}(B^{n+1}A)\Big) = *$.

Remark 5.5 (Failure of $\mathbb{R}$-rationalization to be idempotent). Ex. 5.4 reveals how rationalization over $\mathbb{R}$ (or over any field k strictly containg the rational numbers, Def. 5.7) is not idempotent, hence not a localization (see also [Bousfield and Gugenheim (1976), Rem. 9.7]): Applying (5.40) twice yields

$$L_{\mathbb{R}} \circ L_{\mathbb{R}}(B^{n+1}A) \simeq B^{n+1}(A \otimes_{\mathbb{Z}} \mathbb{R} \otimes_{\mathbb{Z}} \mathbb{R}), \tag{5.41}$$

but $\mathbb{R} \otimes_{\mathbb{Z}} \mathbb{R} \not\simeq \mathbb{R}$, in contrast to $\mathbb{Q} \otimes_{\mathbb{Z}} \mathbb{Q} \simeq \mathbb{Q}$ (reflecting that $\mathbb{Q}$ is a *solid ring* [Bousfield and Kan (1972a), §2.4], while $\mathbb{R}$ is not).

Example 5.5 ($\mathbb{R}$-Rationalization of complex projective spaces). From the defining homotopy fiber sequence for complex projective n-space $\mathbb{C}P^n$, $n \in \mathbb{N}$ (e.g. [Bott and Tu (1982), Ex. 14.22])

$$\begin{array}{ccccc} S^1 & \hookrightarrow & S^{2n+1} & \twoheadrightarrow & \mathbb{C}P^n \\ \| & & \| & & \| \\ \mathrm{U}(1) & \hookrightarrow & \mathrm{U}(n+1)/\mathrm{U}(n) & \twoheadrightarrow & \mathrm{SU}(n+1)/\mathrm{U}(n) \end{array} \tag{5.42}$$

the corresponding long exact sequence of homotopy groups (e.g. [tom Dieck (2008), Thm. 6.1.2]) yields the following homotopy groups:

$$\begin{aligned} \pi_k\big(\mathbb{C}P^n\big) &= \begin{cases} * & \mid k \in \{0,1\} \\ \mathbb{Z} & \mid k \in \{2, 2n+1\} \\ \pi_k\big(S^{2n+1}\big) & \mid k \geq 2n+1 \\ 0 & \mid \text{otherwise}. \end{cases}, \\ H^k\big(\mathbb{C}P^n; \mathbb{R}\big) &= \begin{cases} \mathbb{R} \mid k \in \{0,1,2,\dots,2n\} \\ 0 \mid \text{otherwise}. \end{cases} \end{aligned} \tag{5.43}$$

Since these homotopy groups (5.43) in degrees $\geq 2n+1$ are finite by Ex. 5.3, hence rationally trivial, it follows, with Prop. 5.13 and Prop. 5.4, that the Chevalley-Eilenberg algebra of the Whitehead L_∞-algebra of $\mathbb{C}P^n$ (Prop. 5.11) has exactly one generator f_2 in degree 2 and one generator h_{2k+1} in degree $2k+1$. Moreover, since the cohomology groups are (e.g. [Bott and Tu (1982), Ex. 14.22.1]) as shown on the right of (5.43), the first of these must be the closed generator of the cohomology ring, and the differential of the latter must exhibit the vanishing of its $(n+1)$st cup product in cohomology (see also, e.g., [Félix *et al.* (2001), p. 203][Menichi (2015), §5.3]):

$$\mathrm{CE}\big(\mathfrak{l}\mathbb{C}P^n\big) = \mathbb{R}\begin{bmatrix} h_{2n+1}, \\ f_2 \end{bmatrix} \Big/ \left(\begin{aligned} d\,h_{2n+1} &= \overbrace{f_2 \wedge \cdots \wedge f_2}^{n+1 \text{ factors}} \\ d\,f_2 &= 0 \end{aligned} \right). \tag{5.44}$$

Example 5.6 ($\mathbb{R}$-Rationalization of loop spaces). The minimal Sullivan model (Def. 4.22) of a loop space $A \simeq \Omega A'$ of $\mathbb{Q}$-finite type (which exists by Ex. 5.1) has vanishing differential (e.g. [Félix *et al.* (2001), p. 143]). Therefore, Prop. 5.13 implies that the rational Whitehead L_∞-algebra $\mathfrak{l}A$ (Prop. 5.11) of A is the direct sum of line Lie n-algebras $\mathfrak{b}^n\mathbb{R}$ (Ex. 4.12) and its Chevalley-Eilenberg algebra (Def. 4.10) is the tensor product of those of the summands:

$$\begin{aligned} \mathfrak{l}A &\simeq \bigoplus_{n \in \mathbb{N}} \mathfrak{b}^n\big(\pi_{n+1}(A) \otimes_{\mathbb{Z}} \mathbb{R}\big) && \in L_\infty\mathrm{Algs}^{\geq 0, \mathrm{nil}}_{\mathbb{R}, \mathrm{fin}}, \\ \mathrm{CE}\big(\mathfrak{l}A\big) &\simeq \bigotimes_{n \in \mathbb{N}} \mathrm{CE}\Big(\mathfrak{b}^n\big(\pi_{n+1}(A) \otimes_{\mathbb{Z}} \mathbb{R}\big)\Big) && \in \mathrm{dgcAlgs}^{\geq 0}_{\mathbb{R}}. \end{aligned}$$

Noticing that the tensor product of dgc-algebras is the coproduct in the category of $\mathrm{dgcAlgs}_{\mathbb{R}}^{\geq 0}$ (Ex. 4.8), and hence the Cartesian product in the opposite category, the right adjoint functor $B\exp_{\mathrm{PL}}$ (5.14) preserves this, so that

$$B\exp_{\mathrm{PL}} \circ \mathrm{CE}(\mathfrak{l}A) \simeq \prod_{n\in\mathbb{N}} \Big(B\exp_{\mathrm{PL}} \circ \mathrm{CE}\big(\mathfrak{b}^n(\pi_{n+1}(A)\otimes_{\mathbb{Z}}\mathbb{R})\big)\Big) \quad \in \mathrm{Ho}\big(\mathrm{TopSp}_{\mathrm{Qu}}\big).$$

Finally, by Ex. 5.4, this means that the rationalization over $\mathbb{R}$ (Def. 5.7) of a loop space is a Cartesian product of Eilenberg-MacLane spaces:

$$A \simeq \Omega A' \quad \Rightarrow \quad L_{\mathbb{R}}(A) \simeq \prod_{n\in\mathbb{N}} B^{n+1}\big(\pi_{n+1}(A)\otimes_{\mathbb{Z}}\mathbb{R}\big). \tag{5.45}$$

Example 5.7 ($\mathbb{R}$-Rationalization of spectra). For E a spectrum (Ex. 2.10), Ex. 5.6 says that its degree-wise rationalization (Def. 5.2) and $\mathbb{R}$-rationalization (Def. 5.7) are both direct sums of the same form of rational Eilenberg-MacLane spectra:

$$\begin{aligned} L_{\mathbb{Q}}(E) &\simeq \bigoplus_{k\in\mathbb{Z}} H\big(\pi_\bullet(E)\otimes_{\mathbb{Z}}\mathbb{Q}\big) \\ L_{\mathbb{R}}(E) &\simeq \bigoplus_{k\in\mathbb{Z}} H\big(\pi_\bullet(E)\otimes_{\mathbb{Z}}\mathbb{R}\big), \end{aligned} \tag{5.46}$$

But rationalization of spectra is also known (review in [Lawson (2022), Ex. 8.12][Bauer (2014), Ex. 1.7 (4)]) to be given by forming the smash product of spectra (e.g. [Elmendorf *et al.* (1997)]) with the rational Eilenberg-MacLane spectrum:

$$L_{\mathbb{Q}}(E_n) \simeq \big(E\wedge H\mathbb{Q}\big)_n. \tag{5.47}$$

Observing with (5.46) that

$$L_{\mathbb{R}}(E) \simeq \big(L_{\mathbb{Q}}(E)\big)\wedge_{H\mathbb{Q}} H\mathbb{R}. \tag{5.48}$$

this implies that the componentwise $\mathbb{R}$-rationalization (Def. 5.7) of spectra is analogously given by the smash product with the real Eilenberg-MacLane:

$$\begin{aligned} L_{\mathbb{R}}(E) &\simeq \big(L_{\mathbb{Q}}(E)\big)\wedge_{H\mathbb{Q}} H\mathbb{R} \text{ by (5.47)} \\ &\simeq E\wedge H\mathbb{Q}\wedge_{H\mathbb{Q}} H\mathbb{R} \text{ by (5.48)} \\ &\simeq E\wedge H\mathbb{R}. \end{aligned} \tag{5.49}$$

It is in this smashing form that $\mathbb{R}$-rationalization of spectra traditionally appears in the construction of differential generalized cohomology theories, see Ex. 9.1 below.

Non-abelian real cohomology. Using these $\mathbb{R}$-rational models, we obtain the first key concept formation towards the character map:

Definition 5.14 (Non-abelian real cohomology). Let $X, A \in \mathrm{TopSp}$. Then the *non-abelian real cohomology* of X with coefficients in A is the non-abelian cohomology

(Def. 2.1) of X with coefficients in the $\mathbb{R}$-rationalization $L_{\mathbb{R}}A$ (Def. 5.7)

$$H\big(X; L_{\mathbb{R}}A\big) \;:=\; \mathrm{Ho}\big(\mathrm{TopSp}_{\mathrm{Qu}}\big)\big(X\,,L_{\mathbb{R}}A\big)\,. \tag{5.50}$$

Example 5.8 (Non-abelian real cohomology subsumes ordinary real cohomology). For $n \in \mathbb{N}$, non-abelian real cohomology (Def. 5.14) with coefficients in the $\mathbb{R}$-rationalized classifying space $L_{\mathbb{R}}\big(B^{n+1}\mathbb{Z}\big) \simeq B^{n+1}\mathbb{R}$ (by Ex. 5.4) is naturally equivalent (by Ex. 2.1) to ordinary real cohomology in degree n:

$$H\big(X; L_{\mathbb{R}}B^{n+1}\mathbb{Z}\big) \simeq H\big(X; B^{n+1}\mathbb{R}\big) \simeq H^{n+1}(X;\mathbb{R})\,.$$

More generally:

Proposition 5.15 (Non-abelian real cohomology with coefficients in loop spaces). *Let $A \in \mathrm{Ho}\big(\Delta\mathrm{Sets}_{\mathrm{Qu}}\big)^{\mathrm{fin}_{\mathbb{Q}}}_{\geq 1,\mathrm{nil}}$ (Def. 5.1) such that it admits loop space structure, hence such that there exists A' with*

$$A \simeq \Omega A' \;\in \mathrm{Ho}\big(\mathrm{TopSp}_{\mathrm{Qu}}\big)\,.$$

Then the non-abelian real cohomology (Def. 5.14) with coefficients in $L_{\mathbb{R}}A$ is naturally equivalent to ordinary real cohomology with coefficients in the rationalized homotopy groups of A:

$$H\big(X; L_{\mathbb{R}}A\big) \;\simeq\; \bigoplus_{n\in\mathbb{N}} H^{n+1}\big(X; \pi_{n+1}(A)\otimes_{\mathbb{Z}}\mathbb{R}\big). \tag{5.51}$$

This is the result of the following sequence of natural bijections:

$$\begin{aligned}
H\big(X; L_{\mathbb{R}}A\big) &\simeq H\Big(X;\, \prod_{n\in\mathbb{N}} B^{n+1}\big(\pi_{n+1}(A)\otimes_{\mathbb{Z}}\mathbb{R}\big)\Big) && \text{by Ex. 5.6}\\
&\simeq \prod_{n\in\mathbb{N}} H\Big(X;\, B^{n+1}\big(\pi_{n+1}(A)\otimes_{\mathbb{Z}}\mathbb{R}\big)\Big) && \text{by Def. 2.1 \& Prop. 1.11}\\
&= \prod_{n\in\mathbb{N}} H^{n+1}\big(X;\, \pi_{n+1}(A)\otimes_{\mathbb{Z}}\mathbb{R}\big) && \text{by Ex. 2.1}\\
&= \bigoplus_{n\in\mathbb{N}} H^{n+1}\big(X;\, \pi_{n+1}(A)\otimes_{\mathbb{Z}}\mathbb{R}\big) && \text{by finite type}\,.
\end{aligned}$$

Relative rational Whitehead L_∞-algebras. In parameterized generalization of Prop. 5.11:

Proposition 5.16 (Relative real Whitehead L_∞-algebras).

For $A, B, F \in \mathrm{Ho}\big(\Delta\mathrm{Sets}_{\mathrm{Qu}}\big)^{\mathrm{fin}_{\mathbb{Q}}}_{\geq 1,\mathrm{nil}}$ (Def. 5.1) and p a Serre fibration (Ex. 1.1) from A to B with fiber F

$$\begin{array}{ccc} F & \xrightarrow{\mathrm{fib}(p)} & A \\ & & {\scriptstyle p}\big\downarrow{\scriptstyle \in\,\mathrm{Fib}} \\ & & B \end{array}$$

there exists a nilpotent L_∞-algebra (Def. 4.15)

$$\mathfrak{l}_B A \;\in\; L_\infty\mathrm{Algs}^{\geq 0,\,\mathrm{nil}}_{\mathbb{R},\,\mathrm{fin}}\,, \tag{5.52}$$

unique up to isomorphism, whose Chevalley-Eilenberg algebra (Def. 4.13) is the relative minimal model (Def. 4.22, Prop. 4.24) of the PL de Rham complex of p (Def. 5.3), relative to $\mathrm{CE}(\mathfrak{l}B)$ *(from Prop. 5.11):*

$$\begin{array}{ccc} \mathrm{CE}(\mathfrak{l}_B A) := \big(\Omega^\bullet_{\mathrm{PLdR}}(A)\big)_{\min_{\mathrm{CE}(\mathfrak{l}B)}} & \xrightarrow[\in \mathrm{W}]{p_A^{\min_B}} & \Omega^\bullet_{\mathrm{PLdR}}(A) \\ \big\uparrow{\scriptstyle \text{relative minimal model } \mathrm{CE}(\mathfrak{l}p)} & & \big\uparrow{\scriptstyle \Omega^\bullet_{\mathrm{PLdR}}(p)} \\ \mathrm{CE}(\mathfrak{l}B) & \xrightarrow[\in \mathrm{W}]{p_B^{\min}} & \Omega^\bullet_{\mathrm{PLdR}}(B)\,. \end{array} \tag{5.53}$$

Proof. By the PL de Rham theorem (Prop. 5.4) and the assumption that A and B are connected, it follows that we have $H\Omega^0_{\mathrm{PLdR}}(A) = \mathbb{R}$ and $H\Omega^0_{\mathrm{PLdR}}(B) = \mathbb{R}$. Moreover, by the assumption that p is a Serre fibration with connected fiber, it follows that $H^1(\Omega^\bullet_{\mathrm{PLdR}}(p))$ is injective (e.g. [Félix *et al.* (2001), p. 196]). Therefore, Prop. 4.24 applies and says that $\big(\Omega^\bullet_{\mathrm{PLdR}}(A)\big)_{\min_B} \in \mathrm{SullModels}^{\geq 1}_{\mathbb{R}}$ exists, and is unique up to isomorphism. With this, the equivalence (4.43) says that $\mathfrak{l}_B A$ exists and is unique up to isomorphism. □

In parameterized generalization of Prop. 5.12 we have:

Proposition 5.17 (Relative $\mathbb{R}$-rationalization as integration of relative Whitehead L_∞-algebras). *For a Serre fibration $A \xrightarrow{p} B$ as in Prop. 5.16, its rationalization over the real numbers (Def. 5.7) is equivalently the image under* $B\exp$ (5.14) *of the image under forming CE-algebras* (5.31) *of its relative Whitehead L_∞-algebra* (5.30)*:*

$$L_{\mathbb{R}}\begin{pmatrix} A \\ \downarrow p \\ B \end{pmatrix} \simeq \begin{array}{c} B\exp_{\mathrm{PL}}\mathrm{CE}(\mathfrak{l}_B A) \\ \big\downarrow{\scriptstyle B\exp_{\mathrm{PL}}\mathrm{CE}(\mathfrak{l}p)} \\ B\exp_{\mathrm{PL}}\mathrm{CE}(\mathfrak{l}B) \end{array}$$

Proof. As in Prop. 5.12, now appealing to Prop. 5.16 for the (co)fibrant replacement. □

Lemma 5.4 (Minimal relative Sullivan models preserve homotopy fibers [Félix *et al.* (2001), §15 (a)][Félix *et al.* (2015), Thm. 5.1]**).** *Consider $F, A, B \in \mathrm{Ho}\big(\Delta\mathrm{Sets}_{\mathrm{Qu}}\big)^{\mathrm{fin}_{\mathbb{Q}}}_{\geq 1,\mathrm{nil}}$ (Def. 5.1) and let p be a Serre fibration from A to B (Ex. 1.1) such that the homology groups $H_\bullet(F,\mathbb{R})$ of the fiber are nilpotent as $\pi_1(B)$-modules (for instance in that B is simply-connected or that the fibration is principal). Then the cofiber of the minimal relative Sullivan model for p* (5.53) *is the minimal Sullivan model* (5.31) *for the homotopy fiber F (Def. 1.14):*

$$\begin{array}{ccc} F & \xrightarrow{\mathrm{fib}(p)} & A \\ & & {\scriptstyle p}\big\downarrow{\scriptstyle \in \mathrm{Fib}} \\ & & B \end{array} \qquad\qquad \begin{array}{ccc} \mathrm{CE}(\mathfrak{l}F) & \xleftarrow{\mathrm{cofib}(\mathrm{CE}(\mathfrak{l}p))} & \mathrm{CE}(\mathfrak{l}_B A) \\ & & \big\uparrow{\scriptstyle \mathrm{CE}(\mathfrak{l}p)} \\ & & \mathrm{CE}(\mathfrak{l}B) \end{array} \tag{5.54}$$

Twisted non-abelian real cohomology.

Proposition 5.18 ($\mathbb{R}$-Rationalization of local coefficients – the *fiber lemma* [Bousfield and Kan (1972b), §II]). *Let*

$$\begin{array}{ccc} A & \longrightarrow & A /\!\!/ G \\ & & \big\downarrow{\scriptstyle \rho} \\ & & BG \end{array}$$

be a local coefficient bundle (Def. 3.2) such that all spaces are connected, nilpotent and of $\mathbb{Q}$-finite type: $A, BG, A/\!\!/G \in \mathrm{Ho}\big(\Delta\mathrm{Sets}_{\mathrm{Qu}}\big)^{\mathrm{fin}_{\mathbb{Q}}}_{\geq 1,\mathrm{nil}}$ (Def. 5.1), and such that the action of $\pi_1(BG)$ on $H_\bullet(A,\mathbb{R})$ is nilpotent (for instance in that BG is simply connected). Then: $\mathbb{R}$-Rationalization (Def. 5.7) preserves the homotopy fiber:

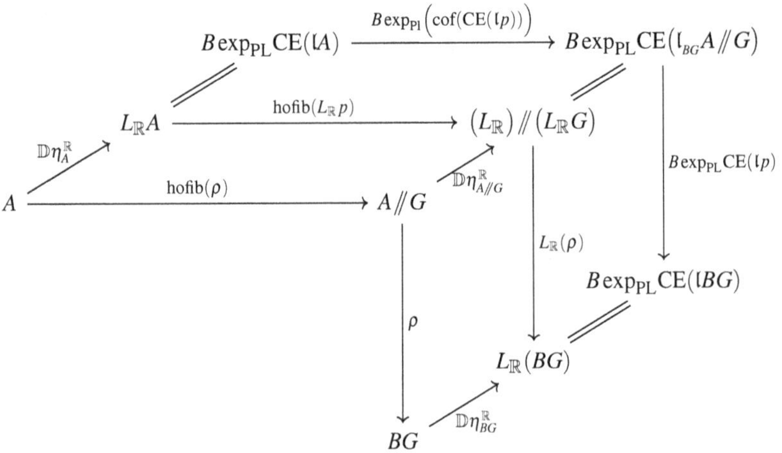

(5.55)

Proof. By Prop. 5.16, Prop. 5.17 and since $B\exp_{\mathrm{PL}}$ preserves fibrations, being a right adjoint, the homotopy fiber of $L_{\mathbb{R}}(p)$ is the image under $B\exp_{\mathrm{PL}}$ of the cofiber of $\mathrm{CE}(\mathfrak{l}p)$. That this is a claimed is the content of Lem. 5.4. □

Due to Prop. 5.18, it makes sense to say, in generalization of Def. 5.14:

Definition 5.19 (Twisted non-abelian real cohomology). Let $X \in \mathrm{TopSp}$ and let $A/\!\!/G \xrightarrow{\rho} BG$ be a local coefficient bundle (Prop. 3.1, Def. 3.2) in $\mathrm{Ho}\big(\Delta\mathrm{Sets}_{\mathrm{Qu}}\big)^{\mathrm{fin}_{\mathbb{Q}}}_{\geq 1,\mathrm{nil}}$ (Def. 5.1). Then the *twisted non-abelian real cohomology* of X with local coefficients ρ is the twisted non-abelian $L_{\mathbb{R}}A$-cohomology (Def. 3.2) of X with coefficients in the rationalized local coefficient bundle $L_{\mathbb{R}}(\rho)$ from Prop. 5.18:

$$H^{\tau}(X; L_{\mathbb{R}}A) \;:=\; \mathrm{Ho}\Big(\big(\Delta\mathrm{Sets}_{\mathrm{Qu}}\big)^{/L_{\mathbb{R}}BG}\Big)\big(\tau, L_{\mathbb{R}}(\rho)\big)\,.$$

Next we discuss how this non-abelian real cohomology is the domain of a non-abelian de Rham isomorphism.

Chapter 6

Non-abelian de Rham theorem

We establish non-abelian de Rham theory for differential forms with values in (nilpotent) L_∞-algebras, following [Sati *et al.* (2009)] [Fiorenza *et al.* (2012)]. The main result is the non-abelian de Rham theorem, Thm. 6.5, and its generalization to the twisted non-abelian de Rham theorem, Thm. 6.15.

L_∞-Algebra valued differential forms.

Definition 6.1 (Flat L_∞-algebra valued differential forms [Sati *et al.* (2009), §6.5] [Fiorenza *et al.* (2012), §4.1]**).**
(i) For $X \in \text{SmoothManifold}$ and $\mathfrak{g} \in L_\infty\text{Algs}^{\geq 0}_{\mathbb{R},\text{fin}}$ (Def. 4.13), a *flat $\mathfrak{g}$-valued differential form on X* is a morphism of dgc-algebras (Def. 4.10)

$$\Omega^\bullet_{\text{dR}}(X) \xleftarrow{\;A\;} \text{CE}(\mathfrak{g}) \qquad \in \text{dgcAlgs}^{\geq 0}_{\mathbb{R}} \tag{6.1}$$

to the smooth de Rham dgc-algebra of X (Ex. 4.9) from the Chevalley-Eilenberg dgc-algebra of $\mathfrak{g}$ (Def. 4.13).
(ii) We write

$$\Omega_{\text{dR}}(X;\mathfrak{g})_{\text{flat}} := \text{dgcAlgs}^{\geq 0}_{\mathbb{R}}\big(\text{CE}(\mathfrak{g}), \Omega^\bullet_{\text{dR}}(X)\big) \tag{6.2}$$

for the set of all flat $\mathfrak{g}$-valued forms on X.

Example 6.1 (Flat Lie algebra valued differential forms). Let $\mathfrak{g} \in \text{LieAlgebras}_{\mathbb{R},\text{fin}}$ be a Lie algebra (4.28) with Lie bracket $[-,-]$. Then a flat $\mathfrak{g}$-valued differential form in the sense of Def. 6.1 is the traditional concept: a $\mathfrak{g}$-valued 1-form satisfying the Maurer-Cartan equation:

$$\Omega^\bullet_{\text{dR}}(X;\mathfrak{g})_{\text{flat}} \simeq \Big\{ A \in \Omega^1_{\text{dR}}(X)\otimes\mathfrak{g} \;\Big|\; dA + [A\wedge A] = 0 \Big\}. \tag{6.3}$$

One way to see this is to appeal to the classical fact that the Chevalley-Eilenberg algebra of a finite-dimensional Lie algebra (Ex. 4.10) is isomorphic to the dgc-algebra of left invariant differential forms on the corresponding Lie group G, which is generated from the Maurer-Cartan form $\theta \in \Omega^1_{\text{dR}}(G)\otimes\mathfrak{g}$ satisfying $\theta_{|T_eG} = \text{id}_{\mathfrak{g}}$ and $d\theta = [\theta\wedge\theta]$. More explicitly, for $\{v_a\}$ a linear basis for $\mathfrak{g}$ (4.17) with structure constants $\{f^c_{ab}\}$ (4.18), we see from (4.19) that a dgc-algebra homomorphims (6.1) has the following components (second line) and constraints (third line):

$$\begin{array}{ccc} \Omega^\bullet_{\mathrm{dR}}(X) & \xleftarrow[A]{\text{flat Lie algebra valued differential form}} & \mathbb{R}\big[\{\theta_1^{(a)}\}\big]\Big/\Big(d\,\theta_1^{(c)} = f^c_{ab}\,\theta_1^{(b)}\wedge\theta_1^{(a)}\Big) \simeq \mathrm{CE}(\mathfrak{g})\,. \\ A^{(c)} & \xleftarrow{\text{components}} & \theta_1^{(c)} \\ d\big\downarrow & & \big\downarrow d \\ dA^{(c)} \overset{\text{constraints}}{=\!=} f^c_{ab}\,A^{(b)}\wedge A^{(a)} & \longleftarrow & f^c_{ab}\,\theta_1^{(a)}\wedge\theta_1^{(b)} \end{array} \qquad (6.4)$$

Example 6.2 (Ordinary closed forms are flat line L_∞-algebra valued forms). For $n \in \mathbb{N}$, consider $\mathfrak{g} = \mathfrak{b}^n\mathbb{R}$ the line Lie $(n+1)$-algebra (Ex. 4.12). Then the corresponding flat $\mathfrak{g}$-valued differential forms (Def. 6.1) are in natural bijection to ordinary closed $(n+1)$-forms:

$$\Omega_{\mathrm{dR}}(X;\mathfrak{b}^n\mathbb{R})_{\mathrm{flat}} \simeq \Omega^{n+1}_{\mathrm{dR}}(X)_{\mathrm{closed}}\,. \qquad (6.5)$$

That is, by (4.30), we see that the elements on the left of (6.5) have the following component (second line) subject to the following constraint (third line):

$$\begin{array}{ccc} \Omega^\bullet_{\mathrm{dR}}(X) & \xrightarrow{\substack{\text{flat}\\ \text{line Lie }(n+1)\text{-algebra-valued}\\ \text{differential form}}} & \mathbb{R}[c_{n+1}]/(d\,c_{n+1}=0) \simeq \mathrm{CE}(\mathfrak{b}^n\mathbb{R})\,. \\ C_{n+1} & \xleftarrow{\text{component}} & c_{n+1} \\ d\big\downarrow & & \big\downarrow d \\ dC_{n+1} \overset{\text{constraint}}{=\!=\!=} 0 & \longleftarrow & 0 \end{array} \qquad (6.6)$$

Example 6.3 (Flat String Lie 2-algebra valued differential forms). Flat L_∞-algebras valued forms (Def. 6.1) with values in a String Lie 2-algebra $\mathfrak{string}_{\mathfrak{g}}$ (Ex. 4.13) are pairs consisting of a flat $\mathfrak{g}$-valued 1-form A_1 (Ex. 6.1) and a coboundary 2-form B_2 for its Chern-Simons form $\mathrm{CS}(A) := c\langle A\wedge[A\wedge A]\rangle$:

$$\Omega_{\mathrm{dR}}\big(X;\mathfrak{string}_{\mathfrak{g}}\big)_{\mathrm{flat}} \simeq \left\{ \begin{array}{l} B_2, \\ A_1 \end{array} \in \Omega^\bullet_{\mathrm{dR}}(X) \;\middle|\; \begin{array}{l} d\,B_2 = \tfrac{1}{c}\mathrm{CS}(A)\,, \\ d\,A_1 = -[A_1\wedge A_1] \end{array} \right\}.$$

Namely, from (4.34) we see that in degree=1 the components of and constraints on such a differential form datum are exactly as in (6.4), while in degree 2 they are as follows:

$$\begin{array}{ccc} \Omega^\bullet_{\mathrm{dR}}(X) & \xleftarrow{\text{flat String Lie 2-algebra valued form}} & \mathbb{R}\left[\begin{array}{c} b_2, \\ \{\theta_1^{(a)}\} \end{array}\right]\Big/\left(\begin{array}{l} d\,b_2 = \mu_{abc}\,\theta_1^{(c)}\wedge\theta_1^{(b)}\wedge\theta_1^{(a)} \\ d\theta_1^{(c)} = f^c_{ab}\,\theta_1^{(b)}\wedge\theta_1^{(a)} \end{array}\right) \simeq \mathrm{CE}\big(\mathfrak{string}_{\mathfrak{g}}\big). \\ B_2 & \xleftarrow{\text{component in degree 2}} & b_2 \\ d\big\downarrow & & \big\downarrow d \\ dB_2 \overset{\text{constraint}}{=\!=\!=} \mu_{abc}A_1^{(c)}\wedge A_1^{(b)}\wedge A_1^{(a)} & \longleftarrow & \mu_{abc}\,\theta_1^{(c)}\wedge\theta_1^{(b)}\wedge\theta_1^{(a)} \end{array} \qquad (6.7)$$

Example 6.4 (Flat sphere-valued differential forms). Flat L_∞-algebras valued forms (Def. 6.1) with values in the rational Whitehead L_∞-algebra (Prop. 5.11) of a sphere (Ex. 5.3) of positive even dimension $2k$ are pairs consisting of a closed differential $2k$-form and a $(4k-1)$-form whose differential equals minus the wedge square of the $2k$-form:

$$\Omega_{\mathrm{dR}}\big(-;\mathfrak{l}S^{2k}\big) \simeq \left\{ \begin{matrix} G_{4k-1}, \\ G_{2k} \end{matrix} \in \Omega^\bullet_{\mathrm{dR}}(X) \,\middle|\, \begin{matrix} d\,G_{4k-1} = -G_{2k}\wedge G_{2k}, \\ d\,G_{2k} \quad = 0 \end{matrix} \right\}.$$

Namely, from (5.38) one sees that the components of and the constraints on an $\mathfrak{l}S^{2k}$-valued form are as follows:

$$\begin{array}{ccccc}
\Omega^\bullet_{\mathrm{dR}}(X) & \xleftarrow{\text{flat } \mathfrak{l}S^{2k}\text{-valued form}} & \mathbb{R}\begin{bmatrix}\omega_{4k-1}, \\ \omega_{2k}\end{bmatrix} \Big/ \begin{pmatrix} d\,\omega_{4k-1} = -\omega_{2k}\wedge\omega_{2k}, \\ d\,\omega_{2k} \quad = 0 \end{pmatrix} & = & \mathrm{CE}\big(\mathfrak{l}S^{2k}\big) \\[2ex]
G_{2k} & \xleftarrow{\text{component in degree } 2k} & \omega_{2k} & & \\
d\big\downarrow & & \big\downarrow d & & \\
d\,G_{2k} \overset{\text{constraint}}{=\!=} 0 & \longleftarrow & 0 & & \\[2ex]
G_{4k-1} & \xleftarrow{\text{component in degree } 4k-1} & \omega_{4k-1} & & \\
d\big\downarrow & & \big\downarrow & & \\
d\,G_{4k-1} \overset{\text{constraint}}{=\!=} -G_{2k}\wedge G_{2k} & \longleftarrow & -\omega_{4k}\wedge\omega_{4k} & &
\end{array} \tag{6.8}$$

For $2k=4$ this is the structure of the equations of motion of the C-field in 11-dimensional supergravity (modulo the Hodge self-duality constraint $G_7 = \star G_4$) [Sati (2018), §2.5].

Example 6.5 (PL de Rham right adjoint via L_∞-algebra valued forms). For $n\in\mathbb{N}$, the right adjoint functor in the PS de Rham adjunction (5.26) sends the Chevalley-Eilenberg algebra (Def. 4.13) of any $\mathfrak{g}\in L_\infty\mathrm{Algs}^{\geq 0,\mathrm{nil}}_{\mathbb{R},\mathrm{fin}}$ (Def. 4.15) to a simplicial set of flat $\mathfrak{g}$-valued differential forms (Def. 6.1):

$$\flat\mathfrak{B}\exp_{\mathrm{PL}}(\mathfrak{g})(\mathbb{R}^n) := B\exp_{\mathrm{PS},n}\big(\mathrm{CE}(\mathfrak{g})\big) \;:\; [k] \longmapsto \Omega_{\mathrm{dR}}\big(\mathbb{R}^n\times\Delta^k;\mathfrak{g}\big)_{\mathrm{flat}} \quad \in \Delta\mathrm{Sets}$$

(by direct comparison of (5.27) with (6.2)). Regarded as a simplicial presheaf over CartSp (Def. 1.22), this construction is the moduli ∞-stack of flat L_∞-algebra valued differential forms (see Chapter 9 below).

Non-abelian de Rham cohomology.

Definition 6.2 (Coboundaries between flat L_∞-algebra valued forms). Let $X\in$ SmthMfds and (from Def. 4.13) $\mathfrak{g}\in L_\infty\mathrm{Algs}^{\geq 0}_{\mathbb{R},\mathrm{fin}}$. For

$$A^{(0)},\, A^{(1)} \in \Omega_{\mathrm{dR}}(X;\mathfrak{g})_{\mathrm{flat}}$$

a pair of flat $\mathfrak{g}$-valued differential forms on X (Def. 6.1), we say that a *coboundary* between them is a flat $\mathfrak{g}$-valued differential form on the cylinder manifold over X (its Cartesian product manifold with the real line):

$$\widetilde{A} \in \Omega(X \times \mathbb{R}; \mathfrak{g})_{\mathrm{flat}} \tag{6.9}$$

such that its restrictions along

$$X \simeq X \times \{0\} \xhookrightarrow{i_0^X} X \times \mathbb{R} \xhookleftarrow{i_1^X} X \times \{1\} \simeq X$$

are equal to $A^{(0)}$ and to $A^{(1)}$, respectively:

$$(i_0^X)^* \widetilde{A} = A^{(0)} \qquad \text{and} \qquad (i_1^X)^* \widetilde{A} = A^{(1)} . \tag{6.10}$$

If such a coboundary exists, we say that $A^{(0)}$ and $A^{(1)}$ are *cohomologous*, to be denoted

$$A^{(0)} \sim A^{(1)} .$$

Definition 6.3 (Non-abelian de Rham cohomology). Let $X \in \mathrm{SmthMfds}$ and $\mathfrak{g} \in L_\infty\mathrm{Algs}^{\geq 0}_{\mathbb{R},\mathrm{fin}}$ (Def. 4.13). Then the *non-abelian de Rham cohomology* of X with coefficients in $\mathfrak{g}$ is the set

$$H_{\mathrm{dR}}(X; \mathfrak{g}) := \big(\Omega_{\mathrm{dR}}(X; \mathfrak{g})_{\mathrm{flat}}\big)_{/\sim} \tag{6.11}$$

of equivalence classes with respect to the coboundary relation from Def. 6.2 on the set of flat $\mathfrak{g}$-valued differential forms on X (Def. 6.1).

We recall the following basic facts (e.g. [Gomi and Terashima (2000), Rem. 3.1]):

Lemma 6.1 (Fiberwise Stokes theorem and Projection formula). *Let X be a smooth manifold and let F be a compact smooth manifold with corners, e.g. $F = \Delta^k$ a standard k-simplex, which for $k = 1$ is the interval $F = [0,1]$.*

Then fiberwise integration over F of differential forms on the Cartesian product manifold $X \times F$

$$\Omega^\bullet_{\mathrm{dR}}(X \times F) \xrightarrow{\quad \int_F \quad} \Omega^{\bullet - \dim(F)}_{\mathrm{dR}}(X) \quad \textit{e.g.} \quad \Omega^\bullet_{\mathrm{dR}}(X \times \mathbb{R}) \xrightarrow{\quad \int_{[0,1]} \quad} \Omega^{\bullet-1}_{\mathrm{dR}}(X)$$

satisfies, for all $\alpha \in \Omega^\bullet_{\mathrm{dR}}(X \times F)$ and $\beta \in \Omega^\bullet_{\mathrm{dR}}(X)$:

(i) *The fiberwise Stokes formula:*

$$\begin{aligned} \textstyle\int_F d\alpha &= (-1)^{\dim(F)} d \textstyle\int_F \alpha + \textstyle\int_{\partial F} \alpha \\ \text{e.g.} \quad d \textstyle\int_{[0,1]} \alpha &= (i_1^X)^* \alpha - (i_0^X)^* \alpha - \textstyle\int_{[0,1]} d\alpha , \end{aligned} \tag{6.12}$$

where

$$X \simeq X \times \{0\} \xhookrightarrow{i_0^X} X \times \mathbb{R} \xhookleftarrow{i_1^X} X \times \{1\} \simeq X$$

are the boundary inclusions.

(ii) *The projection formula*

$$\begin{aligned} & \textstyle\int_F \left(\mathrm{pr}_X^*\beta\right)\wedge\alpha \;=\; (-1)^{\dim(F)\deg(\beta)}\beta\wedge\int_F\alpha\,, \\ & e.g. \quad \textstyle\int_{[0,1]}\left(\mathrm{pr}_X^*\beta\right)\wedge\alpha \;=\; (-1)^{\deg(\beta)}\beta\wedge\int_{[0,1]}\alpha\,, \end{aligned} \tag{6.13}$$

where

$$X\times F \xrightarrow{\;\mathrm{pr}_X\;} X$$

is projection onto the first factor.

Proposition 6.4 (Non-abelian de Rham cohomology subsumes ordinary de Rham cohomology). *For any $n\in\mathbb{N}$, let $\mathfrak{g}=\mathfrak{b}^n\mathbb{R}$ be the line Lie $(n+1)$-algebra (Ex. 4.12). Then the non-abelian de Rham cohomology with coefficients in $\mathfrak{g}$ (Def. 6.3) is naturally equivalent to ordinary de Rham cohomology in degree $n+1$:*

$$H_{\mathrm{dR}}(-;\mathfrak{b}^n\mathbb{R}) \;\simeq\; H^{n+1}_{\mathrm{dR}}(-)\,. \tag{6.14}$$

Proof. From Ex. 6.2, we know that the canonical cocycle sets are in natural bijection

$$\Omega_{\mathrm{dR}}(X;\mathfrak{b}^n\mathbb{R})_{\mathrm{flat}} \;\simeq\; \Omega^{n+1}_{\mathrm{dR}}(X)_{\mathrm{closed}}\,.$$

Therefore, it just remains to see that the coboundary relations in both cases coincide. By the explicit nature (6.6) of the above natural bijection and by the Def. 6.2 of non-abelian coboundaries, we hence need to see that a pair of closed forms

$$C^{(0)}_{n+1},\, C^{(1)}_{n+1} \;\in \Omega^{n+1}_{\mathrm{dR}}(X)_{\mathrm{closed}}$$

has a de Rham coboundary, i.e.,

$$\begin{aligned} & \exists\, h_n \;\in\; \Omega^n_{\mathrm{dR}}(X)\,, \\ & \text{such that } \; C^{(0)}_{n+1} + dh_n \;=\; C^{(1)}_{n+1}\,, \end{aligned} \tag{6.15}$$

precisely if the pair extends to a closed $(n+1)$-form on the cylinder over X, as in (6.9) (6.10):

$$\exists\, \widetilde{C}_{n+1} \;\in\; \Omega^{n+1}_{\mathrm{dR}}(X\times\mathbb{R})_{\mathrm{closed}}\,, \quad \text{such that} \quad \left(i_0^X\right)^*\widetilde{C}_{n+1} \;=\; C^{(0)}_{n+1} \quad\text{and}\quad \left(i_1^X\right)^*\widetilde{C}_{n+1} \;=\; C^{(1)}_{n+1}\,. \tag{6.16}$$

That (6.15) $\Leftrightarrow$ (6.16) is a standard argument: Let t denote the canonical coordinate function on $\mathbb{R}$. In one direction, given h_n as in (6.15), the choice

$$\widetilde{C}_{n+1} \;:=\; (1-t)\,\mathrm{pr}_X^*\left(C^{(0)}_{n+1}\right) + t\,\mathrm{pr}_X^*\left(C^{(1)}_{n+1}\right) + dt\wedge\mathrm{pr}_X^*\left(h_n\right)$$

clearly satisfies (6.16). In the other direction, given $\widetilde{C}_{n+1}$ as in (6.16), the choice

$$h_n := \int_{[0,1]} \widetilde{C}_{n+1}$$

satisfies (6.15), by the fiberwise Stokes theorem (Lem. 6.1). □

The non-abelian de Rham theorem.

Theorem 6.5 (Non-abelian de Rham theorem). *Let $X \in \mathrm{Ho}(\Delta\mathrm{Sets}_{\mathrm{Qu}})$ and $A \in \mathrm{Ho}(\Delta\mathrm{Sets}_{\mathrm{Qu}})^{\mathrm{fin}_{\mathbb{Q}}}_{\geq 1,\mathrm{nil}}$ (Def. 5.1), and let X admit the structure of a smooth manifold. Then the non-abelian de Rham cohomology (Def. 6.3) of X with coefficients in the real Whitehead L_∞-algebra $\mathfrak{l}A$ (Prop. 5.11) is in natural bijection with the non-abelian real cohomology (Def. 5.14) of X with coefficients in $L_{\mathbb{R}}A$ (Def. 5.7):*

$$H(X; L_{\mathbb{R}}A) \simeq H_{\mathrm{dR}}(X; \mathfrak{l}A)\,. \tag{6.17}$$

Proof. This is the result of the following sequence of bijection:

$$\begin{aligned} & H(X; L_{\mathbb{R}}A) \\ & = \mathrm{Ho}(\Delta\mathrm{Sets}_{\mathrm{Qu}})(X, L_{\mathbb{R}}A) && \text{by Def. 5.14} \\ & \simeq \mathrm{Ho}\big((\mathrm{dgcAlgs}_{\mathbb{R}}^{\geq 0})_{\mathrm{trinj}}\big)\big(\Omega^\bullet_{\mathrm{PLdR}}(A), \Omega^\bullet_{\mathrm{PLdR}}(X)\big) && \text{by Def. 5.7 \& Prop. 5.5} \\ & \simeq \mathrm{Ho}\big((\mathrm{dgcAlgs}_{\mathbb{R}}^{\geq 0})_{\mathrm{trinj}}\big)\big(\mathrm{CE}(\mathfrak{l}A), \Omega^\bullet_{\mathrm{dR}}(X)\big) && \text{by Prop. 5.11 \& Lem. 6.4} \\ & \simeq H_{\mathrm{dR}}(X; \mathfrak{l}A) && \text{by Lem. 6.3.} \end{aligned} \tag{6.18}$$

The two lemmas invoked here are proved next. □

Lemma 6.2 (De Rham complex over cylinder of manifold is path space object). *For $X \in \mathrm{SmthMfds}$, consider the following morphisms of dgc-algebras (Def. 4.10)*

$$\Omega^\bullet_{\mathrm{dR}}(X) \xrightarrow{(\mathrm{pr}_X)^*} \Omega^\bullet_{\mathrm{dR}}(X \times \mathbb{R}) \xrightarrow{(i_0^*, i_1^*)} \Omega^\bullet_{\mathrm{dR}}(X) \oplus \Omega^\bullet_{\mathrm{dR}}(X) \tag{6.19}$$

(from the de Rham complex of X (Ex. 4.9) to that of its cylinder manifold $X \times \mathbb{R}$, to its Cartesian product with itself, by Ex. 4.8), given by pullback of differential forms along these smooth functions:

$$X \xleftarrow{\mathrm{pr}_X} X \times \mathbb{R} \xleftarrow{(i_0, i_1)} (X \times \{0\}) \sqcup (X \times \{1\}) \simeq X \sqcup X\,.$$

This is a path space object (Def. 1.5) for $\Omega^\bullet_{\mathrm{dR}}(X)$ in $(\mathrm{dgcAlgs}_{\mathbb{R}}^{\geq 0})_{\mathrm{trinj}}$ (Prop. 4.19).

Proof. **(i)** It is clear by construction that the composite morphism is the diagonal.
(ii) That $(\mathrm{pr}_X)^*$ is a weak equivalence, hence a quasi-isomorphism, follows from the de Rham theorem, using that ordinary cohomology is homotopy invariant: $H^\bullet(X \times \mathbb{R}; \mathbb{R}) \simeq H^\bullet(X; \mathbb{R})$.

(iii) That (i_0^*, i_1^*) is a fibration, namely degreewise surjective, is seen from the fact that any pair of forms on the boundaries $X \times \{0\}$, $X \times \{1\}$ may be smoothly interpolated to zero along any small enough positive parameter length, and then glued to a form on $X \times \mathbb{R}$. □

Lemma 6.3 (Non-abelian de Rham cohomology via the dgc-homotopy category). *Let $X \in \mathrm{SmthMfds}$ and $\mathfrak{g} \in L_\infty\mathrm{Algs}^{\geq 0,\mathrm{nil}}_{\mathbb{R},\mathrm{fin}}$ (Def. 4.15). Then the non-abelian de Rham cohomology of X with coefficients in $\mathfrak{g}$ (Def. 6.3) is in natural bijection with the hom-set in the homotopy category of $\big(\mathrm{dgcAlgs}^{\geq 0}_{\mathbb{R}}\big)_{\mathrm{trinj}}$ (Prop. 4.19) from $\mathrm{CE}(\mathfrak{g})$ (Def. 4.13) to $\Omega^\bullet_{\mathrm{dR}}(X)$ (Ex. 4.9):*

$$H_{\mathrm{dR}}(X;\mathfrak{g}) \;\simeq\; \mathrm{Ho}\Big(\big(\mathrm{dgcAlgs}^{\geq 0}_{\mathbb{R}}\big)_{\mathrm{trinj}}\Big)\big(\mathrm{CE}(\mathfrak{g})\,,\,\Omega^\bullet_{\mathrm{dR}}(X)\big). \tag{6.20}$$

Proof. Consider a pair of dgc-algebra homomorphisms

$$A^{(0)},\, A^{(1)} \;\in\; \mathrm{dgcAlgs}^{\geq 0}_{\mathbb{R}}\big(\mathrm{CE}(\mathfrak{g})\,,\,\Omega^\bullet_{\mathrm{dR}}(X)\big) \tag{6.21}$$

hence of flat $\mathfrak{g}$-valued differential forms, according to Def. 6.1. Observe that:

(i) $\mathrm{CE}(\mathfrak{g})$ is cofibrant in $\big(\mathrm{dgcAlgs}^{\geq 0}_{\mathbb{R}}\big)_{\mathrm{trinj}}$ (4.44). (by Prop. 4.21, and since $\mathfrak{g}$ is assumed to be nilpotent (4.42));

(ii) $\Omega^\bullet_{\mathrm{dR}}(X)$ is fibrant in $\big(\mathrm{dgcAlgs}^{\geq 0}_{\mathbb{R}}\big)_{\mathrm{trinj}}$ (4.44). (by Rem. 4.3);

(iii) A right homotopy (Def. 1.6) between the pair (6.21) of morphisms, with respect to the path space object $\Omega^\bullet_{\mathrm{dR}}(X \times \mathbb{R})$ from Lem. 6.2, namely a morphism $\widetilde{A}$ making the following diagram commute

$$\begin{array}{ccc}
\Omega^\bullet_{\mathrm{dR}}(X) & \xleftarrow{\quad A^{(0)}\quad} & \\
{\scriptstyle i_0^*}\uparrow & & \\
\Omega^\bullet_{\mathrm{dR}}(X \times \mathbb{R}) & \xleftarrow{\quad \widetilde{A}\quad} & \mathrm{CE}(\mathfrak{g}) \\
{\scriptstyle i_1^*}\downarrow & & \\
\Omega^\bullet_{\mathrm{dR}}(X) & \xleftarrow{\quad A^{(1)}\quad} &
\end{array} \tag{6.22}$$

is manifestly the same as a coboundary $\widetilde{A}$ between the corresponding flat $\mathfrak{g}$-valued forms according to Def. 6.2.

Therefore, Prop. 1.10 says that the quotient set (6.11) defining the non-abelian de Rham cohomology is in natural bijection to the hom-set in the homotopy category. □

Lemma 6.4 (PL de Rham complex on smooth manifold is equivalent to smooth de Rham complex). *Let X be a smooth manifold. Then*

(i) *There exists a zig-zag of weak equivalences (Def. 4.18) in $\big(\mathrm{dgcAlgs}^{\geq 0}_{\mathbb{R}}\big)_{\mathrm{trinj}}$ (4.44) between the smooth de Rham complex of X (Ex. 4.9) and the PL de Rham complex of its underlying topological space (Def. 5.3).*

(ii) *In particular, both are isomorphic in the homotopy category:*

$$X \text{ smooth manifold} \quad \Rightarrow \quad \Omega^\bullet_{\mathrm{dR}}(X) \simeq \Omega^\bullet_{\mathrm{PLdR}}(X) \quad \in \mathrm{Ho}\Big(\big(\mathrm{dgcAlgs}^{\geq 0}_{\mathbb{R}}\big)_{\mathrm{trinj}}\Big).$$

Proof. Let $\Omega^\bullet_{\mathrm{PSdR}}(-)$ (for "piecewise smooth") be defined as the PL de Rham complex in Def. 5.3, but with smooth differential forms on each simplex. Notice that this comes with the canonical natural inclusion

$$\Omega^\bullet_{\mathrm{PLdR}}(-) \xhookrightarrow{\;i_{\mathrm{poly}}\;} \Omega^\bullet_{\mathrm{PSdR}}(-)\,.$$

Let then $\mathrm{Tr}(X) \in \Delta\mathrm{Sets}$ be any smooth triangulation of X (Ex. 1.16). This means that we have a homeomorphism out of its geometric realization (1.50), $|\mathrm{Tr}(X)| \xrightarrow[\text{homeo}]{p} X$ (1.58), which restricts on the interior of each simplex to a diffeomorphism onto its image; and that we have an inclusion (1.59)

$$\mathrm{Tr}(X) \xhookrightarrow[\in \mathrm{W}]{\eta_{\mathrm{Tr}(X)}} \mathrm{Sing}\big(|\mathrm{Tr}(X)|\big) \xrightarrow[\in \mathrm{Iso}]{\mathrm{Sing}(p)} \mathrm{Sing}(X)\,, \tag{6.23}$$

which is a weak equivalence (by Ex. 1.15). In summary, this gives us the following zig-zag of dgc-algebra homomorphisms:

$$\begin{array}{ccccc} & \Omega^\bullet_{\mathrm{PLdR}}\big(\mathrm{Tr}(X)\big) & & & \Omega^\bullet_{\mathrm{dR}}(X) \\ {\scriptstyle (\eta_S)^*}\nearrow & & \searrow{\scriptstyle i_{\mathrm{poly}}} & {\scriptstyle p^*}\swarrow & \\ \Omega^\bullet_{\mathrm{PLdR}}(X) = \Omega^\bullet_{\mathrm{PLdR}}\big(\mathrm{Sing}(X)\big) & & & \Omega^\bullet_{\mathrm{PSdR}}\big(\mathrm{Tr}(X)\big) & \end{array}$$

Here the two morphisms on the right are quasi-isomorphisms by [Griffiths and Morgan (2013), Cor. 9.9] (as in Prop. 5.10). The morphism on the left is a quasi-isomorphism because η_S is a weak homotopy equivalence (6.23) which is preserved by $\Omega^\bullet_{\mathrm{PLdR}}$ (using Lem. 1.1), since this is a Quillen left adjoint (by Prop. 5.5) and since every simplicial set is cofibrant (Ex. 1.2). □

Flat twisted L_∞-algebra valued differential forms. We generalize the above discussion to include twistings.

Definition 6.6 (Local L_∞-algebraic coefficients). We say that a *local L_∞-algebraic coefficient bundle* is a fibration

$$\begin{array}{ccc} \mathfrak{g} & \longrightarrow & \widehat{\mathfrak{b}} \\ & & \downarrow{\scriptstyle p} \\ & & \mathfrak{b} \end{array} \tag{6.24}$$

in $L_\infty\mathrm{Algs}^{\geq 0}_{\mathbb{R},\mathrm{fin}}$ (Def. 4.13), hence a morphism such that under passage to Chevalley-Eilenberg algebras (4.27) we have a cofibration

$$\begin{array}{ccc} \mathrm{CE}(\mathfrak{g}) & \xleftarrow{\;\mathrm{cofib}(\mathrm{CE}(\mathfrak{p}))\;} & \mathrm{CE}(\widehat{\mathfrak{b}}) \\ & & \Big\uparrow{\scriptstyle \mathrm{CE}(\mathfrak{p})\,\in\,\mathrm{Cof}} \\ & & \mathrm{CE}(\mathfrak{b}) \end{array} \tag{6.25}$$

in $\left(\mathrm{dgcAlgs}^{\geq 0}_{\mathbb{R}}\right)_{\mathrm{trinj}}$ (Prop. 4.19).

In generalization of Def. 6.1, we say:

Definition 6.7 (Flat twisted L_∞-algebra valued differential forms).
(i) Let $X \in \mathrm{SmthMfds}$ and $\widehat{\mathfrak{b}}$ (6.24) a local L_∞-algebraic coefficient bundle (Def. 6.6). For

$$\tau_{\mathrm{dR}} \in \Omega_{\mathrm{dR}}(X;\mathfrak{b})_{\mathrm{flat}} \tag{6.26}$$

a flat $\mathfrak{b}$-valued differential form on X (Def. 6.1), we say that a *flat τ-twisted $\mathfrak{g}$-valued differential form* on X is a morphism of dgc-algebras (Def. 4.10) in the slice over $\mathrm{CE}(\mathfrak{b})$

$$\Omega^\bullet_{\mathrm{dR}}(X) \xleftarrow[A]{\text{flat } \tau_{\mathrm{dR}}\text{-twisted } \mathfrak{g}\text{-valued differential form}} \mathrm{CE}(\widehat{\mathfrak{b}}), \quad \Omega^\bullet_{\mathrm{dR}}(X) \xleftarrow[\text{twist}]{\tau_{\mathrm{dR}}} \mathrm{CE}(\mathfrak{b}) \xhookrightarrow[\text{local } L_\infty\text{-algebraic coefficients}]{\mathrm{CE}(\mathfrak{p})} \mathrm{CE}(\widehat{\mathfrak{b}}) \tag{6.27}$$

(ii) We write

$$\Omega^{\tau_{\mathrm{dR}}}_{\mathrm{dR}}(X;\mathfrak{g})_{\mathrm{flat}} := \left(\mathrm{dgcAlgs}^{\geq 0}_{\mathbb{R}}\right)_{/\mathrm{CE}(\mathfrak{b})}(\tau_{\mathrm{dR}}, \mathfrak{p})$$

for the set of all flat τ_{dR}-twisted $\mathfrak{g}$-valued differential forms on X.

Remark 6.1 (Underlying flat forms of flat twisted forms). Let $X \in \mathrm{SmthMfds}$, let $\mathfrak{g} \to \widehat{\mathfrak{b}} \xrightarrow{\mathfrak{p}} \mathfrak{b}$ be a local L_∞-algebraic coefficient bundle (Def. 6.6), and let $\tau_{\mathrm{dR}} \in \Omega_{\mathrm{dR}}(X;\mathfrak{b})$. Then there is a canonical forgetful natural transformation

$$\Omega^{\tau_{\mathrm{dR}}}(X;\mathfrak{g})_{\mathrm{flat}} \longrightarrow \Omega\big(X;\widehat{\mathfrak{b}}\big)_{\mathrm{flat}} \tag{6.28}$$

from flat τ_{dR}-twisted $\mathfrak{g}$-valued differential forms (Def. 6.7) to flat $\widehat{\mathfrak{b}}$-valued differential forms (Def. 6.1), given by remembering only the top morphism in (6.27).

Example 6.6 (L_∞-coefficient bundle for H_3-twisted differential forms [Fiorenza *et al.* (2017), §4][Fiorenza *et al.* (2018), §4][Braunack-Mayer *et al.* (2019), Lem. 2.31]**).** Consider the local L_∞-algebraic coefficient bundle (Def. 6.6) given by the following multivariate polynomial dgc-algebras (Def. 4.15):

$$
\begin{array}{ccc}
\mathrm{CE}\big(\mathfrak{l}\mathrm{ku}_1\big) & & \mathrm{CE}\Big(\mathfrak{l}\big(\mathrm{ku}_1 /\!\!/ B\mathrm{U}(1)\big)\Big) \\
\| & & \| \\
\mathbb{R}\left[\begin{array}{c} \vdots \\ f_5, \\ f_3, \\ f_1 \end{array}\right] \Big/ \left(\begin{array}{c} \vdots \\ d\,f_5 = 0 \\ d\,f_3 = 0 \\ d\,f_1 = 0 \end{array}\right) & \xleftarrow{\;\omega_{2k+1} \leftarrow\!\shortmid\, \omega_{2k+1}\;} & \mathbb{R}\left[\begin{array}{c} \vdots \\ f_5, \\ f_3, \\ f_1, \\ h_3 \end{array}\right] \Big/ \left(\begin{array}{l} \vdots \\ d\,f_5 = h_3 \wedge f_3, \\ d\,f_3 = h_3 \wedge f_1, \\ d\,f_1 = 0, \\ d\,h_3 = 0 \end{array}\right) \\
 & & \Big\uparrow {\scriptstyle h_3 \,\leftarrow\!\shortmid\, h_3} \\
 & & \mathbb{R}\big[h_3\big]\big(d\,h_3 = 0\big) = \mathrm{CE}\big(\mathfrak{b}^2\mathbb{R}\big)
\end{array}
$$

Here the rational model of the classifying space ku_1 for complex topological K-theory in degree 1 and for its twisted version is as in [Fiorenza *et al.* (2017), §4][Fiorenza *et al.* (2018), §4][Braunack-Mayer *et al.* (2019), Lem. 2.31]. In this case:

(i) A twist (6.26) is equivalently an ordinary closed 3-form form (by Ex. 6.2):

$$
H_3 \in \Omega_{\mathrm{dR}}\big(X; \mathfrak{b}^2\mathbb{R}\big)_{\mathrm{flat}} \simeq \Omega^3_{\mathrm{dR}}(X)_{\mathrm{closed}}\,. \tag{6.29}
$$

(ii) The flat $\tau_{\mathrm{dR}} \sim H_3$-twisted $\mathfrak{l}\mathrm{ku}_1$-valued differential forms according to Def. 6.7 are equivalently sequences of odd-degree differential forms $F_{2k+1} \in \Omega^{2k+1}_{\mathrm{dR}}(X)$ satisfying the H_3-twisted de Rham closure condition (see [Rohm and Witten (1986), (23)][Grady and Sati (2019b)]):

$$
\Omega^{\tau_{\mathrm{dR}}}\big(X; \mathfrak{l}\mathrm{ku}_1\big)_{\mathrm{flat}} \simeq \left\{ F_{2\bullet+1} \in \Omega^{2\bullet+1}_{\mathrm{dR}} \,\middle|\, d\sum_k F_{2k+1} = H_3 \wedge \sum_k F_{2k-1} \right\} \tag{6.30}
$$

(where we set $F_{2k-1} := 0$ if $2k-1 < 0$, for convenience of notation).

In direct generalization of Ex. 6.6, we have:

Example 6.7 (L_∞-coefficient bundle for higher twisted differential forms [Fiorenza *et al.* (2020a), Def. 2.14]**).** For $r \in \mathbb{N}$, $r \geq 1$, consider the local L_∞-algebraic coefficient bundle

(Def. 6.6) given by the following multivariate polynomial dgc-algebras (Def. 4.15):

$$
\begin{array}{ccc}
\mathrm{CE}\Big(\underset{k\in\mathbb{N}}{\oplus}\mathfrak{b}^{2rk}\mathbb{R}\Big) & & \mathrm{CE}\Big(\big(\underset{k\in\mathbb{N}}{\oplus}\mathfrak{b}^{2rk}\mathbb{R}\big)/\!\!/B^{2r-1}\mathrm{U}(1)\Big) \\
\| & & \| \\
\mathbb{R}\left[\begin{array}{c}\vdots \\ f_{4r+1}, \\ f_{2r+1}, \\ f_1\end{array}\right] \Big/ \left(\begin{array}{rl} & \vdots \\ d\,f_{4r+1} &= 0 \\ d\,f_{2r+1} &= 0 \\ d\,f_1 &= 0\end{array}\right) & \xleftarrow{\;f_{2rk+1}\,\leftarrow\!\shortmid\, f_{2rk+1}\;} & \mathbb{R}\left[\begin{array}{c}\vdots \\ f_{4r+1}, \\ f_{2r+1}, \\ f_1, \\ h_{2r+1}\end{array}\right] \Big/ \left(\begin{array}{rl} & \vdots \\ d\,f_{4r+1} &= h_{2r+1}\wedge f_{2r+1}, \\ d\,f_{2r+1} &= h_{2r+1}\wedge f_1, \\ d\,f_1 &= 0, \\ d\,h_{2r+1} &= 0\end{array}\right) \\
& & \Big\uparrow {\scriptstyle h_{2r+1} \,\leftarrow\!\shortmid\, h_{2r+1}} \\
& & \mathbb{R}\big[h_{2r+1}\big]\big(d\,h_{2r+1} = 0\big) \\
& & \| \\
& & \mathrm{CE}\big(\mathfrak{b}^{2r}\mathbb{R}\big)
\end{array}
\tag{6.31}
$$

In this case:

(i) A twist (6.26) is equivalently an ordinary closed $(2r+1)$-form form (by Ex. 6.2):

$$
H_{2r+1} \in \Omega_{\mathrm{dR}}\big(X;\, \mathfrak{b}^{2r}\mathbb{R}\big)_{\mathrm{flat}} \simeq \Omega^{2r+1}_{\mathrm{dR}}(X)_{\mathrm{closed}}\,. \tag{6.32}
$$

(ii) The flat $\tau_{\mathrm{dR}} \sim H_{2r+1}$-twisted $\underset{k\in\mathbb{N}}{\oplus}\mathfrak{b}^{2rk}\mathbb{R}$-valued differential forms according to Def. 6.7 are equivalently sequences of differential forms $F_{2r\bullet+1} \in \Omega^{2k\bullet+1}_{\mathrm{dR}}(X)$ satisfying the $H_{(2r+1)}$-twisted de Rham closure condition (6.43):

$$
\Omega^{\tau_{\mathrm{dR}}}\big(X;\, \underset{k\in\mathbb{N}}{\oplus}\mathfrak{b}^{2rk}\mathbb{R}\big)_{\mathrm{flat}} \simeq \left\{ F_{2r\bullet+1} \in \Omega^{2r\bullet+1}_{\mathrm{dR}} \;\middle|\; d\sum_k F_{2rk+1} = H_{2r+1}\wedge\sum_k F_{2rk-1} \right\} \tag{6.33}
$$

(where we set $F_{2rk-1} := 0$ if $2rk-1 < 0$, for convenience of notation).

In twisted generalization of Ex. 6.4, we have the following:

Example 6.8 (Flat twisted differential forms with values in Whitehead L_∞-algebras of spheres and twistor space). The L_∞-algebraic local coefficient bundles (Def. 6.6) given as the relative Whitehead L_∞-algebras (Prop. 5.16) of the local coefficient bundles (3.35) for twisted and twistorial Cohomotopy (Ex. 3.11) are as shown on the right of the following diagram [Fiorenza *et al.* (2020b), Lem. 3.19][Fiorenza *et al.* (2022), Thm. 2.14]:

$$\mathrm{CE}\big(l_{BSp(2)}(\mathbb{C}P^3 /\!\!/ \mathrm{Sp}(2))\big) = \mathrm{CE}(lBSp(2)) \begin{bmatrix} h_3, \\ f_2, \\ \omega_7, \\ \omega_4 \end{bmatrix} \Big/ \begin{pmatrix} d\,h_3 = \omega_4 - \tfrac{1}{4}p_1 - f_2 \wedge f_2 \\ d\,f_2 = 0 \\ d\,\omega_7 = -\omega_4 \wedge \omega_4 + (\tfrac{1}{4}p_1)^2 - \chi_8 \\ d\,\omega_4 = 0 \end{pmatrix}$$

$$\Omega^\bullet_{\mathrm{dR}}(X) \xleftarrow[(G_4, 2G_7)]{} \mathrm{CE}\big(l_{BSp(2)}(S^4 /\!\!/ \mathrm{Sp}(2))\big) = \mathrm{CE}(lBSp(2)) \begin{bmatrix} \omega_7, \\ \omega_4 \end{bmatrix} \Big/ \begin{pmatrix} d\,\omega_7 = -\omega_4 \wedge \omega_4 + (\tfrac{1}{4}p_1)^2 - \chi_8 \\ d\,\omega_4 = 0 \end{pmatrix}$$

$$\mathrm{CE}(lBSp(2)) = \mathbb{R}\begin{bmatrix} \chi_8, \\ \tfrac{1}{2}p_1 \end{bmatrix} \Big/ \begin{pmatrix} d\ \chi_8 = 0 \\ d\,\tfrac{1}{2}p_1 = 0 \end{pmatrix}$$

with the maps (G_4, G_7, F_2, H_3), $(t_\mathbb{H} /\!\!/ \mathrm{Sp}(2))^*$, and τ_{dR}.

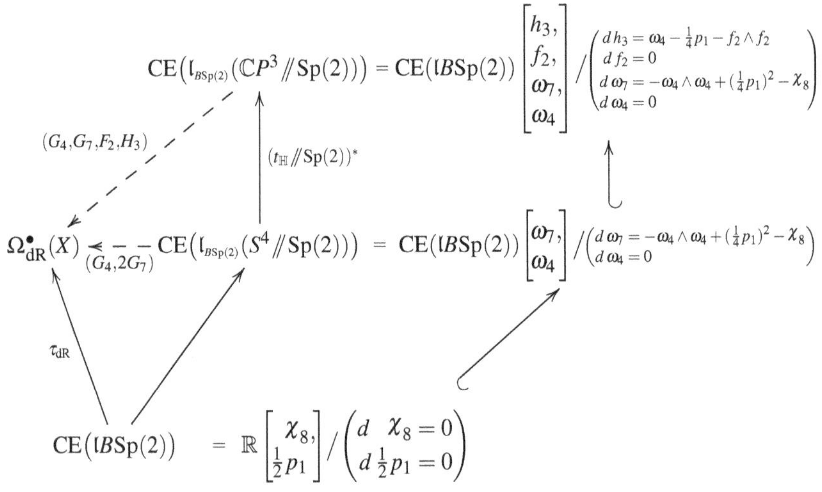

Therefore, given a smooth 8-dimensional spin-manifold X equipped with tangential $\mathrm{Sp}(2)$-structure τ (3.33), the flat τ_{dR}-twisted lS^4- and $l\mathbb{C}P^3$-valued differential forms (Def. 6.7) are of the following form [Fiorenza *et al.* (2020b), Prop. 3.20][Fiorenza *et al.* (2022), Prop. 3.9]:

$$\Omega^{\tau_{\mathrm{dR}}}_{\mathrm{dR}}(X; lS^4) = \left\{ \begin{matrix} 2G_7, \\ G_4 \end{matrix} \in \Omega^\bullet_{\mathrm{dR}}(X) \,\middle|\, \begin{matrix} d\,2G_7 = -\big(G_4 - \tfrac{1}{4}p_1(\nabla)\big) \wedge \big(G_4 + \tfrac{1}{4}p_1(\nabla)\big) - \chi_8(\nabla) \\ d\ G_4 = 0 \end{matrix} \right\}$$

$$\Omega^{\widetilde{\tau}_{\mathrm{dR}}}_{\mathrm{dR}}(X; l\mathbb{C}P^3) = \left\{ \begin{matrix} H_3, \\ F_2, \\ 2G_7, \\ G_4 \end{matrix} \in \Omega^\bullet_{\mathrm{dR}}(X) \,\middle|\, \begin{matrix} d\,H_3 = G_4 - \tfrac{1}{4}p_1(\nabla) - F_2 \wedge F_2, \\ d\,F_2 = 0 \\ d\,2G_7 = -\big(G_4 - \tfrac{1}{4}\big) \wedge \big(G_4 + \tfrac{1}{4}\big) - \chi_8(\nabla) \\ d\ G_4 = 0, \end{matrix} \right\}$$

$$(6.34)$$

Notice:

(a) Here we are using (Ex. 8.2) that the de Rham image τ_{dR} of the rationalization $L_\mathbb{R}\tau$ of the twist τ is given by evaluating characteristic forms (Def. 8.2) on any $\mathrm{Sp}(2)$-connection ∇.

(b) In the second equation of (6.34) we are using the above minimal model for $\mathbb{C}P^3 /\!\!/ \mathrm{Sp}(2)$ relative to $S^4 /\!\!/ \mathrm{Sp}(2)$ (instead of relative to $B\mathrm{Sp}(2)$).

Twisted non-abelian de Rham cohomology. In generalization of Def. 6.2, we set:

Definition 6.8 (Coboundaries between flat twisted L_∞-algebraic forms). Let $X \in$ SmthMfds, let $\mathfrak{g} \twoheadrightarrow \widehat{\mathfrak{b}} \xrightarrow{p} \mathfrak{b}$ be a local L_∞-algebraic coefficient bundle (Def. 6.6), and let $\tau_{\mathrm{dR}} \in \Omega_{\mathrm{dR}}(X; \mathfrak{b})$. Then for

$$A^{(0)}, A^{(1)} \in \Omega^{\tau_{\mathrm{dR}}}_{\mathrm{dR}}(X; \mathfrak{g})$$

a pair of flat τ_{dR}-twisted $\mathfrak{g}$-valued differential forms on X (Def. 6.7), a *coboundary* between them is a coboundary

$$\widetilde{A} \in \Omega_{\mathrm{dR}}\big(X \times \mathbb{R}; \widehat{\mathfrak{b}}\big) \tag{6.35}$$

in the sense of Def. 6.2 between the underlying flat $\widehat{\mathfrak{b}}$-valued forms (via Rem. 6.1), such that the underlying $\mathfrak{b}$-valued form of H equals the pullback of the twist τ_{dR} along $X \times \mathbb{R} \xrightarrow{\mathrm{pr}_X} X$:

$$\mathfrak{p}_*(H) \;=\; \mathrm{pr}_X^*(\tau_{\mathrm{dR}})\,. \tag{6.36}$$

If such a coboundary exists, we say that $A^{(0)}$ and $A^{(1)}$ are *cohomologous*, to be denoted

$$A^{(0)} \;\sim\; A^{(1)}\,.$$

In generalization of Def. 6.3, we set:

Definition 6.9 (Twisted non-abelian de Rham cohomology). Let $X \in \mathrm{SmthMfds}$, let $\mathfrak{g} \longrightarrow \widehat{\mathfrak{b}} \xrightarrow{\mathfrak{p}} \mathfrak{b}$ be a local L_∞-algebraic coefficient bundle (Def. 6.6), and let $\tau_{\mathrm{dR}} \in \Omega_{\mathrm{dR}}\big(X; \mathfrak{b}\big)$. Then the τ_{dR} *-twisted non-abelian de Rham cohomology* of X with coefficients in $\mathfrak{g}$ is the set

$$H^{\tau_{\mathrm{dR}}}_{\mathrm{dR}}(X; \mathfrak{g}) \;:=\; \big(\Omega^{\tau_{\mathrm{dR}}}_{\mathrm{dR}}(X; \mathfrak{g})_{\mathrm{flat}}\big)_{/\sim} \tag{6.37}$$

of equivalence classes with respect to the coboundary relation from Def. 6.8 on the set of flat τ_{dR}-twisted $\mathfrak{g}$-valued differential forms on X (Def. 6.7).

Remark 6.2 (Independence of the representative of the twist). The twisted non-abelian de Rham theorem below (Thm. 6.15) makes manifest that the twisted non-abelian de Rham cohomology in Def. 6.9 depends on the twist τ_{dR} only through its class $[\tau_{\mathrm{dR}}] \in H_{\mathrm{dR}}(X; \mathfrak{b})$ in (un-twisted) non-abelian de Rham cohomology (Def. 6.3).

The example of traditional twisted de Rham cohomology. Twisted de Rham cohomology is traditionally familiar in the form of degree-3 twisted cohomology of even/odd degree differential forms [Rohm and Witten (1986), §III, Appendix][Bouwknegt *et al.* (2002), §9.3][Mathai and Stevenson (2003), §3][Freed *et al.* (2008), §2][Teleman (2004), Prop. 3.7][Cavalcanti (2005), §I.4][Sati (2010)][Mathai and Wu (2011)][Grady and Sati (2019a)] (which is the target of the twisted Chern character in degree-3 twisted K-theory, see Prop. 10.1).

We discuss now how this archetypical example (Def. 6.10) and its higher-degree generalization (Def. 6.12) are subsumed by our general Def. 6.9.

Definition 6.10 (Degree-3 twisted abelian de Rham cohomology). For $X \in \mathrm{SmthMfds}$, and $H_3 \in \Omega^3_{\mathrm{dR}}(X)_{\mathrm{closed}}$ a closed differential 3-form, the H_3*-twisted de Rham cohomology* of X is the cochain cohomology[1]

$$H_{\mathrm{dR}}^{\bullet+H_3}(X) := \frac{\ker^\bullet\big(d - H_3 \wedge (-)\big)}{\mathrm{im}^\bullet\big(d - H_3 \wedge (-)\big)} \tag{6.38}$$

of the following 2-periodic cochain complex:

$$\cdots \longrightarrow \bigoplus_k \Omega_{\mathrm{dR}}^{(n-1)+2k}(X) \xrightarrow{(d-H_3\wedge(-))} \bigoplus_k \Omega_{\mathrm{dR}}^{n+2k}(X) \xrightarrow{(d-H_3\wedge(-))} \bigoplus_k \Omega_{\mathrm{dR}}^{(n+1)+2k}(X) \longrightarrow \cdots . \tag{6.39}$$

We show that this is a special case of twisted non-abelian de Rham cohomology according to Def. 6.9:

Proposition 6.11 (Twisted non-abelian de Rham cohomology subsumes H_3-twisted abelian de Rham cohomology). *Given a twisting 3-form as in* (6.29)

$$\begin{array}{ccc} \tau_{\mathrm{dR}} & \longleftrightarrow & H_3 \\ \pitchfork & & \pitchfork \\ \Omega\big(X; \mathfrak{b}^2\mathbb{R}\big)_{\mathrm{flat}} & \simeq & \Omega^3(X)_{\mathrm{closed}} \end{array}$$

the τ_{dR}-twisted non-abelian de Rham cohomology (Def. 6.9) of flat twisted $\mathfrak{l}\mathrm{ku}_1$*-valued differential forms (Ex. 6.6) is naturally equivalent to H_3-twisted abelian de Rham cohomology (Def. 6.10) in odd degree*[2]

$$\underbrace{H_{\mathrm{dR}}^{\tau_{\mathrm{dR}}}(X; \mathfrak{l}\mathrm{ku}_1)}_{\substack{\mathfrak{b}^2\mathbb{R}\text{-twisted } \mathfrak{l}\mathrm{ku}_1\text{-valued} \\ \text{non-abelian de Rham cohomology}}} \simeq \underbrace{H_{\mathrm{dR}}^{1+H_3}(X)}_{\substack{\text{traditional } H_3\text{-twisted} \\ \text{de Rham cohomology}}}$$

Proof. By (6.30) in Ex. 6.6 the cocycle sets on both sides are in natural bijection. Hence it is sufficient to see that the coboundary relations on the cocycle sets coincide, under this identification. In one direction, consider a coboundary in the sense of twisted non-abelian de Rham cohomology (Def. 6.8) with coefficients as in Ex. 6.6:

$$\widetilde{F}_{2\bullet+1} \in \Omega_{\mathrm{dR}}\big(X \times \mathbb{R}; \mathfrak{l}\mathrm{ku}_1\big).$$

We claim that

$$h_{2\bullet} := \int_{[0,1]} \widetilde{F}_{2\bullet+1} \tag{6.40}$$

[1]The notation "H_3" for the twist (and of "H_{2r+1}" for the higher twists later) originates in the physics literature and has made it as a convention in differential geometry as well. Not to be confused with a third homology group, of course.

[2]The discussion for even degrees is directly analogous; we omit it for brevity.

satisfies the coboundary condition (6.38):

$$\big(d - H_3 \wedge\big)\sum_k h_{2k} \;=\; \sum_k \big(F^{(1)}_{2k+1} - F^{(0)}_{2k+1}\big)\,. \tag{6.41}$$

To see this, we may compute as follows:

$$\begin{aligned} d\sum_k h_{2k} &= \sum_k \Bigg(F^{(1)}_{2k+1} - F^{(0)}_{2k+1} - \int_{[0,1]} d\widetilde{F}_{2k+1} \Bigg) \\ &= \sum_k \Bigg(F^{(1)}_{2k+1} - F^{(0)}_{2k+1} - \int_{[0,1]} (\mathrm{pr}_X^* H_3) \wedge \widetilde{F}_{2k-1} \Bigg) \\ &= \sum_k \Bigg(F^{(1)}_{2k+1} - F^{(0)}_{2k+1} + H_3 \wedge \int_{[0,1]} \widetilde{F}_{2k-1} \Bigg) \\ &= \sum_k \Big(F^{(1)}_{2k+1} - F^{(0)}_{2k+1} + H_3 \wedge h_{2k-2} \Big)\,, \end{aligned}$$

where the first step is the fiberwise Stokes formula (6.12) together with the defining restrictions (6.10) of $\widetilde{F}_{2\bullet+1}$; the second step is the cocycle condition (6.30) on $\widetilde{F}_{2\bullet+1}$ using the constraint (6.36); the third step is the projection formula (6.13); and the last step uses again the definition (6.40).

Conversely, given $h_{2\bullet}$ satisfying (6.41), we claim that

$$\widetilde{F}_{2\bullet+1} \;:=\; (1-t)\,\mathrm{pr}_1^*\big(F^{(0)}_{2\bullet+1}\big) + t\,\mathrm{pr}_1^*\big(F^{(1)}_{2\bullet+1}\big) + dt \wedge \mathrm{pr}_X^*\,(h_{2\bullet}) \tag{6.42}$$

is a coboundary of twisted non-abelian cocycles, in the sense of Def. 6.8: It is immediate that (6.42) has the required restrictions (6.10). We check by direct computation that it satisfies the required differential equation:

$$\begin{aligned} d\sum_k \widetilde{F}_{2k+1} &= \sum_k \Big(- dt \wedge \mathrm{pr}_X^*\big(F^{(0)}_{2k+1}\big) + (1-t)\,\mathrm{pr}_X^*\,(H_3) \wedge \mathrm{pr}_X^*\big(F^{(0)}_{2k-1}\big) \\ &\qquad + dt \wedge \mathrm{pr}_X^*\big(F^{(1)}_{2k+1}\big) + t\,\mathrm{pr}_X^*\,(H_3) \wedge \mathrm{pr}_X^*\big(F^{(1)}_{2k-1}\big) \\ &\qquad - dt \wedge \mathrm{pr}_X^*\big(\underbrace{d\,h_{2k}}_{=F^{(1)}_{2k+1} - F^{(0)}_{2k+1} + H_3 \wedge h_{2k}}\big) \Big) \\ &= \sum_k \Big(\mathrm{pr}_X^*\,(H_3) \wedge \widetilde{F}_{2k-1} \Big)\,. \end{aligned}$$

□

In generalization of Def. 6.10, there are twisted abelian Rham complexes with twist any odd-degree closed form [Teleman (2004), §3][Sati (2009)][Mathai and Wu (2011)][Sati (2010)][Grady and Sati (2019a)] (these serve as the targets for the Chern character [Macdonald *et al.* (2021)] on higher-twisted ordinary K-theory [Teleman (2004)][Guerra (2008)][Dadarlat and Pennig (2015)][Pennig (2016)], see Ex. 10.1 below; and for the LSW-character on twisted higher K-theories [Lind *et al.* (2020), §2.1], see Prop. 10.2 below):

Definition 6.12 (Higher twisted abelian de Rham cohomology). For $X \in \mathrm{SmthMfds}$, $r \in \mathbb{N}$ and $H_{2r+1} \in \Omega^{2r+1}_{\mathrm{dR}}(X)_{\mathrm{closed}}$ a closed differential $(2r+1)$-form, the *H_{2r+1}-twisted de Rham cohomology* of X is the cochain cohomology

$$\Omega^{\bullet + H_{2r+1}}_{\mathrm{dR}}(X) := \frac{\ker^\bullet\big(d - H_{2r+1} \wedge (-)\big)}{\mathrm{im}^\bullet\big(d - H_{2r+1} \wedge (-)\big)} \tag{6.43}$$

of the following $2r$-periodic cochain complex:

$$\cdots \to \bigoplus_k \Omega^{(n-1)+2rk}_{\mathrm{dR}}(X) \xrightarrow{(d - H_{2r+1} \wedge (-))} \bigoplus_k \Omega^{n+2rk}_{\mathrm{dR}}(X) \xrightarrow{(d - H_{2r+1} \wedge (-))} \bigoplus_k \Omega^{(n+1)+2rk}_{\mathrm{dR}}(X) \to \cdots.$$

In direct generalization of Prop. 6.11, we find:

Proposition 6.13 (Twisted non-abelian de Rham cohomology subsumes higher twisted abelian de Rham cohomology). *For $r \in \mathbb{N}$, $r \geq 1$, consider a twisting $(2r+1)$-form as in* (6.32)

$$\begin{array}{ccc} \tau_{\mathrm{dR}} & \longleftrightarrow & H_{2r+1} \\ \pitchfork & & \pitchfork \\ \Omega\big(X; \mathfrak{b}^{2r}\mathbb{R}\big)_{\mathrm{flat}} & \simeq & \Omega^{2r+1}\big(X\big)_{\mathrm{closed}} \end{array}$$

The τ_{dR}-twisted non-abelian de Rham cohomology (Def. 6.9) of flat twisted $\mathfrak{l}K^{2r-2}(\mathrm{ku})_1$-valued differential forms (Ex. 6.7) is naturally equivalent to H_{2r+1}-twisted abelian de Rham cohomology (Def. 6.12) in degree[3] *1 mod $2r$.*

$$\underbrace{H^{\tau_{\mathrm{dR}}}_{\mathrm{dR}}\Big(X; \bigoplus_{k \in \mathbb{N}} \mathfrak{b}^{2rk}\mathbb{R}\Big)}_{\text{twisted non-abelian de Rham cohomology}} \simeq \underbrace{H^{1+H_{2r+1}}_{\mathrm{dR}}(X)}_{\text{higher } H_{2r+1}\text{-twisted de Rham cohomology}}.$$

Proof. By Ex. 6.7, the cocycle sets on both sides are in natural bijection. Hence it remains to see that the coboundary relations correspond to each other, under this identification. This proceeds verbatim, up to degree shifts, as in the proof of Prop. 6.11 (which is the special case of $r = 1$). □

Example 6.9 (Degree-1 twisted non-abelian de Rham cohomology). Def. 6.12 subsumes also the case of a twist in ("lower") degree 1, for $k = 0$. By classical theory of sheaf cohomology for local systems (see e.g. [Chen and Yang (2019), Prop. 2.3] following [Voisin (2007), §II 5.1.1]) the *degree-1 twisted de Rham cohomology* in the sense of Def. 6.12 is equivalently classical sheaf cohomology with coefficients in the flat local sections of a trivial line bundle with flat connection. Beware that for more general local systems of lines (or even of vector spaces) some authors still speak of "twisted de Rham cohomology" (e.g. [Chen and Yang (2019), §2.1]), though the twist itself is then no longer

[3] The discussion for other degrees is directly analogous, and we omit it for brevity.

in real/de Rham cohomology, whence this more general case is, in our terminology, no longer an example of Def. 6.12, but is an example of (torsion-)twisted differential cohomology [Grady and Sati (2018c)].

Example 6.10 (Cohomology operation in (higher-) twisted de Rham cohomology). Degree-3 twisted de Rham cohomology (Def. 6.10) supports the following twisted cohomology operations (Def. 3.6):

(i) *wedge product with* H_3:

$$\begin{array}{ccc} H_{\mathrm{dR}}^{\bullet+H_3}(X) & \longrightarrow & H_{\mathrm{dR}}^{\bullet+3+H_3}(X) \\ \sum\limits_k F_k & \longmapsto & \sum\limits_k F_k \wedge H_3 \end{array}$$

(ii) *wedge square*:

$$\begin{array}{ccc} \bigoplus\limits_r H_{\mathrm{dR}}^{2r+H_3}(X) & \longrightarrow & \bigoplus\limits_r H_{\mathrm{dR}}^{2r+2H_3}(X) \\ \sum\limits_k F_k & \longmapsto & \Big(\sum\limits_k F_k\Big) \wedge \Big(\sum\limits_k F_k\Big) \end{array}$$

(iii) *compositions of these*:

$$\begin{array}{ccc} \bigoplus\limits_r H_{\mathrm{dR}}^{2r+H_3}(X) & \longrightarrow & \bigoplus\limits_r H_{\mathrm{dR}}^{2r+1+2H_3}(X) \\ \sum\limits_k F_k & \longmapsto & \Big(\sum\limits_k F_k\Big) \wedge \Big(\sum\limits_k F_k\Big) \wedge H_3 \end{array}$$

It is noteworthy that terms of the form **(iii)** arise in type IIA string theory, together with terms of the form $I_8 \cup [H_3]$ (8.7), see [Grady and Sati (2019b)].

This evidently generalizes to higher twisted de Rham cohomology (Def. 6.12) and higher twisted real cohomology in the sense of [Grady and Sati (2019a)], with H_3 replaced by H_{2r+1} for $r \in \mathbb{N}$.

Homotopical formulation of twisted non-abelian de Rham cohomology. In preparation of the twisted non-abelian de Rham theorem (Thm. 6.15) we give a homotopy-theoretic reformulation of twisted non-abelian de Rham cohomology (Def. 6.9):

Lemma 6.5 (Pullback to de Rham complex over cylinder of manifold is relative path space object).
Let $X \in \mathrm{SmthMfds}$, *let* $\mathfrak{b} \in L_\infty\mathrm{Algs}^{\geq 0}_{\mathbb{R},\mathrm{fin}}$ *(Ex. 4.10) with Chevalley-Eilenberg algebra* $\mathrm{CE}(\mathfrak{b}) \in \mathrm{dgcAlgs}^{\geq 0}_{\mathbb{R}}$ *(4.26), and let* $\big\{\Omega^\bullet_{\mathrm{dR}}(X) \xleftarrow{\tau^*_{\mathrm{dR}}} \mathrm{CE}(\mathfrak{b})\big\} \in \big(\mathrm{dgcAlgs}^{\geq 0}_{\mathbb{R}}\big)^{\mathrm{CE}(\mathfrak{b})/}_{\mathrm{trinj}}$ *be a morphism of dgc-algebras to the de Rham complex of* X *(Ex. 4.9), regarded as an object in the coslice model category (Ex. 1.5) of* $\big(\mathrm{dgcAlgs}^{\geq 0}_{\mathbb{R}}\big)_{\mathrm{trinj}}$ *(Prop. 4.19) under* $\mathrm{CE}(\mathfrak{b})$.

Then a path space object (Def. 1.5) for τ^*_{dR} is given by this diagram:

$$
\begin{array}{ccccc}
\Omega^\bullet_{\mathrm{dR}}(X) & \xrightarrow{\mathrm{pr}_X^* \in \mathrm{W}} & \Omega^\bullet_{\mathrm{dR}}(X \times \mathbb{R}) & \xrightarrow{(i_0^*, i_1^*) \in \mathrm{Fib}} & \Omega^\bullet_{\mathrm{dR}}(X) \oplus \Omega^\bullet_{\mathrm{dR}}(X) \\
 & {\scriptstyle \tau^*_{\mathrm{dR}}} \nwarrow & \uparrow {\scriptstyle \mathrm{pr}_X^* \circ \tau^*_{\mathrm{dR}}} & \nearrow {\scriptstyle (\tau^*_{\mathrm{dR}}, \tau^*_{\mathrm{dR}})} & \\
 & & \mathrm{CE}(\mathfrak{b}), & &
\end{array}
$$

where the top morphisms are from (6.19).

Proof. It is clear that the diagram commutes, by construction. Moreover, the top morphisms are a weak equivalence followed by a fibration in $\left(\mathrm{dgcAlgs}_{\mathbb{R}}^{\geq 0}\right)_{\mathrm{trinj}}$, by Lem. 6.2. Therefore, by the nature of the coslice model structure (Ex. 1.5) the total diagram constitutes a factorization of the diagonal on τ^*_{dR} through a weak equivalence followed by a fibration, as required (1.20). (To see that the composite really is still the diagonal morphism in the coslice, observe that Cartesian products in any coslice category are reflected in the underlying category.) It only remains to observe that τ^*_{dR} is actually a fibrant object in the coslice model category. But the terminal object in the coslice is clearly the unique morphism from $\mathrm{CE}(\mathfrak{b})$ to the zero-algebra (Ex. 4.7), so that in fact every object in the coslice is still fibrant

$$
\begin{array}{ccc}
\Omega^\bullet_{\mathrm{dR}}(X) & \xrightarrow{\in \mathrm{Fib}} & 0 \\
 {\scriptstyle \tau^*_{\mathrm{dR}}} \nwarrow & & \nearrow \\
 & \mathrm{CE}(\mathfrak{b}) &
\end{array}
\tag{6.44}
$$

as in Rem. 4.3. □

Proposition 6.14 (Twisted non-abelian de Rham cohomology via the coslice dgc-homotopy category). *Consider* $X \in \mathrm{SmthMfds}$, *let*

$$
\begin{array}{ccc}
\mathfrak{g} & \longrightarrow & \widehat{\mathfrak{b}} \\
 & & \downarrow {\scriptstyle p} \\
 & & \mathfrak{b}
\end{array}
\quad \in L_\infty\mathrm{Algs}_{\mathbb{R},\mathrm{fin}}^{\geq 0,\mathrm{nil}}
$$

be an L_∞-algebraic local coefficient bundle (Def. 6.6) of nilpotent L_∞-algebras (Def. 4.15) with Chevalley-Eilenberg algebra $\mathrm{CE}(\widehat{\mathfrak{b}})$, $\mathrm{CE}(\mathfrak{b}) \in \mathrm{dgcAlgs}_{\mathbb{R}}^{\geq 0}$ (4.26), *and let*

$$
\Omega^\bullet_{\mathrm{dR}}(X) \xleftarrow{\tau^*_{\mathrm{dR}}} \mathrm{CE}(\mathfrak{b}) \quad \in \left(\mathrm{dgcAlgs}_{\mathbb{R}}^{\geq 0}\right)_{\mathrm{trinj}}^{\mathrm{CE}(\mathfrak{b})/}
\tag{6.45}
$$

be a morphism of dgc-algebras to the de Rham complex of X (Ex. 4.9), hence a flat $\mathfrak{b}$-valued differential form (Def. 6.1)

$$
\tau_{\mathrm{dR}} \in \Omega_{\mathrm{dR}}(X; \mathfrak{b}),
$$

equivalently regarded as an object in the coslice model category (Ex. 1.5) of $\left(\mathrm{dgcAlgs}_{\mathbb{R}}^{\geq 0}\right)_{\mathrm{trinj}}$ *(Prop. 4.19) under* $\mathrm{CE}(\mathfrak{b})$. *Then the τ_{dR}-twisted non-abelian de Rham cohomology of X with coefficients in $\mathfrak{g}$ (Def. 6.9) is in natural bijection with the hom-set in* *the homotopy category (Def. 1.8) of the coslice model category* $\left(\mathrm{dgcAlgs}^{\geq 0}_{\mathbb{R}}\right)^{\mathrm{CE}(\mathfrak{b})}_{\mathrm{trinj}}$ *(Ex. 1.5) of the model structure on dgc-algebras (Prop. 4.19) from* $\mathrm{CE}(\mathfrak{p})$ (6.25) *to* τ^*_{dR} (6.45)*:*

$$H^{\tau_{\mathrm{dR}}}_{\mathrm{dR}}(X;\mathfrak{g}) \simeq \mathrm{Ho}\Big(\left(\mathrm{dgcAlgs}^{\geq 0}_{\mathbb{R}}\right)^{\mathrm{CE}(\mathfrak{b})/}_{\mathrm{trinj}}\Big)\big(\mathrm{CE}(\mathfrak{p}),\, \tau^*_{\mathrm{dR}}\big). \tag{6.46}$$

Proof. Consider a pair of dgc-algebra homomorphisms in the coslice

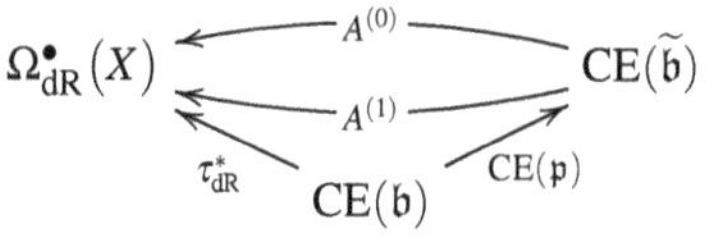

$$\in \left(\mathrm{dgcAlgs}^{\geq 0}_{\mathbb{R}}\right)^{\mathrm{CE}(\mathfrak{b})/}_{\mathrm{trinj}}\big(\mathrm{CE}(\mathfrak{p}),\, \tau^*_{\mathrm{dR}}\big), \tag{6.47}$$

hence of flat τ_{dR}-twisted $\mathfrak{g}$-valued differential forms, according to Def. 6.7. Observe that:

(i) $\mathrm{CE}(\mathfrak{p})$ is cofibrant in $\left(\mathrm{dgcAlgs}^{\geq 0}_{\mathbb{R}}\right)^{\mathrm{CE}(\mathfrak{p})/}_{\mathrm{trinj}}$, since:

(a) the initial object in the coslice is $\mathrm{CE}(\mathfrak{b}) \xleftarrow{\mathrm{id}} \mathrm{CE}(\mathfrak{b})$,

(b) the unique morphism from this object to $\mathrm{CE}(\mathfrak{p})$ is

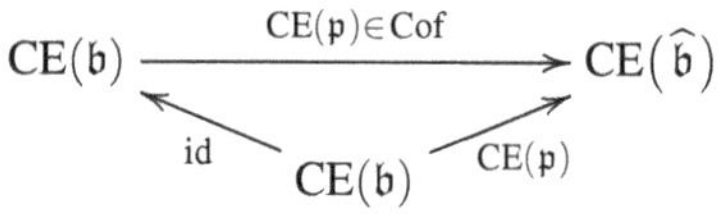

(6.48)

(c) $\mathrm{CE}(\mathfrak{p})$ is a cofibration in $\left(\mathrm{dgcAlgs}^{\geq 0}_{\mathbb{R}}\right)_{\mathrm{trinj}}$, by (6.25), so that the diagram (6.48) is a cofibration in the coslice model category, by Ex. 1.5.

(ii) $\mathrm{pr}^*_X \circ \tau^*_{\mathrm{dR}}$ is fibrant in $\left(\mathrm{dgcAlgs}^{\geq 0}_{\mathbb{R}}\right)^{\mathrm{CE}(\mathfrak{b})/}_{\mathrm{trinj}}$, by (6.44);

(iii) A right homotopy (Def. 1.6) between the pair (6.47) of coslice morphisms, with respect to the path space object from Lem. 6.5, namely a $\widetilde{A}$ that makes the following diagram commute

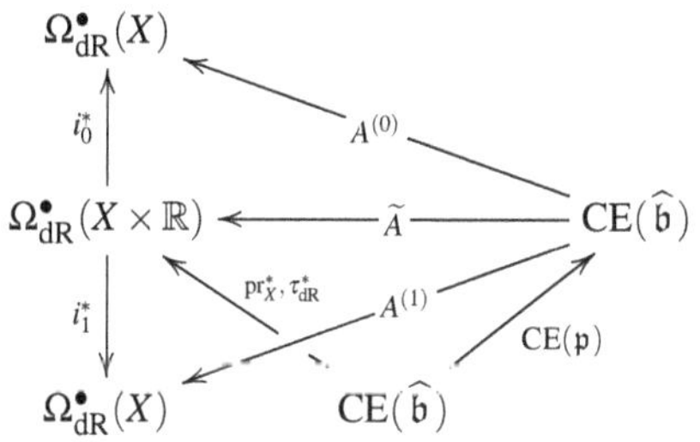

(6.49)

is manifestly the same as a coboundary $\widetilde{A}$ between the corresponding flat twisted $\mathfrak{g}$-valued forms according to Def. 6.8:

(a) The top part of (6.49) is, just as in (6.22), the flat twisted $\widehat{\mathfrak{g}}$-valued form on the cylinder over X that is required by (6.35);

(b) the bottom part of (6.49) is the condition (6.36) on the extension of the twist to the cylinder over X.

Therefore, Prop. 1.10 says that the quotient set (6.37) defining the twisted non-abelian de Rham cohomology is in natural bijection to the hom-set in the coslice homotopy category. □

The twisted non-abelian de Rham theorem. With this in hand we may finally prove the main result in this section, generalizing the non-abelian de Rham theorem (Thm. 6.5) to the twisted case:

Theorem 6.15 (Twisted non-abelian de Rham theorem). *Let $X \in \mathrm{Ho}\big(\Delta\mathrm{Sets}_{\mathrm{Qu}}\big)$ equipped with the structure of a smooth manifold, and let*

$$\begin{array}{ccc} A & \longrightarrow & A /\!\!/ G \\ & {\scriptstyle \text{local coefficient bundle}} & \big\downarrow \rho \\ & & BG \end{array} \quad \in \mathrm{Ho}\big(\Delta\mathrm{Sets}_{\mathrm{Qu}}\big)^{\mathrm{fin}_{\mathbb{Q}}}_{\geq 1,\mathrm{nil}} \tag{6.50}$$

be a local coefficient bundle (3.2) of connected $\mathbb{Q}$-finite nilpotent homotopy types (Def. 5.1) such that the action of $\pi_1(BG) = \pi_0(G)$ on the real homology groups of A is nilpotent. Consider, via Prop. 5.18, the rationalized coefficient bundle $L_{\mathbb{R}}(\rho)$ with corresponding L_∞-algebraic coefficient bundle $\mathfrak{l}\rho$ (Def. 6.6) of the relative real Whitehead L_∞-algebra (Prop. 5.16):

$$\begin{array}{ccc} L_{\mathbb{R}}A & \longrightarrow & (L_{\mathbb{R}}A)/\!\!/(L_{\mathbb{R}}G) \\ & {\scriptstyle \text{rationalized local coefficient bundle}} & \big\downarrow L_{\mathbb{R}}(\rho) \\ & & L_{\mathbb{R}}BG \end{array} \qquad \begin{array}{ccc} \mathfrak{a} & \longrightarrow & \widehat{\mathfrak{b}}\,. \\ & {\scriptstyle L_\infty\text{-algebraic coefficient bundle of Whitehead } L_\infty\text{-algebras}} & \big\downarrow \mathfrak{p} \\ & & \mathfrak{b} \end{array}$$

Then, for

$$X \xrightarrow{\ \tau\ } L_{\mathbb{R}}BG \quad \in \mathrm{Ho}\big(\Delta\mathrm{Sets}_{\mathrm{Qu}}\big)$$

a twist, the τ-twisted non-abelian real cohomology (Def. 5.19) of X with local coefficients in $L_{\mathbb{R}}(\rho)$ (Prop. 5.18) is in natural bijection with the τ_{dR}-twisted non-abelian de Rham cohomology (Def. 6.9) of X with local coefficients in $\mathfrak{p}$,

$$\underbrace{H^{\tau}\big(X; L_{\mathbb{R}}A\big)}_{\tau\text{-twisted non-abelian real cohomology}} \;\simeq\; \underbrace{H^{\tau_{\mathrm{dR}}}_{\mathrm{dR}}(X; \mathfrak{a})}_{\tau_{\mathrm{dR}}\text{-twisted non-abelian de Rham cohomology}}, \tag{6.51}$$

where the twists are related by the plain non-abelian de Rham theorem (Thm. 6.5):

$$\begin{array}{ccc} [\tau] & \longleftrightarrow & [\tau_{\mathrm{dR}}] \\ \cap & & \cap \\ H\big(X; L_{\mathbb{R}}BG\big) & \simeq & H_{\mathrm{dR}}\big(X; \mathfrak{b}\big) \end{array} \tag{6.52}$$

Proof. This is established by the following sequence of natural bijections of hom-sets (where on the right we are illustrating the structure of their elements):

$$
\begin{aligned}
& H^{\tau}(X; L_{\mathbb{R}}A) \\
& = \mathrm{Ho}\big(\Delta\mathrm{Sets}_{\mathrm{Qu}}^{/L_{\mathbb{R}}BG}\big)\big(\tau, L_{\mathbb{R}}(\rho)\big) && = \left\{ \begin{array}{ccc} X & \dashrightarrow & L_{\mathbb{R}}(A /\!\!/ G) \\ & {\scriptstyle\tau}\searrow \quad \swarrow {\scriptstyle L_{\mathbb{R}}(\rho)} & \\ & L_{\mathbb{R}}BG & \end{array} \right\} \\
& = \mathrm{Ho}\big(\Delta\mathrm{Sets}_{\mathrm{Qu}}^{/B\exp_{\mathrm{PL}}\mathrm{CE}(\mathfrak{b})}\big)\big(\tau, B\exp_{\mathrm{PL}}\mathrm{CE}(\mathfrak{p})\big) && = \left\{ \begin{array}{ccc} X & \dashrightarrow & B\exp_{\mathrm{PL}}\mathrm{CE}(\mathfrak{a}) \\ & {\scriptstyle\tau}\searrow \quad \swarrow {\scriptstyle B\exp_{\mathrm{PL}}(\mathrm{CE}(\mathfrak{p}))} & \\ & B\exp_{\mathrm{PL}}\mathrm{CE}(\mathfrak{b}) & \end{array} \right\} \\
& \simeq \mathrm{Ho}\Big(\big((\mathrm{dgcAlgs}_{\mathbb{R}}^{\geq 0})^{\mathrm{op}}_{\mathrm{trinj}}\big)^{/\mathrm{CE}(\mathfrak{g})}\Big)\big(\widetilde{\tau}, \mathrm{CE}(\mathfrak{p})\big) && = \left\{ \begin{array}{ccc} \Omega^{\bullet}_{\mathrm{PLdR}}(X) & \dashleftarrow & \mathrm{CE}(\widehat{\mathfrak{b}}) \\ & {\scriptstyle\widetilde{\tau}}\nwarrow \quad \nearrow {\scriptstyle \mathrm{CE}(\mathfrak{p})} & \\ & \mathrm{CE}(\mathfrak{b}) & \end{array} \right\} \\
& \simeq \mathrm{Ho}\Big(\big((\mathrm{dgcAlgs}_{\mathbb{R}}^{\geq 0})^{\mathrm{op}}_{\mathrm{trinj}}\big)^{/\mathrm{CE}(\mathfrak{g})}\Big)\big(\tau_{\mathrm{dR}}, \mathrm{CE}(\mathfrak{p})\big) && = \left\{ \begin{array}{ccc} \Omega^{\bullet}_{\mathrm{dR}}(X) & \dashleftarrow & \mathrm{CE}(\widehat{\mathfrak{b}}) \\ & {\scriptstyle\tau^{*}_{\mathrm{dR}}}\nwarrow \quad \nearrow {\scriptstyle \mathrm{CE}(\mathfrak{p})} & \\ & \mathrm{CE}(\mathfrak{b}) & \end{array} \right\} \\
& \simeq H^{\tau_{\mathrm{dR}}}(X; \mathfrak{l}A) .
\end{aligned}
$$

Here the first line is the definition of twisted non-abelian real cohomology (Def. 5.19), while the second line inserts the definition of $L_{\mathbb{R}}$ (Def. 5.7), with $\mathrm{CE}(\mathfrak{l}\rho)$ serving as the required (1.37) fibrant resolution (1.19) of $\Omega^{\bullet}_{\mathrm{PLdR}}(\rho)$.

The key step is the third line, which uses the hom-isomorphism (1.2) of the derived adjunction (1.35) of the sliced Quillen adjunction (Ex. 1.8) of the PLdR-adjunction (Prop. 5.5), using the form (1.32) of its left adjoint with the observation that this is already derived (1.36) since τ is necessarily cofibrant, by (1.15) and (1.19).

The fourth step is composition with the slice morphism exhibiting (6.52)

$$
\begin{array}{ccc}
\Omega^{\bullet}_{\mathrm{dR}}(X) & \overset{\in \mathrm{W}}{\dashleftarrow} & \Omega^{\bullet}_{\mathrm{PLdR}}(X) , \\
& {\scriptstyle\tau^{*}_{\mathrm{dR}}}\nwarrow \quad \nearrow {\scriptstyle\widetilde{\tau}} & \\
& \mathrm{CE}(\mathfrak{b}) &
\end{array}
$$

which is an isomorphism in the homotopy category by Lem. 6.4 (as, in the untwisted case, in the last step of (6.18)). The last step is Prop. 6.14. □

PART IV
The (differential) non-abelian character map

We introduce the character map in non-abelian cohomology (Def. IV.2 below) and then discuss how it specializes to:

Chapter 7 – the Chern-Dold character on generalized cohomology;
Chapter 8 – the Chern-Weil homomorphism on degree-1 nonabelian cohomology;
Chapter 9 – the Cheeger-Simons differential character on degree-1 nonabelian cohomology.

Definition IV.1 (Rationalization and realification in non-abelian cohomology). Let $A \in \mathrm{Ho}\big(\Delta\mathrm{Sets}_{\mathrm{Qu}}\big)^{\mathrm{fin}_{\mathbb{Q}}}_{\geq 1,\mathrm{nil}}$ (Def. 5.1).
(i) We write

$$(\eta_A^{\mathbb{Q}})_* \;:\; \underset{\text{cohomology}}{\overset{\text{non-abelian}}{H(-;A)}} \xrightarrow[\text{rationalization}]{H(-;\eta_A^{\mathbb{Q}}) = H(-;\mathbb{D}\eta_A^{\mathrm{P}\mathbb{Q}\mathrm{L}})} \underset{\text{rational cohomology}}{\overset{\text{non-abelian}}{H\big(-;L_{\mathbb{Q}}A\big)}} \qquad \text{(IV.1)}$$

for the cohomology operation (Def. 2.3) from non-abelian A-cohomology (Def. 2.1) to non-abelian rational cohomology (Def. 5.14), which is induced (2.20) by the rationalization map $\eta_A^{\mathbb{Q}}$ (Def. 5.2), or equivalently, via the Fundamental Theorem (Prop. 5.6), by the derived unit of the rational PL de Rham adjunction.
(ii) Analogously, we write

$$(\eta_A^{\mathbb{R}})_* \;:\; \underset{\text{cohomology}}{\overset{\text{non-abelian}}{H(-;A)}} \xrightarrow[\text{real-ification}]{H(-;\mathbb{D}\eta_A^{\mathrm{P}\mathbb{R}\mathrm{L}})} \underset{\text{real cohomology}}{\overset{\text{non-abelian}}{H\big(-;L_{\mathbb{R}}A\big)}} \qquad \text{(IV.2)}$$

for the cohomology operation to non-abelian real cohomology that is induced by the derived PL de Rham adjunction unit over the real numbers into $L_{\mathbb{R}}$ (5.21).
(iii) For, moreover, $X \in \mathrm{Ho}\big(\Delta\mathrm{Sets}_{\mathrm{Qu}}\big)^{\mathrm{fin}_{\mathbb{Q}}}_{\geq 1,\mathrm{nil}}$ (Def. 5.1), we consider the cohomology operation shown by the dashed arrow here:

$$\begin{array}{ccc}
H\big(X;L_{\mathbb{Q}}A\big) & \overset{(-)\otimes_{\mathbb{Q}}\mathbb{R}}{\dashrightarrow} & H\big(X;L_{\mathbb{R}}A\big) \\
\big\downarrow \widetilde{(-)}\ \wr & & \big\uparrow \wr\ \widetilde{(-)} \\
\mathrm{Ho}\Big(\big(\mathrm{dgcAlgs}^{\geq 0}_{\mathbb{Q}}\big)^{\mathrm{op}}_{\mathrm{trinj}}\Big)\Big(\mathbb{D}\Omega^\bullet_{\mathrm{P}\mathbb{Q}\mathrm{LdR}}(X), \mathbb{D}\Omega^\bullet_{\mathrm{P}\mathbb{Q}\mathrm{LdR}}(A)\Big) & \xrightarrow{\mathbb{D}\big((-)\otimes_{\mathbb{Q}}\mathbb{R}\big)} & \mathrm{Ho}\Big(\big(\mathrm{dgcAlgs}^{\geq 0}_{\mathbb{R}}\big)^{\mathrm{op}}_{\mathrm{trinj}}\Big)\Big(\mathbb{D}\Omega^\bullet_{\mathrm{P}\mathbb{R}\mathrm{LdR}}(X), \mathbb{D}\Omega^\bullet_{\mathrm{P}\mathbb{R}\mathrm{LdR}}(A)\Big),
\end{array} \qquad \text{(IV.3)}$$

hence the composition of:
(i) the hom-isomorphisms $\widetilde{(-)}$ (1.2) of the derived (1.35) PL de Rham Quillen adjunction (Prop. 5.5) over the rational and over the real numbers, respectively;

(ii) the corresponding hom-component of the right derived extension-of-scalars functor from Lem. 5.3 (the operation of "tensoring a space with $\mathbb{R}$" from [Deligne *et al.* (1975), Footnote 5]);

While real-ification (IV.2), in constrast to rationalization (IV.1), is not directly induced by a localization of spaces, it is equivalent to rationalization followed by derived extension of scalars:

Proposition IV.1 (Realification is rationalization followed by extension of scalars). *The operation of real-ification* (IV.2) *factors through rationalization* (IV.1) *via extension of scalars* (IV.3) *in that the following diagram commutes:*

$$\begin{array}{ccc} & \xrightarrow{(\eta_A^{\mathbb{Q}})_*} & H(X; L_{\mathbb{Q}}A) \\ H(X;A) & & \big\downarrow{\scriptstyle (-)\otimes_{\mathbb{Q}}\mathbb{R}} \\ & \xrightarrow[(\eta_A^{\mathbb{R}})_*]{} & H(X; L_{\mathbb{R}}A)\,. \end{array} \tag{IV.4}$$

Proof. Consider the following diagram:

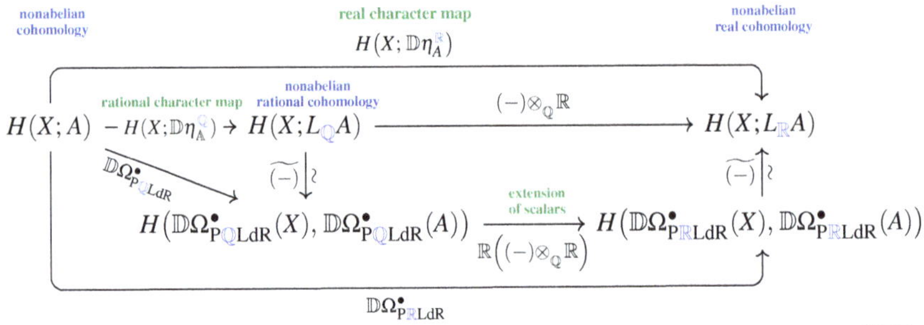

(IV.5)

Here the triangle on the left as well as the outer rectangle commute by general properties of adjunctions (the naturality of the hom-isomorphism (1.2) combined with the definition (1.3) of the adjunction unit). The square on the right commutes by definition (IV.3), and the bottom part commutes by Prop. 5.8. Together, these imply that the top rectangle commutes, which is the statement to be shown. □

Definition IV.2 (Non-abelian character map). Let $X, A \in \mathrm{Ho}\big(\Delta\mathrm{Sets}_{\mathrm{Qu}}\big)^{\mathrm{fin}_{\mathbb{Q}}}_{\geq 1,\mathrm{nil}}$ (Def. 5.1) such that X admits the structure of a smooth manifold. Then we say that the *non-abelian character map* in non-abelian A-cohomology (Def. 2.1) is the cohomology operation (Def. IV.1)

$$\underset{\text{character map}}{\text{non-abelian}}\quad \mathrm{ch}_A \;:\; \underset{\text{cohomology}}{\overset{\text{non-abelian}}{H(X;A)}} \xrightarrow[\mathbb{R}\text{-rationalization}]{(\eta_A^{\mathbb{R}})_*} \underset{\text{real cohomology}}{\overset{\text{non-abelian}}{H(X;L_{\mathbb{R}}A)}} \xrightarrow[\text{de Rham theorem}]{\simeq\;\;\text{non-abelian}} \underset{\text{de Rham cohomology}}{\overset{\text{non-abelian}}{H_{\mathrm{dR}}(X;\mathfrak{l}A)}} \tag{IV.6}$$

from non-abelian A-cohomology (Def. 2.1) to non-abelian de Rham cohomology (Def. 6.3) with coefficients in the rational Whitehead L_∞-algebra $\mathfrak{l}A$ of A (Prop. 5.11), which is the composite of

(i) the operation (IV.2) of real rationalization of coefficients (Def. IV.1),
(ii) the equivalence (6.17) of the non-abelian de Rham theorem (Thm. 6.5).

Unwinding the definitions and theorems that go into Def. IV.2, shows that the non-abelian character map on a non-abelian cohomology theory with classifying space a (connected, nilpotent and $\mathbb{Q}$-finite) homotopy type A assigns flat non-abelian differential form data (Def. 6.1) satisfying the differential relations of the CE-algebra of the Whitehead L_∞-algebra of A (Prop. 5.11):

Example IV.1 (Non-abelian character on Cohomotopy theory). The non-abelian character (Def. IV.2) of

(i) a class $[c] \in \pi^n(X) = H^1\big(X; \Omega S^n\big)$ in Cohomotopy (Ex. 2.7) is (by Ex. 5.3, Ex. 6.4) of this form:

$$\mathrm{ch}_{S^n}(c) = \begin{cases} \big[G_n \in \Omega^n(X) \,\big|\, d\,G_n = 0 \big] & \text{if } n = 2k+1 \text{ is odd} \\ \left[\begin{array}{l|l} G_{2n-1} \in \Omega^{2n-1}_{\mathrm{dR}}(X) & d\,G_{2n-1} = -G_n \wedge G_n \\ G_n \in \Omega^n_{\mathrm{dR}}(X) & d\,G_n = 0 \end{array} \right] & \text{if } n = 2k \text{ is even} \end{cases}$$

(ii) a class $[c] \in H^1\big(X; \Omega \mathbb{C}P^n\big)$ in the non-abelian cohomology theory represented by complex projective n-space is (by Ex. 5.5) of this form:

$$\mathrm{ch}_{\mathbb{C}P^n}(c) = \left[\begin{array}{l|l} H_{2n+1} \in \Omega^{2n+1}_{\mathrm{dR}}(X) & d\,H_{2k+1} = \overbrace{F_2 \wedge \cdots \wedge F_2}^{n+1 \text{ factors}} \\ F_2 \in \Omega^2_{\mathrm{dR}}(X) & d\,F_2 = 0 \end{array} \right]$$

We come back to these new and deeply non-abelian examples in Chapter 12 below. First we now turn attention to verifying that the non-abelian character map of Def. IV.2 correctly subsumes more classical structures of differential topology.

Chapter 7

Chern-Dold character

We prove (Thm. 7.4) that the non-abelian character map reproduces the Chern-Dold character on generalized cohomology theories (recalled as Def. 7.3) and in particular the Chern character on topological K-theory (Ex. 7.2).

Remark 7.1 (Chern-Dold character over the real numbers). In view of Prop. IV.1 and Ex. 5.7 we may and will regard Dold's equivalence (Prop. 7.1) and the Chern-Dold character (Def. 7.3) over the real numbers instead of over the rational numbers. This does not affect the information contained in the character but serves to allow, over smooth manifolds, for composition with the de Rham isomorphism.

Proposition 7.1 (Dold's equivalence [Dold (1972), Cor. 4][Hilton (1971), Thm. 3.18] [Rudyak (1998), §II.3.17]**).** *Let E be a generalized cohomology theory (Ex. 2.10). Then its $\mathbb{R}$-rationalization $E_{\mathbb{R}}$ is equivalent to ordinary cohomology with coefficients in the rationalized stable homotopy groups of E:*

$$E^n_{\mathbb{Q}}(X) \xrightarrow[\simeq]{\mathrm{do}_E} \bigoplus_{k \in \mathbb{Z}} H^{n+k}\big(X; \pi_k(E) \otimes_{\mathbb{Z}} \mathbb{Q}\big).$$

Remark 7.2 (Rational stable homotopy theory). In modern stable homotopy theory, Dold's equivalence (Prop. 7.1) is a direct consequence of the fundamental theorem [Schwede and Shipley (2003b), Thm. 5.1.6] that rational spectra are direct sums of Eilenberg-MacLane spectra with coefficients in the rationalized stable homotopy groups [Braunack-Mayer *et al.* (2019), Prop. 2.17].

But we may explicitly re-derive Dold's equivalence using the unstable rational homotopy theory from Part III and passing to rationalization over the real numbers.

Proposition 7.2 (Dold's equivalence via non-abelian real cohomology). *Let E be a generalized cohomology theory (Ex. 2.10) and let $n \in \mathbb{N}$ such that the nth coefficient space (2.13) is of rational finite homotopy type (Def. 5.1) $E_n \in \mathrm{Ho}\big(\Delta\mathrm{Sets}_{\mathrm{Qu}}\big)^{\mathrm{fin}_{\mathbb{Q}}}$. Then there is a natural equivalence between the non-abelian real cohomology (Def. 5.14) with coefficients in E_n and ordinary cohomology with coefficients in the $\mathbb{R}$-rationalized homotopy groups of E:*

$$H\big(-; L_{\mathbb{R}} E_n\big) \simeq \bigoplus_{k \in \mathbb{N}} H^{n+k}\big(-; \pi_k(E) \otimes_{\mathbb{Z}} \mathbb{R}\big). \tag{7.1}$$

Proof. Since E_n is an infinite-loop space, it is nilpotent (Ex. 5.1). We may assume without restriction that it is also connected, for otherwise we apply the following argument to each connected component (Rem. 5.1). Hence $E_n \in \mathrm{Ho}\big(\Delta\mathrm{Sets}_{\mathrm{Qu}}\big)^{\mathrm{fin}_{\mathbb{Q}}}_{\geq 1,\mathrm{nil}}$ (Def. 5.1) and the discussion in Chapter 5 applies. Again, since E_n is a loop space (2.13), Prop. 5.15 gives $H(-; L_{\mathbb{R}}E_n) \simeq \underset{k\in\mathbb{N}}{\oplus} H^k\big(-; \pi_k(E_n) \otimes_{\mathbb{Z}} \mathbb{R}\big)$. The claim follows from the definition of stable homotopy groups as $\pi_{k-n}(E) = \pi_k(E_n)$ for $k, n \geq 0$, (as E is an "Ω-spectrum" (2.13)). □

Definition 7.3 (Real Chern-Dold character [Buhštaber (1970)][Hilton (1971), p. 50]**).** Let E be a generalized cohomology theory (Ex. 2.10). The real *Chern-Dold character* in E-cohomology theory is the cohomology operation to ordinary cohomology which is the composite of rationalization in E-cohomology with Dold's equivalence (Prop. 7.1):

$$
\begin{array}{ccccc}
\underset{\text{character}}{\overset{\text{Chern-Dold}}{}}\ \mathrm{ch}_E : & E^\bullet(-) & \xrightarrow[\text{in } E\text{-cohomoloy}]{\mathbb{R}\text{-rationalization}} & E^\bullet_{\mathbb{R}}(-) & \xrightarrow[\simeq]{\overset{\text{Dold's equivalence}}{\mathrm{do}_E}} \bigoplus_k H^{\bullet+k}\big(-;\pi_k(E)\otimes_{\mathbb{Z}}\mathbb{R}\big) \\
& {\scriptstyle (2.14)}\downarrow\simeq & & \simeq\downarrow{\scriptstyle (2.14)} & \nearrow_{\simeq\ (7.1)} \\
& H(-;E_\bullet) & \xrightarrow[\text{(IV.2)}]{(\eta^{\mathbb{R}}_{E_\bullet})_*} & H(-;L_{\mathbb{R}}E_\bullet)\,. &
\end{array}
\tag{7.2}
$$

Here the bottom part serves to make the nature of the top maps fully explicit, using Ex. 2.10, Def. IV.1 and Prop. 7.2.

Remark 7.3 (Rationalization in the Chern-Dold character). That the first map in the Dold-Chern character (7.2) is the rationalization localization (here shown exended to the real numbers) is stated somewhat indirectly in the original definition [Buhštaber (1970)] (the concept of rationalization was fully formulated later in [Bousfield and Kan (1972b)]). The role of rationalization in the Chern-Dold character is made fully explicit in [Lind *et al.* (2020), §2.1]. The same rationalization construction of the generalized Chern character, but without attribution to [Buhštaber (1970)] or [Dold (1972)], is considered in [Hopkins and Singer (2005), §4.8] (see also [Bunke and Nikolaus (2019), p. 17]).

We now come to the main result in this section:

Theorem 7.4 (Non-abelian character subsumes Chern-Dold character). *Let E be a generalized cohomology theory (Ex. 2.10) and let $n \in \mathbb{N}$ such that the nth coefficient space (2.13) is of rational finite homotopy type (Def. 5.1). Let moreover X be a smooth manifold.*

Then the non-abelian character (Def. IV.2) coincides with the Chern-Dold character (Def. 7.3) on E-cohomology in degree n, in that the following diagram commutes:

$$
\begin{array}{ccc}
H\big(X;E_n\big) & \xrightarrow[\mathrm{ch}_{E_n}]{} & H_{\mathrm{dR}}\big(\mathfrak{l}E_n\big) \\
{\scriptstyle (2.14)}\uparrow\simeq & & \simeq\downarrow{\scriptstyle (6.17)\,(7.1)} \\
E^n(X) & \xrightarrow[(\mathrm{ch}_E)^n]{} & \bigoplus_k H^{n+k}\big(X;\pi_k(E)\otimes_{\mathbb{Z}}\mathbb{R}\big).
\end{array}
\tag{7.3}
$$

Here the equivalence on the left is from Ex. 2.10, while the equivalence on the right is the inverse non-abelian de Rham theorem (Thm. 6.5) composed with that from Prop. 7.2.

Proof. Since E_n is an infinite-loop space, it is necessarily nilpotent (Ex. 5.1). We may assume without restriction that it is also connected, for otherwise we apply the following argument to each connected component (Rem. 5.1). Hence $E_n \in \mathrm{Ho}\big(\Delta\mathrm{Sets}_{\mathrm{Qu}}\big)^{\mathrm{fin}_{\mathbb{Q}}}_{\geq 1,\mathrm{nil}}$ (Def. 5.1) and the discussion in Chapter 5 and Chapter 6 applies:

The non-abelian de Rham isomorphism (6.17) in the definition (IV.6) of the non-abelian character cancels against its inverse on the right of (7.3). Commutativity of the remaining diagram

$$
\begin{array}{ccc}
H\big(X; E_n\big) & \xrightarrow{\quad (\eta^{\mathbb{R}}_{E_n})_* \quad} & H\big(X; L_{\mathbb{R}} E_n\big) \\
{\scriptstyle (2.14)}\ \big\uparrow \simeq & & \simeq \big\downarrow\ {\scriptstyle (7.1)} \\
E^n(X) & \xrightarrow{\quad \mathrm{ch}_{E_n} \quad} & \bigoplus_k H^{n+k}\big(X; \pi_k(E) \otimes_{\mathbb{Z}} \mathbb{R}\big)
\end{array}
$$

is the very definition of the Chern-Dold character (Def. 7.3). □

Example 7.1 (de Rham homomorphism in ordinary cohomology). On ordinary integral cohomology (Ex. 2.1), the non-abelian character (Def. IV.2) reduces to extension of scalars from the integers to the real numbers, followed by the de Rham isomorphism, in that the following diagram commutes:

$$
\begin{array}{ccccc}
H\big(-; B^{n+1}\mathbb{Z}\big) & & \xrightarrow[\quad \mathrm{ch}_{B^{n+1}\mathbb{Z}} \quad]{\text{non-abelian character on ordinary cohomology}} & & H_{\mathrm{dR}}\big(-; \mathfrak{l}B^{n+1}\mathbb{Z}\big) \\
{\scriptstyle (2.5)}\ \big\uparrow \simeq & & & & \simeq \big\downarrow\ {\scriptstyle (5.39)\,(6.14)} \\
H^{n+1}(-;\mathbb{Z}) & \xrightarrow[\text{extension of scalars}]{} & H^{n+1}(-;\mathbb{R}) & \xrightarrow[\text{ordinary de Rham isomorphism}]{\simeq} & H^{n+1}_{\mathrm{dR}}(-)
\end{array}
$$

Example 7.2 (Chern character on complex K-theory). The spectrum (2.13) representing complex K-theory has 0th component space $\mathrm{KU}_0 \simeq \mathbb{Z} \times B\mathrm{U}$ (2.15). Here the connected components $B\mathrm{U}$, the classifying space of the infinite unitary group (2.16), are clearly of finite rational type (since their rational cohomology is the ring of universal Chern classes, e.g. [Kochman (1996), Thm. 2.3.1]). Therefore, Thm. 7.4 applies and says that the non-abelian character map (Def. IV.2) for coefficients in $\mathbb{Z} \times B\mathrm{U}$ reduces to the Chern-Dold character on complex K-theory. This, in turn, is equivalent (by [Hilton (1971), Thm. 5.8]) to the original Chern character ch on complex K-theory [Hirzebruch (1956), §12.1][Borel and Hirzebruch (1958), §9.1][Atiyah and Hirzebruch (1961), §1.10] (review in [Hilton (1971), §V]):

$$
\begin{array}{ccc}
\overset{\text{non-abelian cohomology with } \mathbb{Z}\times BU\text{-coefficients}}{H(X;\mathbb{Z}\times B\mathrm{U})} & \xrightarrow[\mathrm{ch}_{\mathbb{Z}\times BU}]{\text{non-abelian character map with } \mathbb{Z}\times BU\text{-coefficients}} & \overset{\text{non-abelian de Rham homology with } \mathfrak{l}(\mathbb{Z}\times BU)\text{-coefficients}}{H_{\mathrm{dR}}\big(X;\,\mathfrak{l}(\mathbb{Z}\times B\mathrm{U})\big)} \\
\| & & \| \\
\mathrm{KU}^0(X) & \xrightarrow[\text{traditional Chern character}]{\mathrm{ch}} & \bigoplus\limits_{k\in\mathbb{Z}} H^{2k}_{\mathrm{dR}}(X) \\
[\nabla] & \longmapsto & \left[\mathrm{tr}\circ\exp\left(\frac{iF_\nabla}{2\pi}\right)\right]
\end{array}
\tag{7.4}
$$

On the bottom we are showing the classical component-formula, which to the K-theory class of a complex vector bundle with any choice of connection ∇ assigns the de Rham cohomology class of the trace of the exponential series of its curvature 2-form (8.5). (More on this *Chern-Weil formalism* in Chapter 8).

Example 7.3 (Pontrjagin character on real K-theory). The *Pontrjagin character* ph on real topological K-theory (see [Greub *et al.* (1973), §9.4][Imaoka and Kuwana (1999)][Igusa (2008)][Grady and Sati (2021b), §2.1]) is defined to be the composite

$$
\begin{array}{ccccccc}
\mathrm{KSpin}^\bullet(-) & \longrightarrow & \mathrm{KSO}^\bullet(-) & \longrightarrow & \mathrm{KO}^\bullet(-) & \xrightarrow{\mathrm{ph}^\bullet} & \bigoplus\limits_k H^{\bullet+4k}(-;\mathbb{R}) \\
 & & & & \downarrow{\scriptstyle \mathrm{cplx}} & & \downarrow \\
 & & & & \mathrm{KU}^\bullet(-) & \xrightarrow{\mathrm{ch}^\bullet} & \bigoplus\limits_k H^{\bullet+2k}(-;\mathbb{R})
\end{array}
$$

(with $\widetilde{\mathrm{ph}}^\bullet$ the arrow from $\mathrm{KSO}^\bullet(-)$ to $\bigoplus_k H^{\bullet+4k}(-;\mathbb{R})$)

of the complexification map (on representing virtual vector bundles) with the Chern character ch on complex K-theory (Ex. 7.2).

(i) By naturality of the complexification map and since the complex Chern character is a Chern-Dold character (by [Hilton (1971), Thm. 5.8]), so is the Pontrjagin character, as well as its restriction $\widetilde{\mathrm{ph}}$ to oriented real K-theory KSO and further to ph on KO-theory and to Spin K-theory, etc.

(ii) The connected components $B\mathrm{O}$ of the classifying space KO_0 for real topological K-theory are of finite $\mathbb{R}$-type (since the real cohomology is the ring of universal Pontrjagin classes). Therefore, Thm. 7.4 applies and says that the non-abelian Chern character (Def. IV.2) for coefficients in $\mathbb{Z}\times B\mathrm{SO}$ coincides with the Pontrjagin character $\widetilde{\mathrm{ph}}$ in KSO-theory:

$$
\underset{\text{on oriented real K-theory}}{\overset{\text{Pontrjagin character}}{\widetilde{\mathrm{ph}}}} \;\simeq\; \mathrm{ch}_{\mathbb{Z}\times B\mathrm{SO}}\,.
$$

(iii) By Rem. 5.1, the construction extends to the Pontrjagin character ph on KO-theory.
(iv) The same applies to the further restriction of the Pontrjagin character to KSpin; see [Li and Duan (1991)][Thomas (1962)] for some subtleties involved and [Sati (2008), §7] for interpretation and applications.

Example 7.4 (Chern-Dold character on Topological Modular Forms). The connective ring spectrum tmf of *topological modular forms* [Hopkins (1995), §9][Hopkins (2002), §4] (see [Douglas *et al.* (2014)]) is, essentially by design, such that under rationalization it yields the graded ring of rational modular forms (e.g. [Douglas and Henriques (2011), p. 2]):

$$\underset{\text{topological modular forms}}{\pi_\bullet(\mathrm{tmf})} \xrightarrow{\;(-)\otimes_{\mathbb{Z}}\mathbb{R}\;} \underset{\text{rational modular forms}}{\mathrm{mf}^{\mathbb{R}}_\bullet} \simeq \mathbb{R}\big[\overbrace{c_4}^{\deg=8}, \overbrace{c_6}^{\deg=12}\big]. \tag{7.5}$$

It follows that the Chern-Dold character (Def. 7.3) on tmf takes values in real cohomology with coefficients in modular forms

$$\mathrm{tmf}^\bullet(-) \xrightarrow[\mathrm{ch}^\bullet_{\mathrm{tmf}}]{\text{Chern-Dold character on topological modular forms}} H^\bullet\big(-;\mathrm{mf}^{\mathbb{R}}_\bullet\big). \tag{7.6}$$

(This is often considered over the rational numbers, sometimes over the complex numbers [Berwick-Evans (2013), Fig. 1]; we may just as well stay over the real numbers, by Rem. 5.2, to retain contact to the de Rham theorem.)

By Thm. 7.4, this is an instance of the non-abelian character map:

$$\underset{\text{Chern-Dold character on topological nodular forms}}{\mathrm{ch}^\bullet_{\mathrm{tmf}}} \simeq \mathrm{ch}_{\mathrm{tmf}_\bullet}.$$

Example 7.5 (The Hurewicz/Boardman homomorphism on topological modular forms). The spectrum tmf (Ex. 7.4) carries the structure of an E_∞-ring spectrum (Ex. 2.10) and hence receives an essentially unique homomorphism of ring spectra from the sphere spectrum:

$$\Sigma^\infty S^0 = \mathbb{S} \xrightarrow{\;e_{\mathrm{tmf}}\;} \mathrm{tmf}.$$

This is also known as the *Hurewicz homomorphism* or rather the *Boardman homomorphism* (e.g. [Adams (1974), §II.7][Kochman (1996), §4.3]) for tmf. The Boardman homomorphism on tmf happens to be a stable weak equivalence in degrees ≤ 6, in that it is an isomorphism on stable homotopy groups in these degrees [Hopkins (2002), Prop. 4.6][Douglas *et al.* (2014), §13]:

$$\pi^s_{\bullet\leq 6} = \pi_{\bullet\leq 6}(\mathbb{S}) \xrightarrow[\simeq]{\;\pi_{\bullet\leq 6}(e_{\mathrm{tmf}})\;} \pi_{\bullet\leq 6}(\mathrm{tmf}).$$

Hence (by Prop. 1.20) when X^{10} is a manifold of dimension $\dim(X) \leq 6+4 = 10$, then the Boardman homomorphism identifies the stable Cohomotopy (Ex. 2.13) of X^{10} in degree 4 with $\mathrm{tmf}^4(X^{10})$:

$$\underset{\text{stable 4-Cohomotopy}}{\pi^4_s(X^{10})} = \mathbb{S}^4(X^{10}) \xrightarrow[\simeq]{\;e^4_{\mathrm{tmf}}\;\text{(Boardman homomorphism)}} \underset{\text{tmf-cohomology in degree 4}}{\mathrm{tmf}^4(X^{10})}. \tag{7.7}$$

$$\mathbb{S}^4(X^{10}) \xrightarrow{\;\mathrm{ch}_{\mathbb{S}^4}\;} H^4_{\mathrm{dR}}(X^{10}) \xleftarrow{\;\mathrm{ch}_{\mathrm{tmf}^4}\;} \mathrm{tmf}^4(X^{10})$$

In this situation, the character map from Ex. 7.4 extracts exactly the datum of a real 4-class.

Remark 7.4 (Clarifying the role of tmf **in string theory).** Ever since the famous computation of [Witten (1987)] (following [Schellekens and Warner (1986)] [Schellekens and Warner (1987)]) showed that the partition function of a 2d super-conformal field theory lands in modular forms, and since the theorem of [Ando *et al.* (2001)][M. Ando and Rezk (2010)] showed that, mathematically, this statement lifts through (what we call above) the tmf-Chern-Dold character (7.6), there have been proposals about a possible role of tmf-cohomology theory in controlling elusive aspects of string theory (see [Kriz and Sati (2005)][Sati (2010)][Douglas and Henriques (2011)][Stolz and Teichner (2011)][Sati (2014)][Gaiotto and Johnson-Freyd (2022)][Gukov *et al.* (2021)][Sati (2019)]). While good progress has been made, it might be fair to say that the situation has remained inconclusive.

(i) Non-abelian enhancement of $\mathrm{tmf}^4(X^{10})$**.** But with the non-abelian generalization (Def. IV.2) of the Chern-Dold character in hand, we may ask for a non-abelian enhancement (Ex. 2.19) of tmf-theory on string background spacetimes. By Ex. 7.5, this is, in degree 4, equivalent to asking for a non-abelian enhancement of stable Cohomotopy theory (Ex. 2.20). This exists (not uniquely but) canonically: given by actual Cohomotopy theory (Ex. 2.7). We work out the non-abelian character map on twisted 4-Cohomotopy in Ex. 12.1 below. The concluding Prop. 12.1 shows that this does capture crucial non-linear phenomena of non-perturbative string theory.

(ii) Non-Torsion classes in $\mathrm{tmf}^\bullet$**.** Part of the statement (7.7) is that the higher non-torsion generators (7.5) of $\pi_\bullet(\mathrm{tmf})$ (hence the actual or "non-topological" modular forms) do not contribute to tmf^4 on 10-manifolds: These start to contribute only on manifolds of dimensionl $4+\deg(c_4) = 12$, where, in string theory language, one computes not fluxes of fields but their (Green-Schwarz-)anomaly densities. Indeed, the original computation of what came to be known as the "Witten genus" interprets it as the generating function for just these anomalies [Schellekens and Warner (1986)][Schellekens and Warner (1987)][Lerche *et al.* (1988)][Sati (2011)]. While the character map (7.6) still applies in these higher dimensions, the non-abelian enhancement by Cohomotopy is restricted exactly to dimension 10, and is what makes the character pick up just those non-linear relations, discussed in Chapter 12, that are expected to cancel the anomalies [Fiorenza *et al.* (2020b)][Fiorenza *et al.* (2022)][Sati and Schreiber (2020a)].

(iii) Torsion classes in $\mathrm{tmf}^\bullet$**.** Indeed, the deep motivation behind topological modular forms is the suggestion that these capture mathematical aspects of 2d supersymmetric field theories even in their non-rational torsion elements – and the beauty of (7.7) is to show that in the relevant degrees and dimensions these aspects are equivalently seen in Cohomotopy. Concretely, a famous conjecture orginating with [Stolz and Teichner (2011)][Douglas and Henriques (2011)] and cast in more pronounced form in [Gaiotto and Johnson-Freyd (2022), §5] says that the elements of $\mathrm{tmf}^\bullet(X)$ correspond bijectively to, roughly, the deformation classes of 2-dimensional supersymmetric field theories with target space X. Specifically the torsion elements in $\pi_3(\mathrm{tmf}) \simeq \pi_3(\mathbb{S}) \simeq \mathbb{Z}/24$, have, conjecturally, been identified, with certain supersymmetric SU(2)-WZW models [Gaiotto *et al.* (2021), p. 17][Gaiotto and Johnson-Freyd (2019)][Johnson-Freyd (2020)], whose "meaning", however, has remained somewhat elusive. But under the equivalence (7.7) with Cohomotopy, these same elements could be understood in [Sati and Schreiber (2021a)] in their role in non-perturbative string theory.

Example 7.6 (Chern-Dold character on integral Morava K-theory). We highlight that a particularly interesting example of the Chern-Dold character, which is not widely known, is that on integral Morava K-theory, whose codomain in real cohomology has a rich coefficient system. Morava K-theories $K(n)$ [Johnson and Wilson (1975)] (reviewed in [Würgler (1991)][Rudyak (1998), §IX.7]) form a sequence of spectra labeled by chromatic level $n \in \mathbb{N}$ and by a prime p (notationally left implicit). Their coefficient ring is pure torsion, and hence vanishes upon rationalization. However, there is an integral version $\widetilde{K}(n)$, highlighted in [Kriz and Sati (2004)][Sati (2010)][Buhné (2011)][Sati and Westerland (2015)][Grady and Sati (2017)], which has an integral p-adic coefficient ring:

$$\widetilde{K}(n)_* \;=\; \mathbb{Z}_p[v_n, v_n^{-1}], \qquad \text{with } \deg(v_n) = 2(p^n - 1). \tag{7.8}$$

This theory more closely resembles complex K-theory than is the case for $K(n)$; in fact, for $n = 1$, it coincides with the p-completion of complex K-theory.

Therefore, the Chern-Dold character (Def. 7.3) on integral Morava K-theory [Grady and Sati (2017), p. 53] is of the form

$$\mathrm{ch}_{\mathrm{Mor}} \,:\, \widetilde{K}(n)(-) \longrightarrow H^*\big(-; \mathbb{Q}_p[v_n, v_n^{-1}] \otimes_{\mathbb{Q}} \mathbb{R}\big)\,, \tag{7.9}$$

where we used (7.8) in (7.2) together with the fact that the rationalization of the p-adic integers is the rational (here: real, by Rem. 5.2) p-adic numbers[1] $\mathbb{Z}_p \otimes_{\mathbb{Z}} \mathbb{R} \simeq \mathbb{Q}_p \otimes_{\mathbb{Q}} \mathbb{R}$.

Now $\mathbb{Q}_p$ is not finite-dimensional over $\mathbb{Q}$, whence $\mathbb{Q}_p \otimes \mathbb{R}$ is not finite-dimensional over $\mathbb{R}$, so that the classifying space for integral Morava K-theory is not of $\mathbb{R}$-finite type (Def. 5.1). Therefore, our *proof* of the non-abelian de Rham theorem (Thm. 6.5), being based on the fundamental theorem of dgc-algebraic rational homotopy theory (Prop. 5.6), does not immediately apply to integral Morava K-theory coefficients; and hence the non-abelian character on integral Morava K-theory with de Rham codomain, in the form defined in Def. IV.2, is not established here. While this is a purely technical issue, as discussed in Rem. 5.1, further discussion is beyond the scope of the present article.

[1] Note, parenthetically, that the classical Chern character ch itself can be extended to cohomology theories with values in graded $\mathbb{Q}$-algebras; see, e.g., [Maakestad (2017)].

Chapter 8

Chern-Weil homomorphism

We prove (Thm. 8.6) that the non-abelian character subsumes the Chern-Weil homomorphism (recalled as Prop. 8.3, review in [Chern (1951), §III][Kobayashi and Nomizu (1963), §XII][Chern and Simons (1974), §2][Milnor and Stasheff (1974), §C][Fiorenza *et al.* (2012), §2.1]) in degree-1 non-abelian cohomology.

Chern-Weil theory. For definiteness, we recall the statements of Chern-Weil theory that we need to prove Thm. 8.6 below.

Remark 8.1 (Attributions in Chern-Weil theory).
(i) What came to be known as the *Chern-Weil homomorphism* (recalled as Def. 8.3 below) seems to be first publicly described by H. Cartan (in May 1950), in his prominent *Séminaire* [Cartan (1950), §7], published as [Cartan (1951)]. Later that year at the ICM (in Aug.-Sep. 1950), Chern discusses this construction in a talk [Chern (1952), (10)], including a brief reference to unpublished work by Weil (which remained unpublished until appearance in Weil's collected works [Weil (2014)]) for the proof that the construction is independent of the choice of connection (which is stated with an announcement of a proof in [Cartan (1950), §7]).
(ii) The new result of Chern's talk was the observation [Chern (1952), (15)] – later called the *fundamental theorem* in [Chern (1951), §III.6], recalled as Prop. 8.5 below – that this differential-geometric construction coincides with the topological construction of real characteristic classes (Ex. 2.17). This crucially uses the identification [Chern (1952), (11)] of the real cohomology of classifying space BG with invariant polynomials, later expanded on by Bott [Bott (1973), p. 239]. (Various subsequent authors, e.g. [Freed (2002), (1.14)], suggest to prove Chern's equation (15) by making sense of a connection on the universal G-bundle – which is possible though notoriously subtle, e.g. [Mostow (1979)] – but the proof in [Chern (1952)] simply observes that for any fixed bound $\leq d$ on the dimension of the domain space, the classifying space for G-principal bundles may be truncated to a $d+1$-dimensional sub-complex (as follows by the cellular approximation theorem [Spanier (1966), p. 404]) this carrying a smooth G principal bundle with ordinary connection, which is universal for G-principal bundles over $\leq d$-manifolds. This argument was later worked out in [Narasimhan and Ramanan (1961)][Narasimhan and Ramanan (1963)][Schlafly (1980)]).

(iii) It is this fundamental theorem [Chern (1952), (15)][Chern (1951), §III.6] which allows to identify the Chern-Weil homomorphism as an instance of the non-abelian character, in Thm. 8.6 below.

Notation 8.1 (Principal bundles with connection). For $G \in$ LieGroups and $X \in$ SmthMfds, we write

$$G\mathrm{Connections}(X)_{/\sim} \longrightarrow\!\!\!\!\!\rightarrow G\mathrm{Bundles}(X)_{/\sim} \qquad (8.1)$$

for the forgetful map from the set of isomorphism classes, over X, of G-principal bundles equipped with principal connections (review in [Nakahara (2003), §9][Rudolph and Schmidt (2017), §1]) to the underlying bundles without connection, .

The function (8.1) is surjective and admits sections, corresponding to a choice of the class of a principal connection on any class of G-principal bundles.

Definition 8.1 (Invariant polynomials [Weil (2014)][Cartan (1950), §7]**).** For $\mathfrak{g} \in$ $\mathrm{LieAlgebras}_{\mathbb{R},\mathrm{fin}}$, we write

$$\mathrm{inv}^\bullet(\mathfrak{g}) := \mathrm{Sym}\big(\mathfrak{b}^2\mathfrak{g}^*\big)^G \in \mathrm{gcAlgs}_{\mathbb{R}}^{\geq 0}$$

for the graded sub-algebra (4.9) on those elements in the symmetric algebra (4.12) of the linear dual of $\mathfrak{g}$ shifted up (Def. 4.4) into degree 2, which are invariant under the adjoint action of G on $\mathfrak{g}^*$.

Definition 8.2 (Characteristic forms [Cartan (1950), §7][Chern (1952), (10)]**).** Let G be a finite-dimensional Lie group with Lie algebra $\mathfrak{g}$, and let $P \xrightarrow{p} X$ be G-principal bundle with connection ∇ (Def. 8.1). Then for $\omega \in \mathrm{inv}^{2n}(\mathfrak{g})$ an invariant polynomial (Def. 8.1), its evaluation on the curvature 2-form $F_\nabla \in \Omega^2(P) \otimes \mathfrak{g}$ of the connection yields a differential form

$$\omega(F_\nabla) \in \Omega^{2n}_{\mathrm{dR}}(X) \xrightarrow{p^*} \Omega^{2n}_{\mathrm{dR}}(P)$$

which, by the second condition on an Ehresmann connection, is *basic*, namely in the image of the pullback operation along the bundle projection p, as shown. Regarded as a differential form on X, this is called the *characteristic form* corresponding to ω.

Lemma 8.1 (Characteristic de Rham classes of characteristic forms [Weil (2014)][Chern (1952), p. 401][Chern (1951), §III.4]**).** *The class in de Rham cohomology*

$$\big[\omega(F_\nabla)\big] \in H^{2n}_{\mathrm{dR}}(X)$$

of a characteristic form in Def. 8.2 is independent of the choice of connection ∇ *and depends only on the isomorphism class of the principal bundle* P.

Definition 8.3 (Chern-Weil homomorphism [Cartan (1950), §7][Chern (1952), (10)]**).** Let G be a finite-dimensional Lie group, with classifying space denoted BG. The *Chern-Weil homomorphism* is the composite map

$$\underset{\text{homomorphism}}{\overset{\text{Chern-Weil}}{}} \mathrm{cw}_G : G\mathrm{Bundles}(X)_{/\sim} \longrightarrow G\mathrm{Connections}(X)_{/\sim} \longrightarrow \mathrm{Hom}\big(\mathrm{inv}^\bullet(\mathfrak{g}), H^\bullet_{\mathrm{dR}}(X)\big)$$
$$\underset{[P]}{\text{principal bundle}} \longmapsto \underset{[P,\nabla]}{\text{with connection}} \longmapsto \Big(\underset{\omega}{\text{invariant polynomial}} \mapsto \underset{[\omega(F_\nabla)]}{\text{de Rham class of characteristic form}} \Big), \tag{8.2}$$

where the first map is any section of (8.1), given by choosing any connection on a given principal bundle; and the second map is the construction of characteristic forms according to Def. 8.2. (The Hom on the right is that in $\mathrm{gcAlgs}^{\geq 0}_{\mathbb{R}}$.) By Lem. 8.1 the second map is well-defined (and its composition with the first turns out to be independent of the choices made, by Prop. 8.5 below).

That this construction is useful, in that it produces interesting real characteristic classes of G-principal bundles (Ex. 2.17), is the following statement:

Proposition 8.4 (Abstract Chern-Weil homomorphism [Chern (1952), (11)][Chern (1951), §III.5][Bott (1973), p. 239]**).** *Let G be a finite-dimensional, compact Lie group, with Lie algebra denoted $\mathfrak{g}$. Then the real cohomology algebra of its classifying space BG is isomorphic to the algebra of invariant polynomials (Def. 8.1):*

$$\mathrm{inv}^\bullet(\mathfrak{g}) \simeq H^\bullet(BG;\mathbb{R}) \in \mathrm{gcAlgs}^{\geq 0}_{\mathbb{R}}. \tag{8.3}$$

We can also obtain the following:

Proposition 8.5 (Fundamental theorem of Chern-Weil theory [Chern (1952), (15)][Chern (1951), §III.6] **(Rem. 8.1)).** *Let G be a finite-dimensional compact Lie group. Then the Chern-Weil homomorphism (Def. 8.3) coincides with the operation of pullback of universal characteristic classes along the classifying maps of G-bundles (Ex. 2.17), in that the following diagram commutes:*

$$\begin{array}{ccc} H(X;BG) & \xrightarrow[c \mapsto c^*(-)]{\text{pullback of universal characteristic classes along classifying map (2.22)}} & \mathrm{Hom}\big(H^\bullet(BG;\mathbb{R}), H^\bullet(X;\mathbb{R})\big) \\ {\scriptstyle (2.7)}\uparrow\simeq & & \simeq\downarrow{\scriptstyle (8.3)} \\ G\mathrm{Bundles}(X)_{/\sim} & \xrightarrow[\text{Chern-Weil homomorphism (8.2)}]{\mathrm{cw}_G} & \mathrm{Hom}\big(\mathrm{inv}^\bullet(\mathfrak{g}), H^\bullet_{\mathrm{dR}}(X)\big) \end{array} \tag{8.4}$$

Here the isomorphism on the left is from Ex. 2.2, while that from the right is from Prop. 8.4 and using the de Rham theorem.

Example 8.1 (Characteristic forms of classical Lie groups (e.g. [Nakahara (2003), Ex. 11.5-7])**).** Let $G = \mathrm{SU}(n)$ be the special unitary group, for $n \in \mathbb{N}$. Then the fundamental Chern-Weil theorem Prop. 8.5 identifies, for any connection ∇ (8.2), on a given

$\mathrm{SU}(n)$-principal bundle, with associated curvature differential form

$$\begin{array}{ccc} \Omega^2\big(X;\mathfrak{u}(n)\big) & \hookrightarrow & \Omega^2\big(X;\mathrm{Mat}_{n\times n}(\mathbb{C})\big) \\ F_\nabla & \longmapsto & \big((F_\nabla)_{ab}\big)_{1\le a,b\le n} \end{array} \tag{8.5}$$

the following (de Rham cohomology classes of) classical characteristic forms (Def. 8.2):

(i) Chern forms. The real-cohomology images of the first couple *Chern classes* $c_i \in H^{2i}\big(BU(n);\mathbb{Z}\big)$ are identified with the de Rham cohomology classes $\mathrm{cw}_{\mathrm{SU}(n)}(c_i) = \big[c_i(\nabla)\big] \in H^{2i}_{\mathrm{dR}}(X)$ of the polynomials in the curvature differential form (8.5) which are the homogeneous components of the following *total Chern form*[1]

$$c(\nabla) := \sum_{k\in\mathbb{N}} \underbrace{c_k(\nabla)}_{\deg=2k} := \det\big(1+\tfrac{i}{2\pi}F_\nabla\big)\,.$$

(ii) Pontrjagin forms. When the structure group G is reduced along the canonical inclusion $\mathrm{SO}(n) \overset{\iota}{\hookrightarrow} \mathrm{SU}(n)$ of the special orthogonal group, then the real images of the first couple *Pontrjagin classes* $p_k \in H^{4k}\big(B\mathrm{SO}(n);\mathbb{Z}\big)$ are identified with the de Rham cohomology classes of the corresponding Chern forms (8.6), up to a signs:

$$\mathrm{cw}_{\mathrm{SO}(n)}(p_k) = \big[p_k(\nabla)\big] = (-1)^k\big[c_{2k}(\iota_*\nabla)\big] \in H^{4k}_{\mathrm{dR}}(X)\,.$$

One finds

$$\begin{aligned} p_1(\nabla) &:= -\tfrac{1}{8\pi^2}\,\mathrm{tr}\big(F_\nabla\wedge F_\nabla\big)\,, \\ p_2(\nabla) &= \tfrac{1}{128\pi^4}\Big(\mathrm{tr}\,(F_\nabla\wedge F_\nabla)\wedge\mathrm{tr}\,(F_\nabla\wedge F_\nabla) - 2\cdot\mathrm{tr}\,(F_\nabla\wedge F_\nabla\wedge F_\nabla\wedge F_\nabla)\Big)\,. \end{aligned} \tag{8.6}$$

The following rational combination of these forms plays a central role in Chapter 12:

$$I_8(\nabla) := \tfrac{1}{48}\big(p_2(\nabla) - \tfrac{1}{4}p_1(\nabla)\wedge p_1(\nabla)\big)\,. \tag{8.7}$$

(iii) Euler form. If, moreover, $n = 2k$ is even, then the real image of the *Euler class* $\chi_n \in H^n\big(B\mathrm{SO}(n);\mathbb{Z}\big)$ is identified with the de Rham cohomology class

$$\mathrm{cw}_{\mathrm{SO}(n)}(\chi_n) = \big[\chi_n(\nabla)\big] \in H^n_{\mathrm{dR}}(X)$$

of the *Pfaffian* wedge-product polynomial of the matrix (8.5):

$$\chi_{2k}(\nabla) := \tfrac{(-1)^{n/2}}{(4\pi)^{n/2}\cdot(n/2)!} \sum_{\sigma\in\mathrm{Sym}(n)} \mathrm{sgn}(\sigma)\cdot(F_\nabla)_{\sigma(1)\sigma(2)}\wedge(F_\nabla)_{\sigma(3)\sigma(4)}\wedge\cdots\wedge(F_\nabla)_{\sigma(n-1)\sigma(n)}\,. \tag{8.8}$$

---

[1] The standard normalization factor $i/2\pi$ appearing here results from identifying $\mathrm{U}(1)$ with $\mathbb{R}/h\mathbb{Z}$ for the choice $h = 2\pi$.

Chern-Weil homomorphism as a special case of the non-abelian character.

Lemma 8.2 (Sullivan model of classifying space). *Let G be a finite-dimensional, compact and simply-connected Lie group, with Lie algebra denoted* $\mathfrak{g}$. *Then the minimal Suillvan model (Def. 4.22) of its classifying space BG is the graded algebra of invariant polynomials (Def. 8.1), regarded as a dgc-algebra with vanishing differential:*

$$\big(\mathrm{inv}(\mathfrak{g}), d=0\big) \simeq \mathrm{CE}(\mathfrak{l}BG) \quad \in \mathrm{dgcAlgs}_{\mathbb{R}}^{\geq 0}\,. \tag{8.9}$$

Proof. According to [Félix *et al.* (2008), Ex. 2.42], we have

$$\mathrm{CE}(\mathfrak{l}BG) \simeq \big(H^\bullet(BG;\mathbb{R}), d=0\big). \tag{8.10}$$

The composition of (8.10) with the isomorphism (8.3) from Prop. 8.4 yields the desired (8.9). □

Lemma 8.3 (Non-abelian de Rham cohomology with coefficients in a classifying space). *Let G be a finite-dimensional, compact and simply-connected Lie group, with Lie algebra denoted* $\mathfrak{g}$. *Then the non-abelian de Rham cohomology (Def. 6.3) with coefficients in the rational Whitehead* L_∞*-algebra* $\mathfrak{l}BG$ *(Prop. 5.11) of the classifying space is canonically identified with the codomain of the classical Chern-Weil construction* (8.2)*:*

$$\underset{\text{de Rham cohomology}}{\overset{\text{nonabelian}}{H_{\mathrm{dR}}\big(X;\mathfrak{l}BG\big)}} \simeq \underset{\text{Chern-Weil construction}}{\overset{\text{traditional codomain of}}{\mathrm{Hom}\big(\mathrm{inv}^\bullet(\mathfrak{g}), H^\bullet_{\mathrm{dR}}(X)\big)}}. \tag{8.11}$$

Proof. Consider the following sequence of natural bijections:

$$\begin{aligned}
H_{\mathrm{dR}}\big(X;\mathfrak{l}BG\big) &:= \mathrm{dgcAlgs}_{\mathbb{R}}^{\geq 0}\big(\mathrm{CE}\big(\mathfrak{l}BG\big)\,,\,\Omega^\bullet_{\mathrm{dR}}(X)\big)_{/\sim} \\
&\simeq \mathrm{dgcAlgs}_{\mathbb{R}}^{\geq 0}\Big(\big(\mathrm{inv}^\bullet(\mathfrak{g}), d=0\big)\,,\,\Omega^\bullet_{\mathrm{dR}}(X)\Big)_{/\sim} \\
&\simeq \mathrm{gcAlgs}_{\mathbb{R}}^{\geq 0}\Big(\mathrm{inv}^\bullet(\mathfrak{g})\,,\,\Omega^\bullet_{\mathrm{dR}}(X)_{\mathrm{closed}}\Big)_{/\sim} \\
&\simeq \mathrm{gcAlgs}_{\mathbb{R}}^{\geq 0}\Big(\mathrm{inv}^\bullet(\mathfrak{g})\,,\,\big(\Omega^\bullet_{\mathrm{dR}}(X)_{\mathrm{closed}}\big)_{/\sim}\Big) \\
&\simeq \mathrm{gcAlgs}_{\mathbb{R}}^{\geq 0}\big(\mathrm{inv}^\bullet(\mathfrak{g})\,,\,H^\bullet_{\mathrm{dR}}(X)\big) \\
&=: \mathrm{Hom}\big(\mathrm{inv}^\bullet(\mathfrak{g})\,,\,H^\bullet_{\mathrm{dR}}(X)\big).
\end{aligned}$$

Here the first line is the definition (Def. 6.3). After that, the first step is Lem. 8.3. The second step unwinds what it means to hom out of a dgc-algebra with vanishing differential (which is generator-wise as in Ex. 6.2), while the third and fourth steps unwind what this means for the coboundary relations (which is generator-wise as in Prop. 6.4). The last line just matches the result to the abbreviated notation used in (8.2). □

Theorem 8.6 (Non-abelian character map subsumes Chern-Weil homomorphism). *Let G be a finite-dimensional compact, connected and simply-connected Lie group, with Lie algebra* $\mathfrak{g}$. *Let* $X \in \mathrm{Ho}\big(\Delta\mathrm{Sets}_{\mathrm{Qu}}\big)^{\mathrm{fin}_{\mathbb{Q}}}_{\geq 1,\mathrm{nil}}$ *(Def. 5.1) be equipped with the structure of a* *smooth manifold. Then the non-abelian character* ch_{BG} *(Def. IV.2) on non-abelian cohomology (Def. 2.1) of X with coefficients in BG coincides with the Chern-Weil homomorphism* cw_G *(Def. 8.3) with coefficients in G, in that the following diagram (of cohomology sets) commutes:*

$$
\begin{array}{ccc}
H(X;BG) & \xrightarrow[\mathrm{ch}_{BG}]{\text{non-abelian character}} & H_{\mathrm{dR}}(X;\mathfrak{l}BG) \\
{\scriptstyle (2.7)}\uparrow \simeq & & \simeq \downarrow {\scriptstyle (8.11)} \\
G\mathrm{Bundles}(X)_{/\sim} & \xrightarrow[\text{Chern-Weil homomorphism}]{\mathrm{cw}_G} & \mathrm{Hom}\big(\mathrm{inv}^\bullet(\mathfrak{g}), H^\bullet_{\mathrm{dR}}(X)\big)
\end{array}
\qquad (8.12)
$$

Here the isomorphism on the left is from Ex. 2.2, while that on the right is from Lem. 8.3.

Proof. First, notice that BG is simply connected (hence nilpotent), by the assumption that G is connected, and that it is of finite rational type by Prop. 8.4. Hence, with Def. 5.1,

$$
BG \in \mathrm{Ho}\big(\Delta\mathrm{Sets}_{\mathrm{Qu}}\big)^{\mathrm{fin}_{\mathbb{Q}}}_{\geq 1,\mathrm{nil}} . \qquad (8.13)
$$

Now, by Def. IV.2, the non-abelian character map on the top of (8.4)

$$
\mathrm{ch}_{BG} \;:\; H(X;BG) \xrightarrow{(\eta^{\mathbb{R}}_{BG})} H\big(X;L_{\mathbb{R}}BG\big) \xrightarrow{\simeq} H_{\mathrm{dR}}\big(X;L_{\mathbb{R}}BG\big)
$$

sends a classifying map

$$
X \xrightarrow{c} BG \;\in H(X;BG) \;=\; \mathrm{Ho}\big(\Delta\mathrm{Sets}_{\mathrm{Qu}}\big)(X\,,BG)
$$

first to its composite with the rationalization map (Def. 5.2). By the fundamental theorem (Thm. 5.6(i), using (8.13)), this is given by the derived adjunction unit $\mathbb{D}\eta_{BG}$ of $\mathbb{D}B\exp_{\mathrm{PL}} \dashv \Omega^\bullet_{\mathrm{PLdR}}$ (5.16):

$$
\begin{aligned}
X \xrightarrow{\;c\;} BG & \xrightarrow{\mathbb{D}_{\mathbb{R}}BG \simeq \mathbb{D}\eta_{BG}} \mathbb{D}B\exp_{\mathrm{PL}} \circ\, \Omega^\bullet_{\mathrm{PLdR}}(BG) \\
& \in \mathrm{Ho}\big(\Delta\mathrm{Sets}_{\mathrm{Qu}}\big)\big(X\,,L_{\mathbb{R}}BG\big) \;=\; H\big(X;L_{\mathbb{R}}BG\big) .
\end{aligned}
$$

Moreover, by part (ii) of the fundamental theorem, the adjunct of the morphism $\mathbb{D}\eta_{BG} \circ c$ under (5.16) is

$$
\Omega^\bullet_{\mathrm{PLdR}}(X) \xleftarrow{\;c^*\;} \Omega^\bullet_{\mathrm{PLdR}}(BG) \quad \in \mathrm{Ho}\big(\big(\mathrm{dgcAlgs}^{\geq 0}_{\mathbb{R}}\big)_{\mathrm{trinj}}\big)
$$

(using that $\Omega^\bullet_{\mathrm{PLdR}}(\mathbb{D}\eta^{\mathbb{R}})$ is an equivalence, by reflectivity of rationalization (5.2)). Hence it is the pullback operation of rational cocycles on BG along the classifying map c. Sending this further along the isomorphism to the bottom right in (8.4) (via Thm. 6.5 and Lem. 8.3) gives, by (6.18):

$$\mathrm{ch}_{BG} : c \mapsto \quad \Omega^\bullet_{\mathrm{dR}}(X) \xleftarrow{\;c^*\;} \Omega^\bullet_{\mathrm{PLdR}}(BG) \xleftarrow{\;\simeq\;} \mathrm{inv}^\bullet(\mathfrak{g}) \quad \in \mathrm{Ho}\Big(\big(\mathrm{dgcAlgs}^{\geq 0}_{\mathbb{R}}\big)_{\mathrm{trinj}}\Big)\,. \tag{8.14}$$

In conclusion, we have found that the commutativity of (8.12) is equivalent to the statement that the characteristic forms obtained by the Chern-Weil construction (8.2) represent the pullback (8.14) of the universal real characteristic classes on BG along the classifying map c of the underlying principal bundle (Ex. 2.17). This is the case by the fundamental theorem of Chern-Weil theory, Prop. 8.5. □

Example 8.2 (de Rham representative of tangential Sp(2)**-twist).** For X a smooth 8-dimensional spin manifold equipped with tangential Sp(2)-structure τ (3.33), Thm. 8.6 says that there exists a smooth Sp(2)-principal bundle on X equipped with an Ehresmann connection ∇ such that the $\mathbb{R}$-rationalization (Def. 5.7) of the twist τ corresponds, under the non-abelian de Rham theorem (Thm. 6.5) to a flat $\mathfrak{l}B\mathrm{Sp}(2)$-valued differential form whose components are the characteristic forms of the Sp(2)-principal connection ∇:

$$\begin{array}{ccccc}
H\big(X; B\mathrm{Sp}(2)\big) & \xrightarrow{(\eta^{\mathbb{R}}_{BG})_*} & H\big(X; L_{\mathbb{R}}B\mathrm{Sp}(2)\big) & \simeq & H_{\mathrm{dR}}\big(X; \mathfrak{l}B\mathrm{Sp}(2)\big) \\
\tau & \longmapsto & L_{\mathbb{R}}\tau & \longleftrightarrow & \Omega^\bullet_{\mathrm{dR}}(X) \xleftarrow{\tau_{\mathrm{dR}}} \mathbb{R}\begin{bmatrix} \chi_8, \\ \tfrac{1}{2}p_1 \end{bmatrix} \Big/ \begin{pmatrix} d\,\tfrac{1}{2}p_1 = 0 \\ d\;\;\chi_8 = 0 \end{pmatrix} \\
 & & & & \tfrac{1}{2}p_1(\nabla) \longleftarrow\!\!\shortmid \tfrac{1}{2}p_1 \\
 & & & & \chi_8(\nabla) \longleftarrow\!\!\shortmid \chi_8
\end{array}$$

Here on the right we are using [Čadek and Vanžura (1998), Thm. 8.1] (see [Fiorenza *et al.* (2022), Lem. 2.12]) to identify generating universal characteristic classes on $B\mathrm{Sp}(2)$: $\tfrac{1}{2}p_1$ is the first Pontrjagin class (see Ex. 8.1) and $\chi_8 = \left(\tfrac{1}{2}p_2 - \left(\tfrac{1}{2}p_1\right)^2\right)$ is the Euler 8-class, which here on $B\mathrm{Sp}(2)$ happens to be proportional to I_8 (8.7), see [Fiorenza *et al.* (2020b), Prop. 3.7].

Chapter 9

Cheeger-Simons homomorphism

We show (Thm. 9.9) that the non-abelian character map induces *secondary* non-abelian cohomology operations (Def. 9.7) which subsume the Cheeger-Simons homomorphism, recalled around (9.41) below, with values in ordinary differential cohomology, recalled around (9.27) below. We follow [Fiorenza *et al.* (2012)] [Sati *et al.* (2012)][Schreiber (2013)] where the Cheeger-Simons homomorphism, generalized to higher principal bundles, is called the ∞-*Chern-Weil homomorphism*. Underlying this is a differential enhancement of the non-abelian character map (Def. 9.2), and an induced notion of differential non-abelian cohomology (Def. 9.3) on smooth ∞-stacks (recalled as Prop. 1.24).

The differential non-abelian character map. We introduce (in Def. 9.2 below) the differential refinement of the non-abelian character map; given as before by rationalization, but now followed not by a map to non-abelian de Rham cohomology, but to its refinement by the full cocycle space of flat non-abelian differential forms (Def. 9.1 below). It is this refinement of the codomain of the character map that allows it to be fibered over the smooth space (Ex. 1.26) of actual flat non-abelian differential forms (instead of just their non-abelian de Rham classes), thus producing differential non-abelian cohomology (Def. 9.3 below).

Definition 9.1 (Moduli ∞-stack of flat L_∞-algebra valued forms [Schreiber (2013), 4.4.14.2]**).** Let $A \in \Delta\mathrm{Sets}$ be of connected, nilpotent, $\mathbb{R}$-finite homotopy type (Def. 5.1). By means of the system of sets (Def. 6.1)

$$X \longmapsto \Omega_{\mathrm{dR}}\big(X; \mathfrak{l}A\big) \quad \in \mathrm{Sets}$$

of flat non-abelian differential forms with coefficient in the Whitehead L_∞-algebra $\mathfrak{l}A$ of A (Prop. 5.11), which are contravariantly assigned to smooth manifolds X, we consider in $\mathrm{Ho}(\mathrm{SmthStacks}_\infty)$ (Def. 1.25):

(i) the *smooth space (Ex. 1.26) of flat $\mathfrak{l}A$-valued differential forms*

$$\Omega_{\mathrm{dR}}\big(-;\mathfrak{l}A\big)_{\mathrm{flat}} := \Big(\mathbb{R}^n \mapsto \Big(\Delta[k] \mapsto \Omega_{\mathrm{dR}}\big(\mathbb{R}^n;\mathfrak{l}A\big)_{\mathrm{flat}}\Big)\Big), \tag{9.1}$$

regarded as a simplicially constant simplicial presheaf (1.69);

(ii) the *smooth ∞-stack of flat* $\mathfrak{l}A$*-valued differential forms* (Ex. 6.5)

$$\flat\mathfrak{B}\exp(\mathfrak{l}A) := \Big(\mathbb{R}^n \mapsto \big(\Delta[k] \mapsto \Omega_{\mathrm{dR}}\big(\mathbb{R}^n \times \Delta^k; \mathfrak{l}A\big)_{\mathrm{flat}}\big)\Big) \tag{9.2}$$

which to any Cartesian space assigns the simplicial set that in degree k is the set of flat $\mathfrak{l}A$-valued differential forms on the product manifold of the Cartesian space with the standard smooth k-simplex $\Delta^k \subset \mathbb{R}^k$;

(iii) the canonical inclusion

$$\begin{array}{ccc} \underset{\text{flat } \mathfrak{l}A\text{-valued forms}}{\overset{\text{smooth space of}}{\Omega(-;\mathfrak{l}A)_{\mathrm{flat}}}} & \xrightarrow{\text{atlas}} & \underset{\text{flat } \mathfrak{l}A\text{-valued forms}}{\overset{\text{smooth } \infty\text{-stack of}}{\flat_{\mathfrak{B}}\exp(\mathfrak{l}A)}} \\ \| & & \| \\ \Big(\mathbb{R}^n \mapsto \big(\Delta[k] \mapsto \Omega_{\mathrm{dR}}\big(\mathbb{R}^n; \mathfrak{l}A\big)_{\mathrm{flat}}\big)\Big) & \hookrightarrow & \Big(\mathbb{R}^n \mapsto \big(\Delta[k]a \mapsto \Omega_{\mathrm{dR}}\big(\mathbb{R}^n \times \Delta^k; \mathfrak{l}A\big)_{\mathrm{flat}}\big)\Big) \end{array} \tag{9.3}$$

exhibiting $\Omega(-;\mathfrak{l}A)$ (9.1) as the presheaf of 0-simplices in the simplicial presheaf $\flat\mathbf{B}\exp(\mathfrak{l}A)$ (9.2) (more abstractly: this is the canonical *atlas* of the smooth moduli ∞-stack, see [Sati and Schreiber (2020c), Prop. 2.70]).

Lemma 9.1 (Moduli ∞-stack of flat forms is equivalent to discrete rational ∞-stack). *For* $A \in \mathrm{Ho}\big(\Delta\mathrm{Sets}_{\mathrm{Qu}}\big)^{\mathrm{fin}_{\mathbb{Q}}}_{\geq 1,\mathrm{nil}}$ *(Def. 5.1), the evident inclusion (by inclusion of polynomial forms into smooth differential forms followed by pullback along* pr_{Δ^k}*)*

$$\begin{array}{ccc} \mathrm{Disc}\big(L_{\mathbb{R}}A\big) & & \\ \wr\| & & \\ \mathrm{Disc}\circ \mathbb{D}B\exp_{\mathrm{PL}}\circ\, \mathrm{CE}\big(\mathfrak{l}A\big) & \xrightarrow{\in \mathrm{W}} & \flat\mathfrak{B}\exp\big(\mathfrak{l}A\big) \\ \| & & \| \\ \Big(\mathbb{R}^n \mapsto \big(\Delta[k] \mapsto \Omega_{\mathrm{PLdR}}\big(\Delta^k; \mathfrak{l}A\big)_{\mathrm{flat}}\big)\Big) & \hookrightarrow & \Big(\mathbb{R}^n \mapsto \big(\Delta[k] \mapsto \Omega_{\mathrm{dR}}\big(\mathbb{R}^n \times \Delta^k; \mathfrak{l}A\big)_{\mathrm{flat}}\big)\Big) \end{array} \tag{9.4}$$

of the image under Disc (1.85) *of the dg-algebraic model* (5.17) *for the rationalization of A (Def. 5.2), given by the fundamental theorem (Prop. 5.6), into the moduli ∞-stack of flat* $\mathfrak{l}A$*-valued differential forms (Def. 9.1) is an equivalence in* $\mathrm{Ho}(\mathrm{SmthStacks}_\infty)$ *(Def. 1.25).*

Proof. By Prop. 5.10, the inclusion is for each $\mathbb{R}^n$ a weak equivalence (5.28) in $\Delta\mathrm{Sets}_{\mathrm{Qu}}$ (Ex. 1.2), hence is a weak equivalence already in the global projective model structure on simplicial presheaves, and therefore also in the local projective model structure (Ex. 1.20). □

Lemma 9.2 (Moduli ∞-stack of closed differential forms is shifted de Rham complex). *For* $n \in \mathbb{N}$,

(i) *we have an equivalence in* $\mathrm{Ho}(\mathrm{SmthStacks}_\infty)$ *(Def. 1.25) from the moduli ∞-stack* $\flat\mathfrak{B}\exp\big(\mathfrak{b}^n\mathbb{R}\big)$ *of flat differential forms (Def. 9.1) with values in the line Lie* $(n+1)$*-algebra*

$\mathfrak{b}^n\mathbb{R}$ (Ex. 4.12) to the image under the Dold-Kan construction (Def. 1.30) of the smooth de Rham complex $\Omega^\bullet_{\mathrm{dR}}(-)$ (Ex. 4.9).

(ii) *This is naturally regarded as a presheaf on* CartSp (1.66) *with values in connective chain complexes (Ex. 1.27) (i.e., with de Rham differential lowering the chain degree) shifted up in degree by $n+1$ and then homologically truncated in degree 0, as shown below.*

$$\flat\mathbf{B}\exp(\mathfrak{b}^n\mathbb{R}) \xrightarrow{\simeq} \mathrm{DK}\left(\begin{array}{c} \vdots \\ \downarrow \\ 0 \\ \downarrow \\ 0 \\ \downarrow \\ \Omega^0_{\mathrm{dR}}(-) \\ \downarrow d \\ \Omega^1_{\mathrm{dR}}(-) \\ \downarrow d \\ \vdots \\ \downarrow d \\ \Omega^{n+1}_{\mathrm{dR}}(-)_{\mathrm{clsd}} \end{array}\right) \in \mathrm{Ho}(\mathrm{SmthStacks}_\infty)$$

Proof. This follows by an enhancement of the proof of Prop. 6.4. First observe, with Ex. 6.2, that the simplicial presheaf

$$\flat\mathbf{B}\exp(\mathfrak{b}^n\mathbb{R})(-) = \left(\Delta[k] \mapsto \Omega^{n+1}_{\mathrm{dR}}\big((-)\times\Delta^k\big)_{\mathrm{clsd}}\right) \tag{9.5}$$

naturally carries the structure of a presheaf of simplicial abelian groups, given by addition of differential forms. Therefore, by the Dold-Kan Quillen equivalence (Prop. 1.28), it is sufficient to prove that we have a quasi-isomorphism of presheaves of chain complexes from the corresponding normalized chain complex (1.86) of (9.5) to the shifted and truncated de Rham complex itself:

$$N\left(\Delta[k] \mapsto \Omega^{n+1}_{\mathrm{dR}}\big((-)\times\Delta^k\big)_{\mathrm{clsd}}\right) \xrightarrow[\simeq]{\int_{\Delta^\bullet}} \left(\cdots \to 0 \to 0 \to \Omega^0_{\mathrm{dR}}(-) \xrightarrow{d} \Omega^1_{\mathrm{dR}}(-) \xrightarrow{d} \cdots \xrightarrow{d} \Omega^{n+1}_{\mathrm{dR}}(-)_{\mathrm{clsd}}\right). \tag{9.6}$$

We claim that such is given by fiber integration of differential forms over the simplices Δ^k:

First, to see that fiber integration does constitute a chain map, we compute for $\omega \in \Omega^\bullet_{\mathrm{dR}}\big((-)\times\Delta^k\big)_{\mathrm{clsd}}$ on the left of (9.6):

$$\int_{\Delta^k}\partial\omega = (-1)^k\int_{\partial\Delta^k}\omega = d\int_{\Delta^k}\omega, \tag{9.7}$$

where the first step is the definition of the differential in the normalized chain complex (1.86) and the second step is the fiberwise Stokes formula (6.12).

Finally, to see that $\int_{\Delta^\bullet}$ is a quasi-isomorphism, notice that the chain homology groups on both sides are

$$H_k(-) \;=\; \begin{cases} \mathbb{R} \mid k = n+1 \\ 0 \mid \text{otherwise} \end{cases}$$

over each Cartesian space: For the left hand side this follows via the weak equivalence (5.28) from the fundamental theorem (Prop. 5.6) via Ex. 5.4, while for the right hand side this follows from the Poincaré lemma.

Hence it is sufficient to see that fiber integration over Δ^{n+1} is an isomorphism on the $(n+1)$st chain homology groups. But a generator of this group on the left is clearly given by the pullback $\mathrm{pr}^*_{\Delta^{n+1}}\omega$ of any $\omega \in \Omega^{n+1}_{\mathrm{dR}}(\Delta^{n+1})$ of unit weight and supported in the interior of the simplex. That this is sent under $\int_{\Delta^{n+1}}$ to a generator $\pm 1 \in \mathbb{R} \simeq \Omega^0_{\mathrm{dR}}(-)_{\mathrm{clsd}}$ on the right follows by the projection formula (6.13). □

Remark 9.1 (Moduli of closed forms via stable Dold-Kan correspondence). Expressed in terms of the stable Dold-Kan construction $\mathrm{DK}_{\mathrm{st}}$ (Prop. 1.29) via the derived stabilization adjunction (Ex. 1.19), Lem. 9.2 says, equivalently, that:

$$\flat\mathfrak{B}\exp(\mathfrak{b}^n\mathbb{R}) \;\simeq\; \mathbb{R}\Omega^\infty\Big(\mathrm{DK}_{\mathrm{st}}\big(\,\Omega^\bullet_{\mathrm{dR}}(-)\otimes_{\mathbb{R}}\mathfrak{b}^{n+1}\mathbb{R}\,\big)\Big) \quad \in \mathrm{Ho}(\mathrm{SmthStacks}_\infty), \tag{9.8}$$

where now $\Omega^\bullet_{\mathrm{dR}}(-) \in \mathrm{PSh}\big(\mathrm{CartSp}, \mathrm{ChainComplexes}_{\mathbb{R}}\big)$ is in non-positive degrees, with $\Omega^0_{\mathrm{dR}}(-)$ in degree 0, and where $\mathfrak{b}^{n+1}\mathbb{R}$ (Def. 4.4) is concentrated on $\mathbb{R}$ in degree $n+1$.

Definition 9.2 (Differential non-abelian character map [Fiorenza *et al.* (2015d), §4]**).** Given $A \in \mathrm{Ho}\big(\Delta\mathrm{Sets}_{\mathrm{Qu}}\big)^{\mathrm{fin}_{\mathbb{Q}}}_{\geq 1,\mathrm{nil}}$ (Def. 5.1), the *differential non-abelian character map* in A-cohomology theory, to be denoted $\mathbf{ch}_A$, is the morphism in $\mathrm{Ho}(\mathrm{SmthStacks}_\infty)$ (1.83) from $\mathrm{Disc}(A)$ (1.85) to the moduli ∞-stack of flat $\mathfrak{l}A$-valued forms $\flat\mathbf{B}\exp(\mathfrak{l}A)$ (9.2) given by the composite of

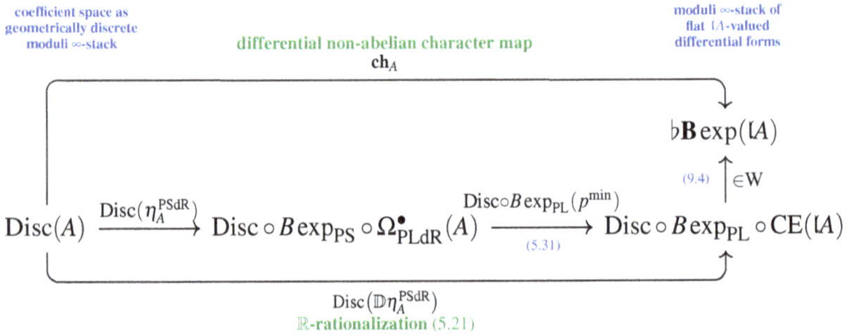

(9.9)

(a) the image under Disc (1.85) of the derived adjunction unit $\mathbb{D}\eta^{\mathrm{PLdR}}_A$ (1.38) of the PS de Rham adjunction (5.26), specifically with (co-)fibrant replacement $p^{\min}$ being the minimal Sullivan model replacement (4.51); (recalling that $B\exp_{\mathrm{PL}}$ is a contravariant functor), with **(b)** the weak equivalence from Lem. 9.1.

Remark 9.2 (Differential non-abelian character map is independent of choices). The differential non-abelian character map (Def. 9.2) is independent, up to equivalence, of the choice of comparison morphism $p^{\min}$ to a minimal model for the coefficients, since, by (4.52) in Prop. 4.23, any two choices factor through each other by an isomorphism of dgc-algebras.

It is this uniqueness which makes minimal models provide canonical form coefficients for non-abelian differential cohomology, see also the second item of Ex. 9.1 below.

Differential non-abelian cohomology.

Definition 9.3 (Differential non-abelian cohomology [Fiorenza *et al.* (2015d), §4]**).** For $A \in \mathrm{Ho}\big(\Delta\mathrm{Sets}_{\mathrm{Qu}}\big)^{\mathrm{fin}_{\mathbb{Q}}}_{\geq 1,\mathrm{nil}}$ (Def. 5.1) we say that:

(i) the *moduli ∞-stack of ΩA-connections* is the object $A_{\mathrm{diff}} \in \mathrm{Ho}(\mathrm{SmthStacks}_\infty)$ in the homotopy category of smooth ∞-stacks (Def. 1.25), which is given by the homotopy pullback (Def. 1.15) of the smooth space of flat non-abelian differential forms $\Omega_{\mathrm{dR}}(-;\mathfrak{l}A)_{\mathrm{flat}}$ (9.3) along the differential non-abelian character map $\mathbf{ch}_A$ (Def. 9.2):

$$
\begin{array}{ccc}
\overset{\substack{\text{moduli }\infty\text{-stack}\\ \text{of }\Omega A\text{-connections}}}{A_{\mathrm{diff}}} & \xrightarrow[F_A]{\substack{\mathfrak{l}A\text{-valued}\\ \text{curvature forms}}} & \overset{\substack{\text{smooth space of}\\ \text{flat }\mathfrak{l}A\text{-valued forms}}}{\Omega_{\mathrm{dR}}(-;\mathfrak{l}A)_{\mathrm{flat}}} \\
{\scriptstyle \substack{\text{universal characteristic class}\\ \text{in non-abelian }A\text{-cohomology}}}\ \Big\downarrow{\scriptstyle c_A} & \text{(hpb)} & \Big\downarrow{\scriptstyle \text{atlas}} \\
\mathrm{Disc}(A) & \xrightarrow[\substack{\text{differential non-abelian}\\ \text{character map}}]{\mathbf{ch}_A} & \underset{\substack{\text{moduli }\infty\text{-stack of}\\ \text{flat }\mathfrak{l}A\text{-valued forms}}}{\flat_{\mathfrak{B}}\exp(\mathfrak{l}A)}
\end{array}
\tag{9.10}
$$

(ii) the *differential non-abelian cohomology* of a smooth ∞-stack $\mathscr{X} \in \mathrm{Ho}(\mathrm{SmthStacks}_\infty)$ (1.83) with coefficients in A is the structured non-abelian cohomology (Rem. 2.3) with coefficients in the moduli ∞-stack A_{diff} of ΩA-connections (9.10), hence the hom-set in the homotopy category of ∞-stacks (Def. 1.25) from $\mathscr{X}$ to A_{diff}

$$
\widehat{H}(\mathscr{X};A) := H(\mathscr{X};A_{\mathrm{diff}}) := \mathrm{Ho}(\mathrm{SmthStacks}_\infty)(\mathscr{X},A_{\mathrm{diff}}). \tag{9.11}
$$

(iii) We call the non-abelian cohomology operations induced from the maps in (9.10) as follows (see (0.16)):

$$
\textbf{(a)}\ \textit{characteristic class:}\quad \widehat{H}(\mathscr{X};A) \xrightarrow{(c_A)_*} H(\mathrm{Shp}(\mathscr{X});A) \qquad (\text{Def. } 2.1) \tag{9.12}
$$

$$
\textbf{(b)}\ \textit{curvature:}\quad \widehat{H}(\mathscr{X};A) \xrightarrow{(F_A)_*} \Omega_{\mathrm{dR}}(\mathscr{X};\mathfrak{l}A)_{\mathrm{flat}} \qquad (\text{Def. } 6.1) \tag{9.13}
$$

$$
\textbf{(c)}\ \textit{differential character:}\quad \widehat{H}(\mathscr{X};A) \xrightarrow{(\mathbf{ch}_A \circ c_A)_*} H_{\mathrm{dR}}(\mathscr{X};\mathfrak{l}A) \qquad (\text{Def. } 6.3) \tag{9.14}
$$

In differential enhancement of Ex. 2.10, we have the following:

Differential generalized cohomology.

Example 9.1 (Differential Whitehead-generalized cohomology). Let $E^\bullet$ be a generalized cohomology theory (Ex. 2.10) with representing spectrum E (2.13) which is connective and whose component spaces E_n are of finite $\mathbb{R}$-type, so that their connected components are, by Ex. 5.1, in $\mathrm{Ho}\big(\Delta\mathrm{Sets}_{\mathrm{Qu}}\big)^{\mathrm{fin}_{\mathbb{Q}}}_{\geq 1,\mathrm{nil}}$ (Def. 5.1).

(i) Then differential non-abelian cohomology, in the sense of Def. 9.3, with coefficients in the component spaces $E_\bullet$, coincides with canonical differential generalized E-cohomology in the traditional sense of [Hopkins and Singer (2005), §4.1][Bunke (2013), Def. 4.53][Bunke and Gepner (2012), §2.2][Bunke *et al.* (2016), §4.4]:

$$\underset{\substack{\text{generalized}\\ \text{differential cohomology}}}{\widehat{E}^n(-)} \;\simeq\; \widehat{H}(-;E_n)\,. \tag{9.15}$$

(ii) Here "canonical", in the sense of [Bunke (2013), Def. 4.46], refers to choosing the curvature differential form coefficients to be $\pi_\bullet(E)\otimes\mathbb{R}$ (instead of some chain complex quasi-isomorphic to this). By Ex. 5.6, this choice corresponds in our Def. 9.3 to the *minimality* (Def. 4.22) of the minimal Sullivan model $\mathrm{CE}(\mathfrak{l}E_n)$ for E_n (Prop. 5.11) that controls the flat L_∞-algebra valued differential forms $\Omega_{\mathrm{dR}}(-;\mathfrak{l}E_n)_{\mathrm{flat}}$ (Def. 6.1) in the top right of (9.32).

(iii) Hence for canonical/minimal curvature coefficients, we have from Ex. 5.6, Lem. 9.2 and Rem. 9.1 that

$$\flat\mathfrak{B}\exp\big(\mathfrak{l}E_n\big) \simeq \mathbb{R}\Omega^\infty\Big(\mathrm{DK}_{\mathrm{st}}\big(\Omega^\bullet_{\mathrm{dR}}(-)\otimes_{\mathbb{Z}}\pi_\bullet(E_n)\big)\Big) \;\in\; \mathrm{Ho}(\mathrm{SmthStacks}_\infty) \tag{9.16}$$

$$\Omega_{\mathrm{dR}}\big(-;\mathfrak{l}E_n\big)_{\mathrm{flat}} \simeq \mathbb{R}\Omega^\infty\Big(\mathrm{DK}_{\mathrm{st}}\big(\Omega^\bullet_{\mathrm{dR}}(-)\otimes_{\mathbb{Z}}\pi_\bullet(E_n)\big)_{\leq 0}\Big) \;\in \mathrm{Ho}(\mathrm{SmthStacks}_\infty)\,. \tag{9.17}$$

(iv) With this, the equivalence 9.15 follows by Ex. 5.7 and observing that the defining homotopy pullback diagram (9.10) for differential non-abelian cohomology with coefficients in $A := E_n$ (1.65) is the image under $\mathbb{R}\Omega^\infty$ (1.64) of the defining homotopy pullback diagram for canonical differential E-cohomology according to [Hopkins and Singer (2005), (4.12)] [Bunke (2013), Def. 4.51][Bunke *et al.* (2016), (24)], and using that right adjoints preserve homotopy pullbacks:

$$\begin{array}{ccc}
(E_0)_{\mathrm{diff}} & \xrightarrow{\;F_{E_0}\;} & \Omega_{\mathrm{dR}}(-;\mathfrak{l}E_0)_{\mathrm{flat}} \\
{\scriptstyle c_{E_0}}\big\downarrow & \text{(hpb)} & \big\downarrow{\scriptstyle \mathrm{atlas}} \\
\mathrm{Disc}(E_0) & \xrightarrow[\;\mathbf{ch}_{E_0}\;]{} & \flat B\exp(\mathfrak{l}E_0)
\end{array}$$

$$\simeq\; \mathbb{R}\Omega^\infty\left(\begin{array}{ccc}
\underset{\substack{\text{moduli }\infty\text{-stack}\\ \text{of }\Omega E_0\text{-connections}}}{\mathrm{Diff}(E,\mathrm{can})} & \longrightarrow & \big(\Omega^\bullet_{\mathrm{dR}}(-)\otimes_{\mathbb{Z}}\pi_\bullet(E)\big)_{\leq 0} \\
\big\downarrow & \text{(hpb)} & \big\downarrow \\
\mathrm{Disc}(E) & \xrightarrow[\;H\mathbb{R}\wedge(-)\;]{} & \Omega^\bullet_{\mathrm{dR}}(-)\otimes_{\mathbb{Z}}\pi_\bullet(E)
\end{array}\right) \tag{9.18}$$

"differential function spectrum" of differential generalized E-cohomology

The same applies to $(E_n)_{\mathrm{diff}}$, by replacing E with $\mathbb{D}\Sigma^n E$ (1.64) on the right of (9.18).

Remark 9.3 (The canonical atlas for the moduli stack of connections). The operation $(-)_{\leq 0}$ in (9.17) is the naive truncation functor on the category of chain complexes

$$\begin{array}{ccc} \mathrm{ChainComplexes}_{\mathbb{Z}} & \xrightarrow{\quad (-)_{\leq 0} \quad} & \mathrm{ChainComplexes}_{\mathbb{Z}}^{\leq 0} \\ \left(\cdots \xrightarrow{\partial_1} V_1 \xrightarrow{\partial_0} V_0 \xrightarrow{\partial_{-1}} V_{-1} \xrightarrow{\partial_{-2}} V_{-1} \to \cdots\right) & \longmapsto & \left(V_0 \xrightarrow{\partial_{-1}} V_{-1} \xrightarrow{\partial_{-2}} V_{-1} \to \cdots\right). \end{array}$$

In contrast to the homological truncation involved in Ω^∞ (1.93), this naive truncation is not homotopy-invariant and does not have a derived functor. Instead, as seen from (9.17) and (9.3), once regarded in differential non-abelian cohomology, this operation serves to construct the canonical *atlas* [Sati and Schreiber (2020c), Prop. 2.70] of the moduli ∞-stack of flat $\mathfrak{l}E_n$-valued differential forms. Via the defining homotopy pullback (9.10), (9.18) this becomes hallmark of differential cohomology: Differential cohomology is the universal solution to lifting the values of the character map from cohomology classes to cochain representatives, namely to curvature forms.

In differential enhancement of Ex. 2.11 and Ex. 7.2, we have:

Example 9.2 (Differential complex K-theory). With the coefficient space $A := \mathrm{KU}_0 = \mathbb{Z} \times B\mathrm{U}$ (2.15) for topological complex K-theory (Ex. 2.11), the corresponding differential non-abelian cohomology theory (Def. 9.3) is, by Ex. 9.1, differential K-theory, whose diagram (0.16) of cohomology operations is of this form (see [Hopkins and Singer (2005)][Bunke and Schick (2009)][Bunke and Schick (2012)][Grady and Sati (2017)])

$$\begin{array}{ccc} \widehat{H}(\mathscr{X}; \mathrm{KU}_0) \simeq \widehat{\mathrm{KU}}^0(\mathscr{X}) & \xrightarrow{\quad F_{\mathrm{KU}_0} \quad} & \left\{ \{F_{2k} \in \Omega^{2k}_{\mathrm{dR}}(\mathscr{X})\}_{k\in\mathbb{N}} \,\middle|\, dF_{2k} = 0 \right\} \\ {\scriptstyle c_{\mathrm{KU}_0}}\big\downarrow & & \big\downarrow \\ \mathrm{KU}^0(\mathscr{X}) & \xrightarrow{\quad \mathrm{ch} \quad} & \bigoplus_{k\in\mathbb{N}} H^{2k}_{\mathrm{dR}}(\mathscr{X}), \end{array} \tag{9.19}$$

where the bottom map is the ordinary Chern character from Ex. 7.2, and the curvature differential forms are identified as in Ex. 6.6.

Remark 9.4 (Differential K-theory via equivalence classes of principal connections). In our context of non-abelian cohomology it is worth highlighting the well-known fact that differential K-theory classes (Ex. 9.2) may equivalently be expressed ([Karoubi (1987)][Lott (1994)][Simons and Sullivan (2010)][Bunke *et al.* (2016), §6], brief review in [Bunke and Schick (2012), §4.1]) in terms of equivalence classes of *vector bundles with connection*, hence equipped with principal connections (Ntn. 8.1) on the underlying $\mathrm{U}(n)$-principal bundles.

Examples of differential non-abelian cohomology. In differential enhancement of Ex. 2.2, we have:

Proposition 9.4 (Differential non-abelian cohomology of principal connections). *Let G be a compact Lie group with classifying space BG (2.8). Then there is a natural map over* *smooth manifolds X, shown dashed in* (9.20), *from equivalence classes of G-principal connections (Ntn. 8.1) to differential non-abelian cohomology with coefficients in BG (Def. 9.3) which covers the classification of G-principal bundles by plain non-abelian cohomology with coefficients in BG (Ex. 2.2), in that the following diagram commutes:*

$$
\begin{array}{ccc}
G\mathrm{Connections}(X)_{/\sim} & \overset{\text{differential non-abelian cohomology}}{\dashrightarrow} & \widehat{H}(X;BG) \\
{\scriptstyle \text{forget connection}}\downarrow & & \downarrow{\scriptstyle c_{BG}} \\
G\mathrm{Bundles}(X)_{/\sim} & \xrightarrow[\simeq]{} & \underset{\text{non-abelian cohomology}}{H(X;BG)}
\end{array}
\tag{9.20}
$$

Proof. By Lem. 8.3, the differential form coefficient in the given case is

$$
\Omega_{\mathrm{dR}}(-;\flat BG)_{\mathrm{flat}} \;\simeq\; \mathrm{Hom}_{\mathbb{R}}\Big(\mathrm{inv}^\bullet(\mathfrak{g})\,,\,\Omega^\bullet_{\mathrm{dR}}(-)_{\mathrm{clsd}}\Big)\,.
$$

Therefore, with Ex. 5.4, we find that

$$
\Big(\Delta[k] \mapsto \mathrm{Hom}_{\mathbb{R}}\big(\mathrm{inv}^\bullet(\mathfrak{g})\,,\,\Omega^\bullet_{\mathrm{dR}}(\Delta^k)_{\mathrm{clsd}}\big)\Big) \;\simeq\; \prod_k K\big(\mathrm{inv}^n(\mathfrak{g}),n\big) \quad \in \mathrm{Ho}\big(\Delta\mathrm{Sets}_{\mathrm{Qu}}\big)
$$

is a product of Eilenberg-MacLane spaces (2.6) for real coefficient groups spanned by the invariant polynomials, and so the defining homotopy pullback (9.10) is here of the following form:

$$
\begin{array}{ccc}
BG_{\mathrm{diff}} & \longrightarrow & \mathrm{Hom}_{\mathbb{R}}\big(\mathrm{inv}^\bullet(\mathfrak{g})\,,\,\Omega^\bullet_{\mathrm{dR}}(-)_{\mathrm{clsd}}\big) \\
\downarrow & \text{(hpb)} & \downarrow \\
\mathrm{Disc}(BG) & \xrightarrow[(c_k)_{k\in\mathbb{N}}]{} & \mathrm{Disc}\Big(\prod\limits_{k\in\mathbb{N}} K\big(\mathrm{inv}^n(\mathfrak{g}),n\big)\Big)\,,
\end{array}
\tag{9.21}
$$

where the bottom map classifies the real characteristic classes of BG via Ex. 2.1. It follows by Ex. 1.12 that maps into BG_{diff} are equivalence classes of triples

$$
\widehat{H}(X;BG) \;\simeq\; \left\{ \big(f,\phi,(\alpha_k)\big) \;\middle|\;
\begin{array}{ccc}
X & \overset{(\alpha_k)}{\dashrightarrow} & \mathrm{Hom}_{\mathbb{R}}\big(\mathrm{inv}^\bullet(\mathfrak{g})\,,\,\Omega^\bullet_{\mathrm{dR}}(-)_{\mathrm{clsd}}\big) \\
{\scriptstyle f}\downarrow & \Leftarrow_{\phi} & \downarrow \\
BG & \longrightarrow & \mathrm{Disc}\Big(\prod\limits_{k\in\mathbb{N}} K\big(\mathrm{inv}^n(\mathfrak{g}),n\big)\Big)
\end{array}
\right\}
\tag{9.22}
$$

consisting of **(a)** a classifying map f for a G-principal bundle (Ex. 2.2), **(b)** a set of closed differential forms α labeled by the invariant polynomials, and **(c)** a set of coboundaries ϕ in real cohomology between these differential forms and the pullbacks f^*c_k.

Now, given a G-connection ∇ on a G-principal bundle f^*EG over X, we obtain such a triple by **(a)** taking f to be the classifying map of the underlying G-principal bundle, **(b)** taking $\alpha_k := \omega_k(F_\nabla)$ to be the characteristic forms (Def. 8.2) of the connection, and **(c)** taking ϕ to be given by the relative Chern-Simons forms [Chern and Simons (1974)] between the given connection and the pullback along f of the universal connection (see Rem. 8.1). This construction is an invariant of the isomorphism class of the connection (see [Hopkins and Singer (2005), p. 28]) and hence defines the desired map (9.20):

$$\begin{array}{ccc} G\mathrm{Connections}(X)_{/\sim} & \longrightarrow & \widehat{H}(X;BG) \\ \big[f^*EG,\nabla\big] & \longmapsto & \big[f,\,\big(\mathrm{cs}_k(\nabla,f^*\nabla_{\mathrm{univ}})\big),\,\big(\omega_k(F_\nabla)\big)\big] \end{array} \tag{9.23}$$

□

Remark 9.5 (The role of principal connections in non-abelian differential cohomology).
(i) It seems unlikely that the map (9.20) in Prop. 9.4 would not be a bijection, but we do not have a proof that it is, in general. A notable case where it is known to be a bijection is the abelian case of the circle group $G = \mathrm{U}(1)$; this case is Prop. 9.5 below.
(ii) However, Thm. 9.9 below shows that the image of G-principal connections in differential non-abelian cohomology $\widehat{H}\big(X;BG\big)$, under this map (9.20), supports the construction of all the secondary characteristic classes of G-principal bundles, hence retains all the relevant information extractable from G-principal connections.
(iii) On the other hand, for each Lie group G with Lie algebra denoted $\mathfrak{g}$, there exists a smooth stack (Prop. 1.24)

$$\mathbf{B}G_{\mathrm{conn}} \simeq \Omega^1(-;\mathfrak{g})/\!\!/G \;\in\; \mathrm{Ho}(\mathrm{SmthStacks}_\infty) \tag{9.24}$$

which is the *moduli stack* of smooth G-principal connections ([Fiorenza *et al.* (2012), Def. 3.2.4][Freed and Hopkins (2013)], exposition in [Fiorenza *et al.* (2015b), §2.4]) in that it not only makes the analogue of the map (9.20) provably a bijection

$$G\mathrm{Connections}(X)_{/\sim} \xrightarrow{\;\sim\;} H\big(X;,\mathbf{B}G_{\mathrm{conn}}\big) \quad \in \mathrm{Sets}$$

but even such that the full mapping space (1.79) into it is equivalent ([Fiorenza *et al.* (2012), Prop. 3.2.5]) to the groupoid (via Ex. 1.3) of gauge transformations between G-principal connections:

$$G\mathrm{Connections}(X) \xrightarrow{\;\sim\;} \mathrm{Maps}\big(X;,\mathbf{B}G_{\mathrm{conn}}\big) \quad \in \mathrm{Ho}\big(\Delta\mathrm{Sets}_{\mathrm{Qu}}\big)\,.$$

(iv) But, while $\mathbf{B}G_{\mathrm{conn}}$ can explicitly be defined as in (9.24), it seems to lack (unless G is abelian, see Prop. 9.5) a more general abstract characterization of the kind that defines BG_{diff} in (9.21), via the systematic Def. 9.3. In particular, it is the construction principle of BG_{diff} – but apparently not that of $\mathbf{B}G_{\mathrm{conn}}$ – which properly generalizes from ordinary non-abelian Lie groups to higher non-abelian groups [Fiorenza *et al.* (2012), §4.3] such as the String 2-group (Ex. 2.4), again for the fact that BG_{diff} canonically supports the secondary characteristic classes: see [Fiorenza *et al.* (2014b), §3-4].

In differential enhancement of Ex. 2.7, we have:

Example 9.3 (Differential Cohomotopy [Fiorenza *et al.* (2015d)]**).** The canonical differential enhancement of (unstable) Cohomotopy theory (Ex. 2.7) in degree n is differential non-abelian cohomology (Def. 9.3) with coefficients in S^n:

$$\underset{\text{differential Cohomotopy}}{\widehat{\pi}^n(-)} := \widehat{H}\big(-;S^n\big).$$

(i) By Ex. 6.4, a cocycle $\widehat{C}_3 \in \widehat{\pi}^4(X)$ in differential 4-Cohomotopy has as curvature (9.10) a pair consisting of a differential 4-form G_4 and a differential 7-form G_7, satisfying the *Cohomotopical Bianchi identity* shown here:

$$\begin{array}{ccc}
\underset{\text{differential 4-Cohomotopy}}{\widehat{\pi}^4(X)} & \xrightarrow[\text{cohomotopical curvature}]{F_{S^4}} & \Omega\big(X;\mathfrak{l}S^4\big)_{\text{flat}} \\
\underset{\substack{\text{cohomotopically} \\ \text{charge-quantized} \\ C_3\text{-field}}}{\widehat{C}_3} & \longmapsto & \left\{ \begin{matrix} G_7(\widehat{C}_3), \\ G_4(\widehat{C}_3) \end{matrix} \in \Omega^\bullet_{\text{dR}}(X) \,\middle|\, \begin{matrix} d\,G_7(\widehat{C}_3) = -G_4(\widehat{C}_3)\wedge G_4(\widehat{C}_3) \\ d\,G_4(\widehat{C}_3) = 0 \end{matrix} \right\}.
\end{array} \tag{9.25}$$

Such differential form data is exactly what characterizes the flux densities of the C_3-field in 11-dimensional supergravity (up to the self-duality constraint $G_7 = \star G_4$). By comparison with Dirac's charge quantization (0.2), we thus see that a natural candidate for charge quantization of the supergravity C_3-field is (nonabelian/unstable) 4-Cohomotopy theory π^4 [Sati (2018), §2.5][Fiorenza *et al.* (2017), §2][Braunack-Mayer *et al.* (2019), §3] (review in [Fiorenza *et al.* (2019), §7]) or rather: differential 4-Cohomotopy theory $\widehat{\pi}^4$ [Fiorenza *et al.* (2015d), p. 9][Grady and Sati (2021a), §3.1].
(ii) The consequence of this Cohomotopical charge quantization is readily seen from the Hurewicz operation on Cohomotopy theory (Ex. 2.21): The de Rham class of the 4-flux density is constrained to be integral, hence to be in the image of the de Rham homomorphism (Ex. 7.1) and its cup square is forced to vanish

$$\big[G_4(\widehat{C}_3)\big] \in H^4(X;\mathbb{Z}) \longrightarrow H^4_{\text{dR}}(X), \qquad \big[G_4(\widehat{C}_3)\big] \cup \big[G_4(\widehat{C}_3)\big] = 0. \tag{9.26}$$

This innocent-looking but *non-linear* cup-square relation is the source of the "quadratic functions in M-theory" [Hopkins and Singer (2005)], revealed here as originating from a deep phenomenon in unstable, hence "non-abelian", homotopy theory, revolving around Hopf maps and Massey products [Kriz and Sati (2005), §4.4][Sati and Schreiber (2021a)] (see [Grady and Sati (2018a)] for differential refinement).
(iii) Passing from 11-dimensional supergravity to M-theory, the curvature data in (9.25) is expected (see [Fiorenza *et al.* (2020b), Table 1]) to be subjected to more refined topological constraints, forcing the class of G_4 to be integral up to a fractional shift by the first Pontrjagin class of the tangent bundle, and deforming its cup square to a quadratic function with non-trivial "background charge". We see, in Prop. 12.1 below, that these more subtle M-theoretic constraints on the C_3-field flux densities are, once more, imposed by charge

quantization in – hence lifting through the non-abelian character map of – the corresponding *twisted* non-abelian cohomology theory, namely: *tangentially twisted* 4-Cohomotopy [Fiorenza *et al.* (2020b)][Fiorenza *et al.* (2022)] (Ex. 12.1 below).

Ordinary differential cohomology. The *ordinary differential cohomology* $\widehat{H}^\bullet(X)$ [Simons and Sullivan (2008)] of a smooth manifold X combines ordinary integral cohomology classes (Ex. 2.1) with closed differential forms that represent the same class in real cohomology, in that it makes a diagram of the following form commute:

$$
\begin{array}{ccc}
\overset{\text{ordinary differential cohomology}}{\widehat{H}^\bullet(X)} & \xrightarrow{\text{curvature}} & \Omega^\bullet_{\mathrm{dR}}(X)_{\mathrm{clsd}} \\
{\scriptstyle \text{underlying integral class}}\big\downarrow & & \big\downarrow{\scriptstyle \text{via de Rham theorem}} \\
H^\bullet(X;\mathbb{Z}) & \xrightarrow{\text{rationalization}} & H^\bullet(X;\mathbb{R})
\end{array}
\qquad (9.27)
$$

In fact, differential cohomology is universal with this property, but not at the coarse level of cohomology sets shown above (where the universal property is shallow) but at the fine level of complexes of sheaves of coefficients (i.e. of moduli ∞-stacks), as made precise in Prop. 9.5 below (see Rem. 9.6).

In degree 2, ordinary differential cohomology classifies ordinary $\mathrm{U}(1)$-principal bundles (equivalently: complex line bundles) with connection [Brylinski (1993), §II], and the curvature map in (9.27) assigns their traditional curvature 2-form. In degree 3, ordinary differential cohomology classifies bundle gerbes with connection [Murray (1996)][Schweigert and Waldorf (2011)] with their curvature 3-form. In general degree, it classifies higher bundle gerbes with connection [Gajer (1997)], or equivalently higher $\mathrm{U}(1)$-principal bundles with connection [Fiorenza *et al.* (2013), 2.6].

One construction of ordinary differential cohomology over smooth manifolds is given in [Cheeger and Simons (1985), §1], now known as *Cheeger-Simons characters*. An earlier construction over schemes, now known as *Deligne cohomology* (Ex. 9.4), due independently to [Deligne (1971), §2.2][Mazur and Messing (1974), §3.1.7][Artin and Mazur (1977), §III.1] and brought to seminal application in [Beĭlinson (1984)] (review in [Esnault and Viehweg (1988)]), is readily adapted to smooth manifolds [Brylinski (1993), §I.5][Gajer (1997)]. The advantage of Deligne cohomology over Cheeger-Simons characters is that is immediately generalizes from smooth manifolds to smooth ∞-stacks, [Fiorenza *et al.* (2012), §3.2.3][Fiorenza *et al.* (2013), §2.5], such as to orbifolds [Sati and Schreiber (2021b)] and to moduli ∞-stacks of higher principal connections where it yields higher Chern-Simons functionals [Sati *et al.* (2012)][Fiorenza *et al.* (2014b)][Fiorenza *et al.* (2015b)][Fiorenza *et al.* (2015a)], as well as allowing for twists in a systematic manner [Grady and Sati (2018c)][Grady and Sati (2019a)].

In differential enhancement of Ex. 2.9, we have:

Example 9.4 (Ordinary differential cohomology on smooth ∞-stacks [Fiorenza *et al.* (2012), §3.2.3][Fiorenza *et al.* (2013), §2.5]**).** Let $n \in \mathbb{N}$.

(i) The smooth *Deligne-Beilinson complex* in degree $n+1$ is the presheaf of connective chain complexes (Ex. 1.27) over CartSp (1.66) given by the truncated and shifted smooth de Rham complex (Ex. 4.9) with a copy of the integers included in degree $n+1$ (as integer valued 0-forms, hence as smooth real-valued functions constant on an integer):

$$\mathrm{DB}^{n+1}_{\bullet} := \Big(\cdots \to 0 \to 0 \to \mathbb{Z} \hookrightarrow \Omega^0_{\mathrm{dR}}(-) \xrightarrow{d} \Omega^1_{\mathrm{dR}}(-) \xrightarrow{d} \cdots \xrightarrow{d} \Omega^n_{\mathrm{dR}}(-) \Big). \tag{9.28}$$

(ii) The de Rham differential in degree 0 gives a morphism of presheaves of complexes

$$\mathrm{DB}^{n+1}_{\bullet} \xrightarrow{(0,0,\ldots,0,d)} \Omega^{n+1}_{\mathrm{dR}}(-)_{\mathrm{clsd}} \tag{9.29}$$

from the Deligne-Beilinson complex (9.28) to the presheaf of closed $(n+1)$-forms, regarded as a presheaf of chain complexes in degree 0.

(iii) *Ordinary differential cohomology* is sheaf hypercohomology with coefficients in the Deligne complex. This means that if we look at the Deligne-Beilinson complex (9.28) as a smooth ∞-stack (Prop. 1.24) by first applying the Dold-Kan construction from Ex. 1.30 and then ∞-stackifying the resulting simplicial presheaf, then ordinary differential cohomology is stacky non-abelian cohomology (Rem. 2.3) with coefficients in the Deligne-Beilinson complex:

$$\overset{\substack{\text{ordinary} \\ \text{differential cohomology}}}{\widehat{H}^{n+1}(\mathscr{X})} := \mathrm{Ho}(\mathrm{SmthStacks}_{\infty})\Big(\mathscr{X}\ ,\ \overset{\substack{\text{Dold-Kan} \\ \text{correspondence}}}{L^{\mathrm{loc}} \circ \mathrm{DK}}\big(\overset{\substack{\text{Deligne-Beilinson} \\ \text{complex}}}{\mathrm{DB}^{n+1}_{\bullet}}\big)\Big). \tag{9.30}$$

(iv) The *curvature map* on ordinary differential cohomology is the cohomology operation induced by (9.29):

$$\begin{array}{ccc}
\overset{\substack{\text{ordinary} \\ \text{differential cohomology}}}{\widehat{H}^{n+1}(\mathscr{X})} & \xrightarrow[\ F\ ]{\text{curvature}} & \Omega^{n+1}_{\mathrm{dR}}(\mathscr{X})_{\mathrm{clsd}} \\
\| & & \| \\
\mathrm{Ho}(\mathrm{SmthStacks}_{\infty})\big(\mathscr{X}, L_{\mathrm{loc}} \circ \mathrm{DK}(\mathrm{DB}^{n+1}_{\bullet})\big) & \xrightarrow{\mathrm{Ho}(\mathrm{SmthStacks}_{\infty})(\mathscr{X}, L_{\mathrm{loc}} \circ \mathrm{DK}(d))} & \mathrm{Ho}(\mathrm{SmthStacks}_{\infty})\big(\mathscr{X}, L_{\mathrm{loc}} \circ \mathrm{DK}(\Omega^{n+1}_{\mathrm{dR}}(-)_{\mathrm{clsd}})\big).
\end{array} \tag{9.31}$$

Proposition 9.5 (Differential non-abelian cohomology subsumes differential ordinary cohomology [Fiorenza *et al.* (2012), Prop. 3.2.26]**).**
Let $n \in \mathbb{N}$ and consider $A = B^n\mathrm{U}(1) \simeq K(\mathbb{Z}, n+1)$ (Ex. 2.9). Then:

(i) *Differential non-abelian A-cohomology (Def. 9.3) coincides with ordinary differential cohomology (Def. 9.4):*

$$\overset{\text{ordinary differential cohomology}}{\widehat{H}^{n+1}(\mathscr{X})} \simeq \widehat{H}\big(\mathscr{X}; B^n\mathrm{U}(1)\big). \tag{9.32}$$

(ii) *The abstract curvature map in differential A-cohomology* (9.10) *reproduces the ordinary curvature map* (9.31).

Proof. In order to compute the defining homotopy pullback (9.10), we use the Dold-Kan correspondence (Prop. 1.27) to obtain a convenient presentation of the differential character map along which to pull back:

(a) Since the Dold-Kan construction DK (Def. 1.30) realizes homotopy groups from homology groups (1.88), and since Eilenberg-MacLane spaces are characterized by their homotopy groups (2.6), we have the vertical identifications on the left of the following diagram:

$$\begin{array}{ccccc}
\mathrm{Disc}(B^{n+1}\mathbb{Z}) & \xrightarrow{\eta^{\mathbb{R}}_{B^{n+1}\mathbb{Z}}} & \mathrm{Disc}(B^{n+1}\mathbb{R}) & \xrightarrow{\simeq} & \flat\mathbf{B}\exp(\mathfrak{b}^n\mathbb{R}) \\
\| & & \| & & \downarrow{\scriptstyle \int_{\Delta^\bullet}}\ \simeq \\
\mathrm{DK}\left(\begin{array}{c}\mathbb{Z}\\ \downarrow\\ 0\\ \downarrow\\ \vdots\\ \downarrow\\ 0\end{array}\right) & \hookrightarrow & \mathrm{DK}\left(\begin{array}{c}\mathbb{R}\\ \downarrow\\ 0\\ \downarrow\\ \vdots\\ \downarrow\\ 0\end{array}\right) & \hookrightarrow & \mathrm{DK}\left(\begin{array}{c}\Omega^0_{\mathrm{dR}}(-)\\ \downarrow d\\ \Omega^1_{\mathrm{dR}}(-)\\ \downarrow d\\ \vdots\\ \downarrow d\\ \Omega^{n+1}_{\mathrm{dR}}(-)_{\mathrm{clsd}}\end{array}\right)
\end{array}
\qquad \text{with } \mathbf{ch}_{B^n\mathrm{U}(1)} \text{ the top composite} \tag{9.33}$$

Under this identification, it is clear that the rationalization map $\eta^{\mathbb{R}}_{B^{n+1}\mathbb{Z}}$ (Def. 5.2) is presented by the canonical inclusion of the integers into the real numbers, as on the bottom left of (9.33).

Moreover, the right vertical equivalence in (9.33) is that from Lem. 9.2.

(b) Since the differential character (9.9) in the present case evidently comes from a morphism of (presheaves of) simplicial abelian groups, with group structure given by addition of ordinary differential forms (Ex. 6.2), we may, using the Dold-Kan correspondence (Prop. 1.27), analyze the remainder of the diagram on normalized chain complexes $N(-)$ (1.87).

Using this, it follows by inspection of the bottom map in (9.9) that the bottom right square in (9.33) commutes, with the bottom morphism on the right being the canonical inclusion of (presheaves of) chain complexes.

Now to use this presentation for identifying the resulting homotopy fiber product (9.10):

(i) Since the DK-construction (Def. 1.30), applied objectwise over CartSp, is a right Quillen functor into the global model structure from Ex. 1.20, and since ∞-stackification preserves homotopy pullbacks (Lem. 1.5), it is now sufficient to show, by definition (9.30), that the homotopy pullback (Def. 1.15) along the bottom map in (9.33), formed in presheaves of chain complexes is the Deligne-Beilinson complex $\mathrm{DB}^{n+1}_\bullet$ (9.28).

For this it is sufficient, by (1.42), to find a fibration replacement of the bottom map in (9.33) whose ordinary fiber product with $\Omega^{n+1}_{\mathrm{dR}}(-)_{\mathrm{clsd}}$ is the Deligne-Beilinson complex. This is provided by a mapping cylinder construction (e.g. [Weibel (1994), §1.5.5]) shown here:

$$
\begin{array}{ccccc}
 & \mathrm{DB}^{n+1}_\bullet = \left(\begin{array}{c} \mathbb{Z} \\ \downarrow^{i} \\ \Omega^0_{\mathrm{dR}}(-) \\ \downarrow^{d} \\ \Omega^1_{\mathrm{dR}}(-) \\ \downarrow^{d} \\ \vdots \\ \downarrow^{d} \\ \Omega^{n-1}_{\mathrm{dR}}(-) \\ \downarrow^{d} \\ \Omega^n_{\mathrm{dR}}(-) \end{array}\right) & \xrightarrow{\left(\begin{smallmatrix} 0 \\ 0 \\ \vdots \\ 0 \\ d \end{smallmatrix}\right)} & \left(\begin{array}{c} 0 \\ \downarrow \\ 0 \\ \downarrow \\ 0 \\ \downarrow \\ \vdots \\ \downarrow \\ 0 \\ \downarrow \\ \Omega^{n+1}_{\mathrm{dR}}(-)_{\mathrm{clsd}} \end{array}\right) \\
 & {}^{i_1}\big\downarrow & \text{(pb)} & \big\downarrow{}^{i} \\
\left(\begin{array}{c} \mathbb{Z} \\ \downarrow \\ 0 \\ \downarrow \\ 0 \\ \downarrow \\ \vdots \\ \downarrow \\ 0 \\ \downarrow \\ 0 \end{array}\right) \xrightarrow[\in \mathrm{W}]{n \mapsto (n,n)} & \left(\begin{array}{ccc} \mathbb{Z} & \oplus & \Omega^0_{\mathrm{dR}}(-) \\ \downarrow^{i} & \swarrow_{-\mathrm{id}} & \downarrow^{d} \\ \Omega^0_{\mathrm{dR}}(-) & \oplus & \Omega^1_{\mathrm{dR}}(-) \\ \downarrow^{d} & \swarrow_{+\mathrm{id}} & \downarrow^{d} \\ \Omega^1_{\mathrm{dR}}(-) & \oplus & \Omega^2_{\mathrm{dR}}(-) \\ \downarrow^{d} & \swarrow_{-\mathrm{id}} & \downarrow^{d} \\ \vdots & \vdots & \vdots \\ \downarrow^{d} & \swarrow & \downarrow^{d} \\ \Omega^{n-1}_{\mathrm{dR}}(-) & \oplus & \Omega^n_{\mathrm{dR}}(-) \\ \downarrow^{d} & \swarrow & \\ \Omega^n_{\mathrm{dR}}(-) & & \end{array}\right) & \xrightarrow[\in \mathrm{Fib}]{\left(\begin{smallmatrix} \mathrm{pr}_2 \\ \mathrm{pr}_2 \\ \vdots \\ \mathrm{pr}_2 \\ d \end{smallmatrix}\right)} & \left(\begin{array}{c} \Omega^0_{\mathrm{dR}}(-) \\ \downarrow^{d} \\ \Omega^1_{\mathrm{dR}}(-) \\ \downarrow^{d} \\ \Omega^2_{\mathrm{dR}}(-) \\ \downarrow^{d} \\ \vdots \\ \downarrow^{d} \\ \Omega^n_{\mathrm{dR}}(-) \\ \downarrow^{d} \\ \Omega^{n+1}_{\mathrm{dR}}(-)_{\mathrm{clsd}} \end{array}\right)
\end{array}
\tag{9.34}
$$

By direct inspection, we see in this diagram that:

- the total bottom morphism is the total bottom morphism from (9.33), factored as a weak equivalence (quasi-isomorphism) followed by a fibration (positive degreewise surjection);
- the ordinary pullback of this fibration is the Deligne-Beilinson complex $\mathrm{DB}^{n+1}_\bullet$ (9.28), as shown, which therefore represents the homotopy pullback (since all chain complexes are projectively fibrant), by Def. 1.15.
- the top morphism out of this (homotopy-)pullback coincides with the curvature map (9.29) on the Deligne complex – which, under the following implication of claim (i), implies claim (ii).

(ii) The image of this homotopy pullback (9.34) under $L^{\mathrm{loc}} \circ \mathrm{DK}$ is still a homotopy pullback (because DK is a right Quillen functor by construction (1.91) and using Lem. 1.5) and hence exhibits the Deligne coefficients (9.30) for ordinary differential cohomology as a model for the differential $B^{n+1}\mathbb{Z}$-cohomology according to Def. 9.3:

$$\begin{array}{ccc} \overset{\text{Deligne complex as smooth }\infty\text{-stack}}{L^{\mathrm{loc}} \circ \mathrm{DK}\big(\mathrm{DB}^{n+1}_{\bullet}\big)} & \xrightarrow{\quad F_{B^{n+1}\mathbb{Z}} \quad} & \Omega_{\mathrm{dR}}\big(-; \mathfrak{l}B^{n+1}\mathbb{Z}\big)_{\mathrm{flat}} \\ \Big\downarrow{\scriptstyle c_{B^{n+1}\mathbb{Z}}} & \text{(hpb)} & \Big\downarrow \\ \mathrm{Disc}\big(B^{n+1}\mathbb{Z}\big) & \xrightarrow[\quad \mathbf{ch}_{B^{n+1}\mathbb{Z}} \quad]{} & \flat\mathfrak{B}\exp\big(\mathfrak{l}B^{n+1}\mathbb{Z}\big) \end{array} \tag{9.35}$$

This implies claim (i), by the definitions. □

Remark 9.6 (The commuting square of ordinary (differential) cohomology groups). The image of the homotopy-pullback square (9.35) under the hom-functor $\mathrm{Ho}(\mathrm{SmthStacks}_\infty)(X,-)$ out of a smooth manifold X gives the commuting square of ordinary (differential) cohomology groups shown in (9.27). Since the hom-functor of a homotopy category does not preserve homotopy pullbacks, in general (only the mapping space functor (1.79) does), the square (9.27) in cohomology is not itself a pullback, in general.

Secondary non-abelian cohomology operations. We define secondary non-abelian cohomology operations (Def. 9.7 below) which generalize the classical notion of secondary characteristic classes (Thm. 9.9, see Rem. 9.7 for the terminology) to higher non-abelian cohomology. To formulate the concept in this generality, we need a technical condition (Def. 9.6) which happens to be trivially satisfied in the classical case (Lem. 9.3 below):

Definition 9.6 (Absolute minimal model). For $A_1, A_2 \in \mathrm{Ho}\big(\Delta\mathrm{Sets}_{\mathrm{Qu}}\big)^{\mathrm{fin}_{\mathbb{Q}}}_{\geq 1,\mathrm{nil}}$ (Def. 5.1), we say that an *absolute minimal model* for a morphism $A_1 \overset{c}{\dashrightarrow} A_2$ in $\Delta\mathrm{Sets}$ is a morphism $\mathfrak{l}A_1 \overset{\mathfrak{c}}{\dashrightarrow} \mathfrak{l}A_2$ between the respective Whitehead L_∞-algebras (Prop. 5.11) which makes this square

$$\begin{array}{ccc} \Omega^\bullet_{\mathrm{dRPL}}(A_1) & \xleftarrow{\;p^{\min}_{A_1}\;} & \mathrm{CE}\big(\mathfrak{l}A_1\big) \\ \Big\uparrow & & \Big\uparrow{\scriptstyle \mathfrak{c}} \\ \Omega^\bullet_{\mathrm{dRPL}}(A_2) & \xleftarrow[\;p^{\min}_{A_2}\;]{} & \mathrm{CE}\big(\mathfrak{l}A_2\big) \end{array} \quad \in \; \mathrm{dgcAlgs}^{\geq 0}_{\mathbb{R}}$$

and hence the square on the right of the following diagram commute:

$$\begin{array}{ccccc} & & \mathbb{D}\eta^{\mathrm{PLdR}}_{A_1} & & \\ A_1 & \xrightarrow{\;\eta^{\mathrm{PLdR}}_{A_1}\;} & B\exp_{\mathrm{PL}} \circ \Omega^\bullet_{\mathrm{PLdR}}(A_1) & \xrightarrow{\;B\exp_{\mathrm{PL}}(p^{\min}_{A_1})\;} & B\exp_{\mathrm{PL}} \circ \mathrm{CE}(\mathfrak{l}A_1) \\ \Big\downarrow{\scriptstyle c} & & \Big\downarrow{\scriptstyle B\exp_{\mathrm{PL}} \circ \Omega^\bullet_{\mathrm{PLdR}}(c)} & & \Big\downarrow{\scriptstyle B\exp_{\mathrm{PL}} \circ \mathrm{CE}(\mathfrak{c})} \\ A_2 & \xrightarrow{\;\eta^{\mathrm{PLdR}}_{A_2}\;} & B\exp_{\mathrm{PL}} \circ \Omega^\bullet_{\mathrm{PLdR}}(A_2) & \xrightarrow{\;B\exp_{\mathrm{PL}}(p^{\min}_{A_2})\;} & B\exp_{\mathrm{PL}} \circ \mathrm{CE}(\mathfrak{l}A_2), \\ & & \mathbb{D}\eta^{\mathrm{PLdR}}_{A_2} & & \end{array}$$

$$\in \; \Delta\mathrm{Sets} \tag{9.36}$$

hence a morphism that yields a transformation between exactly those derived adjunction units $\mathbb{D}\eta^{\mathrm{PLdR}}$ (1.38) of the PL-de Rham adjunction (5.14) that are given by *minimal* fibrant replacement.[1] In this case, the commuting diagram (9.36) evidently extends to a strict transformation between the differential non-abelian characters (9.9) on the A_i (Def. 9.2), in that the following diagram of simplicial presheaves (Def. 1.22) commutes:

$$\begin{array}{ccc} \mathrm{Disc}(A_1) & \xrightarrow{\ \mathbf{ch}_{A_1}\ } & \flat\mathbf{B}\exp(\mathfrak{l}A_1) \\ {\scriptstyle \mathrm{Disc}(c)}\big\downarrow & & \big\downarrow{\scriptstyle \flat\mathbf{B}\exp(\mathfrak{c})} \\ \mathrm{Disc}(A_2) & \xrightarrow{\ \mathbf{ch}_{A_2}\ } & \flat\mathbf{B}\exp(\mathfrak{l}A_2) \end{array} \quad \in \mathrm{PSh}\big(\mathrm{CartSp}, \Delta\mathrm{Sets}\big). \tag{9.37}$$

In differential enhancement of Def. 2.3, we have:

Definition 9.7 (Secondary non-abelian cohomology operation). Let $A_1 \xrightarrow{c} A_2$ in ΔSets, with induced cohomology operation (Def. 2.3)

$$H(-;A_1) \xrightarrow{\ c_*\ } H(-;A_2),$$

have an absolute minimal model $\mathfrak{c}$ (Def. 9.6). Then the corresponding *secondary non-abelian cohomology operation* is the structured cohomology operation (Rem. 2.3)

$$\widehat{H}(-;A_1) \xrightarrow[\text{secondary non-abelian character}]{(c_{\mathrm{diff}})_*} \widehat{H}(-;A_2) \tag{9.38}$$

on differential non-abelian cohomology (Def. 9.3) which is induced, as in (2.25), by the dashed morphism c_{diff} in the following diagram, which in turn is induced from c and $\mathfrak{c}$ (9.37) by the universal property of the defining homotopy pullback operation (9.9):

$$\begin{array}{ccccccc}
\text{secondary/differential cohomology operation} & (A_1)_{\mathrm{diff}} & \overset{c_{\mathrm{diff}}}{\dashrightarrow} & & (A_2)_{\mathrm{diff}} & & \\
& & \searrow^{F_{A_1}} & & & \searrow^{F_{A_2}} & \\
& {\scriptstyle c_{A_1}}\downarrow & & \Omega_{\mathrm{dR}}(-;\mathfrak{l}A_1)_{\mathrm{flat}} & \xrightarrow{\ \mathfrak{c}_*\ } {\scriptstyle c_{A_2}}\downarrow & & \Omega_{\mathrm{dR}}(-;\mathfrak{l}A_2)_{\mathrm{flat}} \\
\text{plain/primary cohomology operation} & \mathrm{Disc}(A_1) & \xrightarrow{\ \mathrm{Disc}(c)\ } & \downarrow & \mathrm{Disc}(A_2) & & \downarrow \\
\text{transformation of differential characters} & & \searrow^{\mathbf{ch}_{A_1}} & & & \searrow^{\mathbf{ch}_{A_2}} & \\
& & & \flat\mathbf{B}\exp(\mathfrak{l}A_1) & \xrightarrow{\ \mathfrak{c}_*\ } & & \flat\mathbf{B}\exp(\mathfrak{l}A_2).
\end{array} \tag{9.39}$$

[1]Notice that the existence of morphisms $\mathfrak{c}$ making this diagram commute is not guaranteed; it is only the existence of the *relative* minimal morphism $\mathfrak{l}_{A_2}(c)$ from Prop. 5.16 which is guaranteed to make the square (5.53) commute.

The left and right squares are the homotopy pullback squares defining differential non-abelian cohomology (Def. 9.3) while the bottom square is the transformation of differential non-abelian characters (Def. 9.2) from (9.37).

In differential enhancement of Ex. 2.21, 7.5, we have:

Example 9.5 (Secondary non-abelian Hurewicz/Boardman homomorphism to differential K-theory). Consider the map

$$S^4 \xrightarrow{\; e^4_{BU} \;} BU \qquad \in \mathrm{Ho}\big(\Delta\mathrm{Sets}_{\mathrm{Qu}}\big)$$

from the 4-sphere to the classifying space of the infinite unitary group (2.16) which classifies a generator in $\pi_4(BU) \simeq \mathbb{Z}$. By Ex. 5.3 and Ex. 5.6, 6.6 the corresponding Whitehead L_∞-algebras (Prop. 5.11) are as shown here:

$$\begin{array}{ccc}
\mathrm{CE}\big(\mathfrak{l}S^4\big) & \longleftarrow & \mathrm{CE}\big(\mathfrak{l}BU\big) \simeq \bigotimes_{k\in\mathbb{N}} \mathrm{CE}\big(\mathfrak{l}K(\mathbb{Z},2k)\big) \\
\Vert & & \Vert \\
\mathbb{R}\left[\begin{smallmatrix}\omega_7, \\ \omega_4\end{smallmatrix}\right] \Big/ \left(\begin{array}{l} d\,\omega_7 = -\omega_4 \wedge \omega_4 \\ d\,\omega_4 = 0 \end{array}\right) & \xleftarrow{\;\left.\begin{smallmatrix}\omega_4 & 2k=4 \\ 0 & \text{else}\end{smallmatrix}\right\} \mapsfrom f_{2k}\;} & \mathbb{R}\left[\begin{smallmatrix}\vdots \\ f_4, \\ f_2\end{smallmatrix}\right] \Big/ \left(\begin{array}{c} \vdots \\ d f_4 = 0 \\ d f_2 = 0 \end{array}\right)
\end{array} \tag{9.40}$$

The morphism shown in (9.40) evidently restricts to the relative rational Whitehead L_∞-algebra inclusion (Prop. 5.16) on the factor $K(\mathbb{R},4) \subset L_{\mathbb{R}}BU$ and is zero elsewhere, hence fits into the required diagram (9.36) exhibiting it as an absolute minimal model (Def. 9.6) for e^4_{BU} (by the commuting diagram in Prop. 4.24).

Cheeger-Simons homomorphism. Where the construction of the Chern-Weil homomorphism (Def. 8.3) invokes connections on principal bundles without actually being sensitive to this choice (by Prop. 8.5), the *Cheeger-Simons homomorphism* [Cheeger and Simons (1985), §2][Hopkins and Singer (2005), §3.3] (based on [Chern and Simons (1974)]) is a refinement of the Chern-Weil homomorphism, now taking values in differential ordinary cohomology (Ex. 9.4), that does detect connection data (hence "differential" data):

$$\begin{array}{ccc}
G\mathrm{Connections}(X)_{/\sim} & \xrightarrow[\;\mathrm{cs}_G\;]{\text{Cheeger-Simons homomorphism}} & \mathrm{Hom}_{\mathbb{Z}}\Big(H^\bullet(BG;\mathbb{Z}),\, \underbrace{\widehat{H}^\bullet(X)}_{\text{differential cohomology}}\Big) \\
{\scriptstyle\text{forget connection}}\big\downarrow & & \big\downarrow{\scriptstyle\text{curvature map}} \\
G\mathrm{Bundles}(X)_{/\sim} & \xrightarrow[\;\text{Chern-Weil homomorphism}\;]{\mathrm{cw}_G} & \mathrm{Hom}_{\mathbb{R}}\Big(\mathrm{inv}^\bullet(\mathfrak{g}),\, \underbrace{H^\bullet_{\mathrm{dR}}(X)}_{\text{de Rham cohomology}}\Big)
\end{array} \tag{9.41}$$

We discuss how the general notion of secondary non-abelian cohomology operations (Def. 9.7) specializes on ordinary principal bundles to the Cheeger-Simons homomorphism, and hence generalizes it to higher non-abelian cohomology:

Lemma 9.3 (Characteristic classes of G-principal bundles have absolute minimal models). *Let G be a connected compact Lie group with classifying space BG (2.8). For $n \in \mathbb{N}$, let $[c] \in H^{n+1}(BG;\mathbb{Z})$ be an indecomposable universal integral characteristic class for G-principal bundles (Ex. 2.3). Then every representative classifying map*

$$BG \xrightarrow{\;c\;} B^{n+1}\mathbb{Z}$$

has an absolute minimal model in the sense of Def. 9.6.

Proof. By Lem. 8.2, the minimal Sullivan model for BG has vanishing differential, while the minimal Sullivan model of $B^{n+1}\mathbb{Z}$ is a polynomial algebra on a single degree $n+1$ generator (by Ex. 5.4), whose inclusion is already the relative minimal Sullivan model $\mathfrak{l}_{B^{n+1}\mathbb{Z}}(c)$ (Prop. 5.16) of c. Therefore, setting

$$\mathrm{CE}(\mathfrak{c}) := \mathrm{CE}\big(\mathfrak{l}_{B^{n+1}\mathbb{Z}}(c)\big) \;:\; \mathbb{R}[c]/(d\,c=0) \hookrightarrow \mathrm{inv}^{\bullet}(\mathfrak{g}) \tag{9.42}$$

gives the required morphism of minimal models that makes makes the square (9.36) commute, by (5.53). □

In differential enhancement of Ex. 2.14 we have:

Definition 9.8 (Secondary characteristic classes of differential non-abelian G-cohomology). Let G be a connected compact Lie group with classifying space BG (2.8). By Lem. 9.3), the construction of secondary characteristic classes (Def. 9.7, on differential non-abelian G-cohomology (Ex. 9.4) yields a $\mathbb{Z}$-linear map of the form

$$H^{\bullet}(BG,;\mathbb{Z}) \simeq H\big(BG;B^{\bullet}\mathbb{Z}\big) \xrightarrow{(-)_{\mathrm{diff}}} \widehat{H}\big(BG_{\mathrm{diff}};B^{\bullet}\mathbb{Z}\big) = H\big(BG_{\mathrm{diff}};B^{\bullet}\mathbb{Z}_{\mathrm{diff}}\big),$$

where on the right we have the ordinary differential non-abelian cohomology (Prop. 9.5) *of* the moduli ∞-stack BG_{diff} (9.10). Combined with the composition operation in $\mathrm{Ho}(\mathrm{SmthStacks}_{\infty})$ (Def. 1.25) this gives a map

$$\begin{array}{ccc} \widehat{H}\big(X;BG\big)\times H\big(BG;B^{\bullet}\mathbb{Z}\big) \xrightarrow{\mathrm{id}\times(-)_{\mathrm{diff}}} & H\big(X;BG_{\mathrm{diff}}\big)\times H\big(BG_{\mathrm{diff}};B^{\bullet}\mathbb{Z}_{\mathrm{diff}}\big) & \\ & \big\downarrow \circ & \\ & H\big(X;B^{\bullet}\mathbb{Z}_{\mathrm{diff}}\big) = & \widehat{H}\big(X;B^{\bullet}\mathbb{Z}\big) \end{array}$$

which is $\mathbb{Z}$-linear in its second argument, and whose hom-adjunct is

$$\widehat{H}(X;BG) \xrightarrow{\nabla \mapsto (c \mapsto c_{\mathrm{diff}}(\nabla))} \mathrm{Hom}_{\mathbb{Z}}\big(H(BG;B^{\bullet}\mathbb{Z}),\widehat{H}(X;B^{\bullet}\mathbb{Z})\big). \tag{9.43}$$

Theorem 9.9 (Secondary non-abelian cohomology operations subsume Cheeger-Simons homomorphism). *Let G be a connected compact Lie group, with classifying* *space denoted BG (2.8). Then the canonical construction (9.43) of secondary characteristic classes on differential non-abelian G-cohomology (Def. 9.8) coincides with the Cheeger-Simons homomorphim (9.41), in that the following diagram commutes:*

$$\begin{array}{ccc} G\mathrm{Connections}(X)_{/\sim} & \xrightarrow[\mathrm{cs}_G]{\text{Cheeger-Simons homomorphism}} & \mathrm{Hom}_{\mathbb{Z}}\Big(H^\bullet(BG;\mathbb{Z}),\, \underset{\text{differential ordinary cohomology}}{\widehat{H}^\bullet(X)}\Big) \\ {\scriptstyle (9.20)}\big\downarrow & & \big\uparrow{\scriptstyle \simeq\ (9.32)} \\ \underset{\text{differential non-abelian cohomology}}{\widehat{H}(X;BG)} & \xrightarrow[\text{secondary non-abelian cohomology operations}]{\nabla \mapsto (c \mapsto c_{\mathrm{diff}}(\nabla))} & \mathrm{Hom}_{\mathbb{Z}}\Big(H(BG;B^\bullet\mathbb{Z}),\, \widehat{H}(X;B^\bullet\mathbb{Z})\Big)\,, \end{array} \tag{9.44}$$

where on the left we have the map from G-connections to differential non-abelian G-cohomology from Prop. 9.4, and on the right the identification of ordinary differential cohomology from Prop. 9.5.

Proof. Let $c \in H\big(BG;\, B^\bullet\mathbb{Z}\big)$ be a characteristic class, and let (f^*EG, ∇) be a G-principal bundle equipped with a G-connection. By Prop. 9.4, its image in differential non-abelian cohomology is given by the first map in the following diagram

$$\begin{array}{ccccc} G\mathrm{Connections}(X)_{/\sim} & \longrightarrow & \widehat{H}(X;BG) & \xrightarrow{(c_{\mathrm{diff}})_*} & \widehat{H}\big(X;B^{n+1}\mathbb{Z}\big) \simeq \widehat{H}^{n+1}(X) \\ \big[f^*EG,\nabla\big] & \longmapsto & \big[f,\, \big(\mathrm{cs}_k(\nabla, f^*\nabla_{\mathrm{univ}})\big)\big(\omega_k(F_\nabla)\big)\big] & \longmapsto & \big[f^*c,\, \mathrm{cs}_c(\nabla, f^*\nabla_{\mathrm{univ}}),\, c(F_\nabla)\big] \end{array} \tag{9.45}$$

Here the triple of data are the three components (Ex. 1.12) of a map into the defining homotopy pullback of differential non-abelian cohomology (9.22). Therefore, the secondary operation induced by the transformation (9.39) of these homotopy pullbacks, which in the present case is of this form:

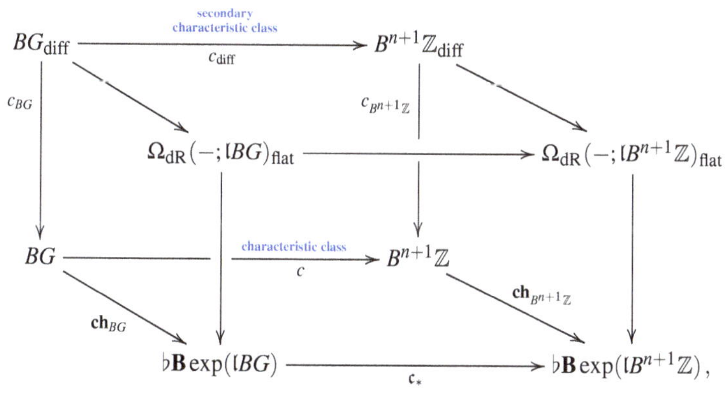

(9.46)

acts **(a)** on the first component in the triple by postcomposition with c, hence as

$$f \mapsto f^*c := c \circ f$$

and **(b)** on the other two components by composition with $\mathfrak{c}$, which by (9.42) corresponds to projecting out the Chern-Simons form and characteristic form corresponding to c, respectively. This is shown as the second map in (9.45). Hence we are reduced to showing that the total map in (9.45) gives the Cheeger-Simons homomorphism. This statement is the content of [Hopkins and Singer (2005), §3.3]. □

Remark 9.7 (Secondary characteristic classes of G-connections). The traditional reason for referring to the Cheeger-Simons homomorphism (9.44) as producing *secondary* invariants is that Cheeger-Simons classes $\mathrm{cs}_G(P,\nabla) \in \widehat{H}(X)$ may be non-trivial even if the underlying characteristic class $\mathrm{cw}_G(P)$ (the "primary" class) vanishes. In this case the $\mathrm{cs}_G(P,\nabla)$ are also called *Chern-Simons invariants*.

(i) This happens, in particular, when the G-connection ∇ is flat, $F(\nabla) = 0$ (by Def. 8.2). Such secondary Chern-Simons invariants exhibit some subtle phenomena ([Reznikov (1995)][Reznikov (1996)][Iyer and Simpson (2007)][Esnault (2009)]).

(ii) In fact, the proof of Thm. 9.44, via the triples (9.22) of homotopy data, shows that, in this case, $\mathrm{cs}_G(P,\nabla)$ measures *how* (or "why") $\mathrm{cw}_G(P)$ vanishes, namely by which class of homotopies.

(iii) Here we may understand secondary classes more abstractly, and explicitly related to the non-abelian character map: Where a (primary) non-abelian cohomology operation, according to Def. 2.3, is induced by a morphism of coefficient spaces (2.20), a secondary non-abelian cohomology operation, according to Def. 9.7, is induced (9.38) by a morphism of non-abelian character maps (9.37) – hence by a morphism of morphisms – on these coefficient spaces.

(iv) Note that classical secondary cohomology operations themselves admit differential refinements. For instance, for the case of Massey products as secondary operations for the cup product this is worked out in [Grady and Sati (2018a)]. While these can also fit into our context on general grounds, we will not demonstrate that explicitly here.

PART V

The twisted (differential) non-abelian character map

We introduce the character map in twisted non-abelian cohomology (Def. V.3) and then discuss how it specializes to:

Chapter 10 – the twisted Chern character on (higher) K-theory;
Chapter 12 – the twisted character on Cohomotopy theory.

Rationalization in twisted non-abelian cohomology. In generalization of Def. IV.1 we now define rationalization of local coefficient bundles (3.2). This operation is transparent in the language of ∞-category theory (Rem. 1.2), where it simply amounts to forming the pasting composite with the homotopy-coherent naturality square of the $\mathbb{R}$-rationalization unit $\eta^{\mathbb{R}}$ (from Def. 5.7):

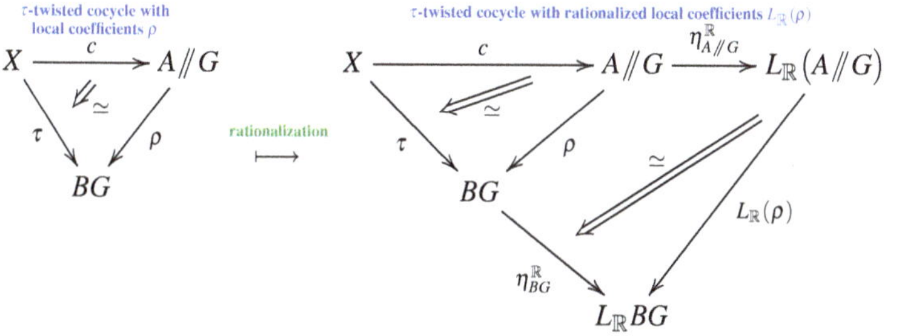

(V.1)

Slightly less directly but equivalently, this is the composite of (**a**) derived base change (Ex. 1.7) along $\eta^{\mathbb{R}}_{BG}$ from the slice over BG to the slice over $L_{\mathbb{R}}BG$, (**b**) followed by the composition with its derived naturality square, now regarded as a morphism in the slice over $L_{\mathbb{R}}BG$:

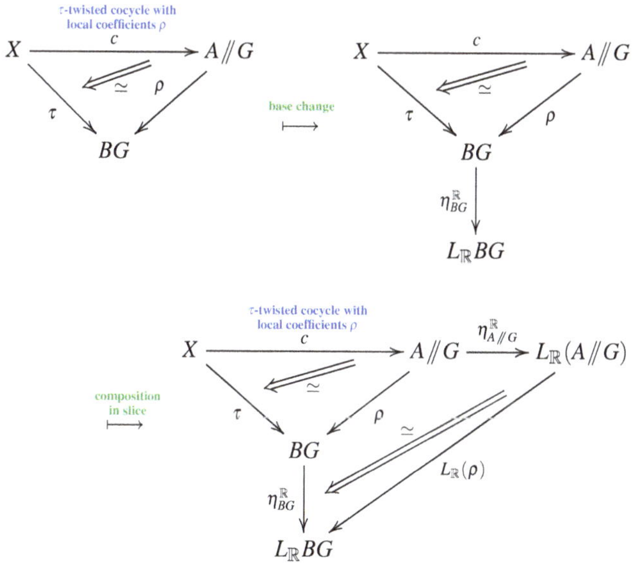

(V.2)

It is in this second form that the operation lends itself to formulation in model category theory (Def. V.2 below). For that we just need to produce a rectified (strictly commuting) model of the $\eta^{\mathbb{R}}$-naturality square:

Definition V.1 (Rectified rationalization unit on coefficient bundle). Consider a local coefficient bundle (3.2) in $\mathrm{Ho}\big(\Delta\mathrm{Sets}_{\mathrm{Qu}}\big)^{\mathrm{fin}_{\mathbb{Q}}}_{\geq 1,\mathrm{nil}}$ (Def. 5.1) with its minimal relative Sullivan model (5.53), (given by Prop. 5.16)

$$
\begin{array}{ccc}
A & \longrightarrow & A/\!\!/G \\
 & {\scriptstyle \text{local coefficient bundle}} & \Big\downarrow{\scriptstyle \rho} \\
 & & BG,
\end{array}
\qquad
\begin{array}{ccc}
\Omega^\bullet_{\mathrm{PLdR}}\big(A/\!\!/G\big) & \xleftarrow[\in \mathrm{W}]{\;p^{\mathrm{min}_{BG}}_{A/\!\!/G}\;} & \mathrm{CE}\big(\mathfrak{l}_{BG}(A/\!\!/G)\big) \\
{\scriptstyle \Omega^\bullet_{\mathrm{PLdR}}(\rho)}\Big\uparrow & & \Big\uparrow{\scriptstyle \mathrm{CE}(\mathfrak{l}p)} \\
\Omega^\bullet_{\mathrm{PLdR}}\big(BG\big) & \xleftarrow[\in \mathrm{W}]{\;p^{\mathrm{min}}_{BG}\;} & \mathrm{CE}\big(\mathfrak{l}(BG)\big)
\end{array}
\tag{V.3}
$$

Then the composite of the image of (V.3) under $B\exp_{\mathrm{PL}}$ with the $\Omega^\bullet_{\mathrm{PLdR}} \dashv B\exp_{\mathrm{PL}}$-adjunction unit (from Prop. 5.5):

$$
\mathbb{D}\eta^{\mathrm{PLdR}}_{\rho} \;:=\;
\begin{array}{ccccc}
 & & \xrightarrow{\;\mathbb{D}\eta^{\mathrm{PLdR}}_{A/\!\!/G} \,\simeq\, \eta^{\mathbb{R}}_{A/\!\!/G}\;} & & \\
A/\!\!/G & \xrightarrow{\;\eta^{\mathrm{PLdR}}_{A/\!\!/G}\;} & B\exp_{\mathrm{PL}} \circ \Omega^\bullet_{\mathrm{PLdR}}\big(A/\!\!/G\big) & \xrightarrow{\;B\exp_{\mathrm{PL}}(p^{\mathrm{min}_{BG}}_{A/\!\!/G})\;} & B\exp_{\mathrm{PL}} \circ \mathrm{CE}\big(\mathfrak{l}_{BG}(A/\!\!/G)\big) \\
{\scriptstyle \rho}\Big\downarrow & & \Big\downarrow{\scriptstyle B\exp_{\mathrm{PL}} \circ \Omega^\bullet_{\mathrm{PLdR}}(\rho)} & & \Big\downarrow{\scriptstyle B\exp_{\mathrm{PL}} \circ \mathrm{CE}(\mathfrak{l}p)} \\
BG & \xrightarrow{\;\eta^{\mathrm{PLdR}}_{BG}\;} & B\exp_{\mathrm{PL}} \circ \Omega^\bullet_{\mathrm{PLdR}}\big(BG\big) & \xrightarrow{\;B\exp_{\mathrm{PL}}(p^{\mathrm{min}}_{BG})\;} & B\exp_{\mathrm{PL}} \circ \mathrm{CE}\big(\mathfrak{l}(BG)\big) \\
 & & \xrightarrow[\;\mathbb{D}\eta^{\mathrm{PLdR}}_{BG} \,\simeq\, \eta^{\mathbb{R}}_{BG}\;]{} & &
\end{array}
\tag{V.4}
$$

is, after passage (1.24) to the classical homotopy category (Ex. 1.14), the naturality square of the rationalization unit on ρ (5.1), namely of the derived adjunction unit (1.38) $\eta^{\mathbb{R}} = \mathbb{D}^{\mathrm{PL}\mathbb{R}\mathrm{dR}}$ (using, with Prop. 4.21, that the right part of (V.4) is the image under $B\exp_{\mathrm{PL}}$ of a fibrant replacement morphism.)

Definition V.2 (Rationalization in twisted non-abelian cohomology). Given a local coefficient bundle ρ and its rectified rationalization unit $\mathbb{D}\eta^{\mathrm{PLdR}}_{\rho}$ (Def. V.1) we say that *rationalization* in twisted non-abelian cohomology with local coefficients ρ (Def. 3.2) is the twisted non-abelian cohomology operation (Def. 3.6)

$$
\big(\eta^{\mathbb{R}}_{\rho}\big)_* \;:\; H^{\tau}(X;A) \xrightarrow{\;(\mathbb{D}\eta^{\mathrm{PLdR}}_{\rho} \circ (-)) \,\circ\, \mathbb{D}(\eta^{\mathbb{R}}_{BG})_!\;} H^{L_{\mathbb{R}}\tau}\big(X; L_{\mathbb{R}}A\big)
\tag{V.5}
$$

given by the composite of

(a) derived left base change $\mathbb{D}(\eta^{\mathbb{R}}_{BG})_!$ (Ex. 1.7) along the rationalization unit (5.1) on the classifying space of twists,
(b) with the rectified rationalization unit (V.4) on the coefficient bundle, regarded as a morphism in the homotopy category (1.24) of the slice model category (Ex. 1.5) of $\Delta\mathrm{Sets}_{\mathrm{Qu}}$ (Ex. 1.2) over $B\exp_{\mathrm{PL}} \circ \mathrm{CE}(\mathfrak{l}BG)$).

Remark V.1 (Rationalization of coefficients and/or of twists). Def. V.2 rationalizes both the coefficients as well as their twist. This is of interest because:

(a) the joint rationalization is defined canonically, in fact functorically, as highlighted around (V.1);
(b) the rationalized twisting appears in the archetypical examples (such as the twisted Chern character on degree-3 twisted K-theory, Chapter 10) and gives the *Bianchi identities* on higher form field/flux data relevant in applications to physics (see Chapter 12).

One may also consider rationalization of just the coefficients, keeping non-rationalized twists; but, in general, this requires making a choice, namely a choice of dashed morphisms in the following transformation diagram of local coefficient bundles:

$$
\begin{array}{ccc}
\underset{\text{local coefficient bundle}}{A} & \xrightarrow{\ \eta_A^{\mathbb{R}}\ } & \underset{\mathbb{R}\text{-rationalized local coefficients}}{L_{\mathbb{R}}(A)} \\
\big\downarrow{\scriptstyle \mathrm{hofib}(\rho)} & & \big\downarrow{\scriptstyle \mathrm{hofib}(\rho^{\mathbb{R}})} \\
A /\!\!/ G & \dashrightarrow & (L_{\mathbb{R}}A) /\!\!/ G \\
\big\downarrow{\scriptstyle \rho} & & \big\downarrow{\scriptstyle \rho^{\mathbb{R}}} \\
BG & = & \underset{\text{non-rationalized twist}}{BG}
\end{array}
\quad \in \mathrm{Ho}\big(\Delta\mathrm{Sets}_{\mathrm{Qu}}\big). \qquad (V.6)
$$

The homotopy-commutativity of the bottom square expresses that and how *rationalization commutes with twisting*.

Given such a choice, then using the bottom square (V.6) in place of the rationalization unit's naturality square on the right of (V.1) produces a definition, directly analogous to Def. V.2, of rationalization of just the A-coefficients in twisted A-cohomology. This is also of interest (see for instance the case of twisted KO-theory in [Grady and Sati (2019d), Prop. 4]), but currently we do not further expand on this generalization here.

Twisted non-abelian character map. In generalization of Def. IV.2, we set:

Definition V.3 (Twisted non-abelian character map). Let $X \in \mathrm{Ho}\big(\Delta\mathrm{Sets}_{\mathrm{Qu}}\big)^{\mathrm{fin}_{\mathbb{Q}}}_{\geq 1,\mathrm{nil}}$ (Def. 5.1) equipped with the structure of a smooth manifold, and

$$
\begin{array}{ccc}
A & \longrightarrow & A /\!\!/ G \\
 & {\scriptstyle \text{local coefficient bundle}} & \big\downarrow{\scriptstyle \rho} \\
 & & BG
\end{array}
\qquad (V.7)
$$

be a local coefficient bundle (3.2) in $\mathrm{Ho}\big(\Delta\mathrm{Sets}_{\mathrm{Qu}}\big)^{\mathrm{fin}_{\mathbb{Q}}}_{\geq 1,\mathrm{nil}}$ (Def. 5.1). Then the *twisted non-abelian character map* in twisted non-abelian cohomology is the twisted cohomology operation

$$
\underset{\substack{\text{twisted}\\ \text{non-abelian}\\ \text{character map}}}{\mathrm{ch}_{\rho}} : \underset{\substack{\text{twisted non-abelian}\\ \text{cohomology}}}{H^{\tau}(X; A)} \xrightarrow[\text{rationalization}]{(\eta_{\rho}^{\mathbb{R}})_*} \underset{\substack{\text{twisted non-abelian}\\ \text{real cohomology}}}{H^{L_{\mathbb{R}}\tau}\big(X; L_{\mathbb{R}}A\big)} \xrightarrow[\substack{\text{twisted non-abelian}\\ \text{de Rham theorem}}]{\simeq} \underset{\substack{\text{twisted non-abelian}\\ \text{de Rham cohomology}}}{H_{\mathrm{dR}}^{\tau_{\mathrm{dR}}}(X; \mathfrak{l}A)} \qquad (V.8)
$$

from twisted non-abelian A-cohomology (Def. 3.2) to twisted non-abelian de Rham cohomology (Def. 6.9) with local coefficients in the rational relative Whitehead L_∞-algebra $\mathfrak{l}\rho$ of ρ (Prop. 5.18) which is the composite of

(i) the operation (V.5) of rationalization of local coefficients (Def. V.2),
(ii) the equivalence (6.51) of the twisted non-abelian de Rham theorem (Thm. 6.15).

Chapter 10

Twisted Chern character on higher K-theory

We discuss (Prop. 10.1) how the twisted non-abelian character map reproduces the twisted Chern character in twisted topological K-theory [Bouwknegt *et al.* (2002), §6.3][Mathai and Stevenson (2003)][Atiyah and Segal (2006), §7] – see also [Tu and Xu (2006)][Mathai and Stevenson (2006), §6][Freed *et al.* (2008), §2][Bressler *et al.* (2008)] [Gomi and Terashima (2010), §4][Karoubi (2012), §8.3][Grady and Sati (2019c), §3.2][Grady and Sati (2019b)]. Then we also consider (Prop. 10.2) the twisted character on twisted iterated K-theory [Lind *et al.* (2020), §2.2].

Character maps on higher-twisted ordinary K-theories.

Remark 10.1 (Twisted Chern character via twisted characteristic forms). The twisted Chern character on twisted K-theory was first proposed in [Bouwknegt *et al.* (2002)] via a natural twisted generalization of the component-wise construction (7.4) of the ordinary Chern character in terms of characteristic curvature forms. Briefly, given a degree-3 twist $\tau_3 \in H^3(X;\mathbb{Z})$ on a (compact) smooth manifold, then every class $[(V_1,V_2)] \in \mathrm{KU}^{\tau_3}(X)$ in τ_3-twisted KU-theory (Ex. 3.4) may be represented by a pair (V_1,V_2) of τ_3-twisted complex vector bundles ([Lupercio and Uribe (2004), §7.2]), generally of infinite rank ("bundle gerbe modules" [Bouwknegt *et al.* (2002), §4]). Now given a choice of lift of τ_3 through the characteristic class map (9.27) to a Deligne cocycle $[h_0,A_1,B_2,H_3]$ (Ex. 9.4) with respect to some open cover $p : (\sqcup_i U_i) \twoheadrightarrow X$ (Ex. 1.24), hence in particular including a choice of "local B-field" B_2, then one may further choose B_2-twisted connections ∇_i [Mackaay (2003)] on the twisted vector bundles, inducing curvature 2-forms F_i:

$$B_2 \in \Omega^2(U), \qquad F_i \in \Omega^2(U;\mathrm{End}(V_i)). \tag{10.1}$$

Now it turns out [Bouwknegt *et al.* (2002), Prop. 9.1] that

- the following trace (10.2) of differences of wedge-product exponentials of these 2-forms (10.1) is well defined (i.e., the trace exists, which is non-trivial since the twisted vector bundles in general have infinite rank)

$$\exp(B_2)\wedge \mathrm{tr}\big(\exp(F_1)-\exp(F_2)\big) \;=\; p^*\mathrm{ch}^{B_2}(\nabla_1,\nabla_2) \quad \in \Omega^{2\bullet}(U) \tag{10.2}$$

  and equals the pullback $p^*(-)$ to the given cover of an even-degree differential form on the base space X,

- which is closed in the H_3-twisted de Rham complex (6.39)

$$(d - H_3 \wedge)\mathrm{ch}^{B_2}(\nabla_1, \nabla_2) \; = \; 0,.$$

- and whose resulting twisted de Rham cohomology class (Def. 6.10) is independent of the choices made:

$$\mathrm{ch}^{\tau_3}(V_1, V_2) \; := \; \big[\mathrm{ch}^{B_2}(\nabla_1, \nabla_2)\big] \; \in H_{\mathrm{dR}}^{3+H_3}(X) \,. \tag{10.3}$$

This class (10.3) was proposed [Bouwknegt *et al.* (2002), p. 26] to be the *twisted Chern character* of the twisted K-theory class $[(V_1, V_2)]$.

A more intrinsic characterization of the twisted Chern character was later found in [Freed *et al.* (2008), §2]. This is the form in which one recognizes the twisted Chern character as an example of the twisted non-abelian character map (Def. V.3), in twisted enhancement of Ex. 7.2:

Proposition 10.1 (Twisted Chern character in twisted topological K-theory). *Consider twisted complex topological K-theory* $\mathrm{KU}^{\tau}(-)$ *(Ex. 3.4), for degree-3 twists given (via Ex. 2.8) by*

$$\tau \in H\big(-; B^2\mathrm{U}(1)\big) \simeq H^3(-;, \mathbb{Z}),$$

and regarded, via (3.17)*, as twisted non-abelian cohomology with local coefficients in* $\mathbb{Z} \times BU /\!/ B^2\mathrm{U}(1)$ (3.16)*. Then the twisted non-abelian character map (Def. V.3)* $\mathrm{ch}^{\tau}_{\mathbb{Z}\times BU}$ *is equivalent to the traditional twisted Chern character* ch^{τ} *on twisted K-theory with values in* H_3*-twisted de Rham cohomology (Def. 6.10):*

$$\underset{\text{character map}}{\overset{\text{twisted non-abelian}}{\mathrm{ch}^{\tau}_{\mathbb{Z}\times BU}}} \;\simeq\; \underset{\text{Chern character}}{\overset{\text{twisted}}{\mathrm{ch}^{\tau}}}.$$

Proof. That the codomain of the twisted non-abelian character map $\mathrm{ch}^{\tau}_{\mathbb{Z}\times BU}$ is indeed H_3-twisted de Rham cohomology is the content of Prop. 6.11. With this, and due to the twisted non-abelian de Rham theorem (Thm. 6.15), it is sufficient to see that the general rationalization map of local non-abelian coefficients from Def. V.2 reproduces the rationalization map underlying the twisted Chern character. This is manifest from comparing the rationalization operation (V.1), that is made formally precise by Def. V.2, to the description of the twisted Chern character as given in [Freed *et al.* (2008), (2.8)-(2.9)]. □

Remark 10.2 (Twisted Pontrjagin character in twisted KO-theory). Similarly, an analogous statement holds for the twisted Pontrjagin character (as in Ex. 7.3) on twisted real K-theory [Grady and Sati (2019d), Prop. 2].

Example 10.1 (Twisted Chern character on higher Cohomotopy-twisted K-theory). For $k \in \mathbb{N}_+$, consider the cohomotopically-twisted complex K-theory from Ex. 3.6.

(i) For $\lambda \in \pi^{2k+1}(X)$ a Cohomotopy class (Ex. 2.7) regarded now as a twist, the corresponding twisted non-abelian character map (Def. V.3) lands, by Thm. 6.5, 6.9 and Ex. 5.4,

5.3, in $\lambda_{\mathrm{dR}} =: H_{2k+1}$-twisted de Rham cohomology (Ex. 6.7, Prop. 6.13):

$$\underset{\text{twist in Cohomotopy}}{\lambda \in \pi^{2k+1}(X)} \;\vdash\; \underset{\text{higher Cohomotopy-twisted K-theory}}{\mathrm{KU}^{\lambda}(X) = H^{\lambda}(X;\mathrm{KU}_0)} \xrightarrow[\text{twisted character map}]{\mathrm{ch}^{\lambda}_{\mathrm{KU}_0}} \underset{\text{higher twisted de Rham cohomology}}{H^{\lambda_{\mathrm{dR}}}_{\mathrm{dR}}(X;\mathfrak{l}\mathrm{KU}_0) \simeq H^{\bullet+H_{2k+1}}(X)}. \tag{10.4}$$

(ii) A map of this form has been defined in [Macdonald *et al.* (2021), §2.2], by direct construction on form representatives. However, [Macdonald *et al.* (2021), Thm. 4.19] implies that this component construction coincides with the rationalization map (Def. V.2) on the local coefficient bundle (3.25), up to application of the de Rham theorem. Therefore, the twisted character map (10.4) obtained as a special case of Def. V.3, reproduces the MMS-Character [Macdonald *et al.* (2021), §2.2] on higher Cohomotopy-twisted K-theory.

Remark 10.3 (Charge quantization of spherical T-duality in M-theory). For $k = 3$, the character map (10.4) on 7-Cohomotopy-twisted K-theory is a candidate for charge quantization (0.2) of the super-rational M-theory fields participating in 3-spherical T-duality over 11-dimensional super-spacetime, as derived in [Fiorenza *et al.* (2020a), Prop. 4.17, Rem. 4.18] (review in [Sati and Schreiber (2018), (8), (19)]). However, the 7-Cohomotopy-twisted K-theory character has some spurious fields of 2-periodic degree in its image, which are not seen in the physics application, where the field degrees are 6-periodic [Sati (2009), §3][Fiorenza *et al.* (2020a), (65)]. Another candidate for charge-quantization of the super-rational M-theory fields participating in 3-spherical T-duality, possibly more accurately reflecting the physics, is the character map on twisted higher K-theory [Lind *et al.* (2020)], which we turn to next (Prop. 10.2).

Character map on twisted higher K-theory.

Remark 10.4 (Higher twisted de Rham coefficients inside rational twisted iterated K-theory). There is a non-trivial twisted cohomology operation (Def. 3.6) from **(a)** twisted non-abelian de Rham cohomology (Def. 6.9) with coefficients in the relative rational Whitehead L_∞-algebra (Prop. 5.16) of the coefficient bundle (3.27) of twisted iterated K-theory (Ex. 3.7) to **(b)** higher twisted de Rham cohomology (Def. 6.12) regarded as twisted non-abelian de Rham cohomology via Prop. 6.13):

$$H^{\tau_{\mathrm{dR}}}_{\mathrm{dR}}\Big(-;\mathfrak{l}K^{\circ_{2r-2}}(\mathrm{ku})_1\Big) \xrightarrow{\;\;\phi_*\;\;} H^{\tau_{\mathrm{dR}}}_{\mathrm{dR}}\Big(-;\bigoplus_{k\in\mathbb{N}}\mathfrak{b}^{2rk}\mathbb{R}\Big), \tag{10.5}$$

given, under the twisted non-abelian de Rham theorem (Thm. 6.15) by the LSW-character from [Lind *et al.* (2020), §2.2] applied to rational coefficients.

Proposition 10.2 (Twisted Chern character in twisted iterated K-theory). *For $r \in \mathbb{N}$, $r \geq 1$, consider twisted iterated K-theory $\big(K^{\circ_{2r-2}}(\mathrm{ku})\big)^{\tau}$ (Ex. 3.7), for degree-$(2r+1)$ twists given (via Ex. 2.9) by*

$$\tau \in H\big(-;B^{2r}\mathrm{U}(1)\big) \simeq H^{2r+1}(-;,\mathbb{Z}),$$

and regarded, via Ex. 3.7, as twisted non-abelian cohomology with local coefficients in $\big(K^{\circ_{2r-2}}(\mathrm{ku})\big)_0$. *Then the twisted non-abelian character map (Def. V.3)* $\mathrm{ch}^{\tau}_{K^{\circ_{2r-2}}(\mathrm{ku})_0}$ *composed with the projection operation* (10.5) *onto higher twisted de Rham cohomology, (Def. 6.12) from Lem. 10.4, is equivalent to the LSW character map* ch_{2r-1} *[Lind et al. (2020), Def. 2.20] restricted along the connective inclusion*

$$\overset{\text{twisted LSW character}}{\mathrm{ch}^{\tau}_{2r-1}} \;\simeq\; \underset{\substack{\text{projection onto}\\ \text{higher twisted}\\ \text{de Rham cohomology}}}{\phi_*} \circ \overset{\substack{\text{twisted non-abelian}\\ \text{character map}}}{\mathrm{ch}^{\tau}_{K^{\circ_{2r-2}}(\mathrm{ku})_0}}\,.$$

Proof. After unwinding the definitions, the statement reduces to the commutativity of the square diagram in [Lind *et al.* (2020), p. 15]: The top morphism there is the plain rationalization map (Def. V.2), the right vertical morphism is ϕ_* from Lem. 10.4 before passing from real to de Rham cohomology, the left morphism is restriction to the connective part and the bottom morphism is the LSW character. □

Chapter 11

Twisted differential non-abelian character

We introduce twisted differential non-abelian cohomology (Def. 11.2 below) and discuss how the corresponding twisted differential non-abelian character subsumes existing constructions on twisted differential K-theory (Ex. 11.2 and 11.3 below).

Twisted differential non-abelian cohomology. From the perspective of structured non-abelian cohomology (Rem. 2.3) that we have developed, it is now evident how to canonically combine

(a) twisted non-abelian cohomology (Def. 3.2) with
(b) differential non-abelian cohomology (Def. 9.3) to get
twisted differential non-abelian cohomology:

Definition 11.1 (Differential non-abelian local coefficient bundles). Let

$$\begin{array}{ccc} A & \longrightarrow & A /\!\!/ G \\ & \text{local coefficient bundle} & \big\downarrow \rho \\ & & BG \end{array}$$

be a local coefficient bundle (3.2) in $\mathrm{Ho}\big(\Delta\mathrm{Sets}_{\mathrm{Qu}}\big)^{\mathrm{fin}_{\mathbb{Q}}}_{\geq 1,\mathrm{nil}}$ (Def. 5.1).

(i) By Lem. 5.4, with Def. V.1, and using that $B\exp_{\mathrm{PL}}$ preserves fibrations (Prop. 5.10), this induces a homotopy fibering (Def. 1.14) in $\mathrm{Ho}(\mathrm{SmthStacks}_\infty)$ (Def. 1.25) of differential non-abelian character maps (Def. 9.2) of this form:

$$\begin{array}{ccccc}
\mathrm{Disc}(A) & \xrightarrow[\mathbf{ch}_A]{\text{differential non-abelian character map with coefficients in fiber space}} & \flat B\exp(\mathfrak{l}A) & \xleftarrow{\text{atlas}} & \Omega_{\mathrm{dR}}(-;\mathfrak{l}A)_{\mathrm{flat}} \\
\big\downarrow{\scriptstyle \mathrm{hofib}(\mathrm{Disc}(\rho))} & & \big\downarrow{\scriptstyle \mathrm{hofib}((\mathfrak{l}\rho)_*)} & & \big\downarrow{\scriptstyle \mathrm{hofib}((\mathfrak{l}\rho)_*)} \\
\mathrm{Disc}(A /\!\!/ G) & \xrightarrow[\text{twisted differential non-abelian character map}]{\mathbf{ch}^{BG}_{A /\!\!/ G}} & \flat B\exp\big(\mathfrak{l}(A /\!\!/ G)\big) & \xleftarrow[\text{atlas}]{} & \Omega_{\mathrm{dR}}\big(-;\mathfrak{l}_{BG}(A /\!\!/ G)\big)_{\mathrm{flat}} \\
\big\downarrow{\scriptstyle \mathrm{Disc}(\rho)} & & \big\downarrow{\scriptstyle (\mathfrak{l}\rho)_*} & & \big\downarrow{\scriptstyle (\mathfrak{l}\rho)_*} \\
\mathrm{Disc}(BG) & \xrightarrow[\text{differential non-abelian character map with coefficients in space of twists}]{\mathbf{ch}_{BG}} & \flat B\exp(\mathfrak{l}BG) & \xleftarrow[\text{atlas}]{} & \Omega_{\mathrm{dR}}(-;\mathfrak{l}BG)_{\mathrm{flat}}
\end{array} \tag{11.1}$$

(ii) Here the *twisted differential non-abelian character map* $\mathbf{ch}^{BG}_{A/\!/G}$ is defined just as in Def. 9.2, but with coefficients the relative Whitehead L_∞-algebra $\mathfrak{l}_{BG}(A/\!/G)$ (Prop. 5.16), as opposed to the absolute Whitehead L_∞-algebra $\mathfrak{l}(A/\!/G)$ (Prop. 5.11).

Remark 11.1 (Differential local coefficient bundles). Since homotopy limits commute over each other, passage to the homotopy fiber products (Def. 1.15) formed from the horizontal stages of (11.1) yields a homotopy fibering (Def. 1.14) of moduli ∞-stacks of ∞-connections (9.10) of this form:

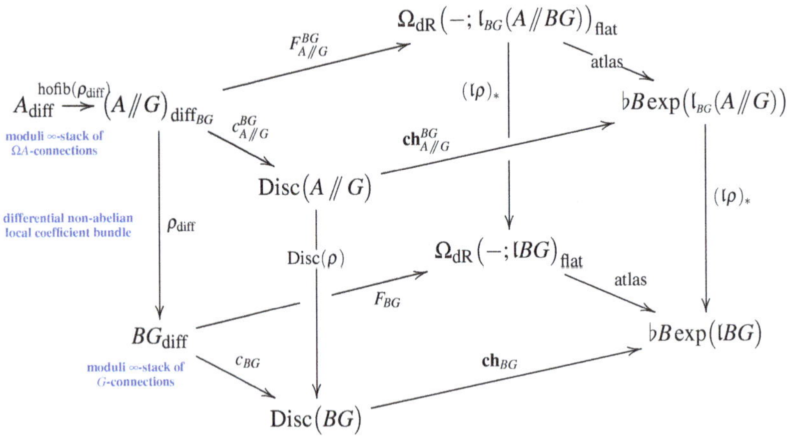

(11.2)

Definition 11.2 (Twisted differential non-abelian cohomology). Given a differential non-abelian local coefficient bundle ρ_{diff} (11.2) according to Def. 11.1, we say that:

(i) A *differential twist* on a $\mathscr{X} \in \mathrm{Ho}(\mathrm{SmthStacks}_\infty)$ (Def. 1.25) is a cocycle τ_{diff} in differential non-abelian cohomology with coefficients in BG (Def. 9.3)

$$\big[\tau_{\mathrm{diff}}\big] \in \widehat{H}\big(\mathscr{X}; BG\big). \tag{11.3}$$

(ii) The τ_{diff}-*twisted differential non-abelian cohomology* with local coefficients in ρ_{diff} is the structured (Rem. 2.3) τ_{diff}-twisted non-abelian cohomology (Def. 3.2) with coefficients in ρ_{diff}, hence the hom-set in the homotopy category (Def. 1.8) of the slice model structure (Def. 1.5) of the local projective model structure $\mathrm{SmoothStacks}_\infty$ on simplicial presheaves over CartSp (Ex. 1.20) from τ_{diff} (11.3) to ρ_{diff} (11.2):

twisted differential non-abelian cohomology

$$\widehat{H}^{\tau_{\mathrm{diff}}}\big(\mathscr{X}; A\big) := \mathrm{Ho}\big(\mathrm{SmthStacks}_\infty^{/BG_{\mathrm{diff}}}\big)\big(\tau_{\mathrm{diff}}, \rho_{\mathrm{diff}}\big)$$

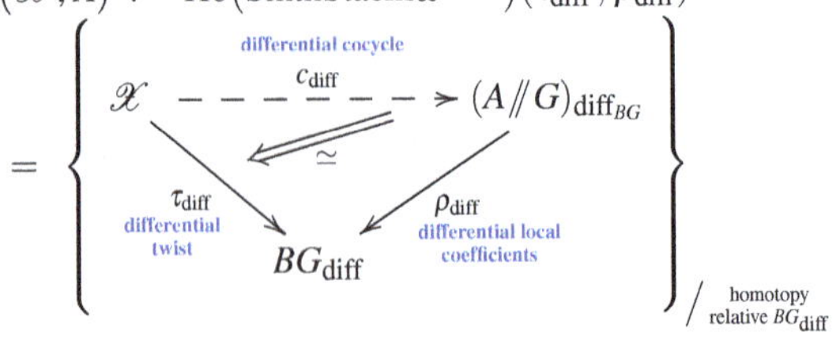

(11.4)

(iii) The twisted non-abelian cohomology operations induced from the maps in (11.2) we call (see (0.16)):

$$\text{(a) } \textit{characteristic class:} \quad \widehat{H}^{\tau_{\mathrm{diff}}}\big(\mathscr{X};A\big) \xrightarrow{\;c_A^{\tau} := \left(c^{BG}_{A/\!\!/G}\right)_*\;} H^{\tau}\big(\mathrm{Shp}(\mathscr{X});A\big) \qquad \text{(Def. 3.2)} \tag{11.5}$$

$$\text{(b) } \textit{curvature:} \quad \widehat{H}^{\tau_{\mathrm{diff}}}\big(\mathscr{X};A\big) \xrightarrow{\;F_A^{\tau_{\mathrm{dR}}} := \left(F^{BG}_{A/\!\!/G}\right)_*\;} \Omega^{\tau_{\mathrm{dR}}}_{\mathrm{dR}}\big(\mathscr{X};\mathfrak{l}A\big)_{\mathrm{flat}} \qquad \text{(Def. 6.7)} \tag{11.6}$$

$$\text{(c) } \textit{differential character:} \quad \widehat{H}^{\tau_{\mathrm{diff}}}\big(\mathscr{X};A\big) \xrightarrow{\;\mathrm{ch}_A^{\tau} := \left(\mathbf{ch}^{BG}_{A/\!\!/G} \circ c^{BG}_{A/\!\!/G}\right)_*\;} H^{\tau_{\mathrm{dR}}}_{\mathrm{dR}}\big(\mathscr{X};\mathfrak{l}A\big) \qquad \text{(Def. 6.9)} \tag{11.7}$$

Twisted differential non-abelian cohomology as non-abelian ∞-sheaf hypercohomology. While the formulation of twisted differential non-abelian cohomology as hom-sets in a slice of $\mathrm{SmoothStacks}_\infty$ (Def. 11.2) is natural and useful, we indicate how this is equivalently incarnated as a non-abelian sheaf hypercohomology over $\mathscr{X}$. This serves to make the connection to existing literature (in Ex. 11.1 below), but is not otherwise needed for the development here. We shall be brief, referring to [Sati and Schreiber (2020c)] for some technical background that is beyond the scope of our presentation here.

Proposition 11.3 (Étale ∞-topos over ∞-stacks [Sati and Schreiber (2020c), Prop. 3.33, Rem. 3.34]**).** *For $\mathscr{X} \in \mathrm{Ho}(\mathrm{SmthStacks}_\infty)$ (Def. 1.25) let*

$$\mathrm{Ho}\big(\mathbf{\acute{E}t}_{\mathscr{X}}\big) \xhookrightarrow{\;\mathbb{D}\mathrm{i}_{\mathscr{X}}\;} \mathrm{Ho}\big(\mathrm{SmthStacks}_\infty^{/\mathscr{X}}\big)$$

be the full subcategory of the homotopy category (Def. 1.8) of the slice model structure over $\mathscr{X}$ (Ex. 1.5) of the local projective model structure on simplicial presheaves (Ex. 1.20) on those $\mathscr{E} \to \mathscr{X}$ which are local diffeomorphisms ([Sati and Schreiber (2020c), Def. 3.26]).

(i) *The inclusion $\mathbb{D}\mathrm{i}_{\mathscr{X}}$ is a left-exact homotopy co-reflection, in that it preserves finite homotopy limits and has a derived right adjoint $\mathbb{R}\mathrm{Et}$ (sending ∞-bundles to their ∞-sheaves of ∞-sections).*
(ii) *There is a global section functor $\mathbb{R}\Gamma_{\mathscr{X}}$ from $\mathrm{Ho}\big(\mathbf{\acute{E}t}_{\mathscr{X}}\big)$ to $\mathrm{Ho}\big(\Delta\mathrm{Sets}_{\mathrm{Qu}}\big)$ (Ex. 1.14) which also admits a left exact left adjoint:*

$$\underset{\infty\text{-bundles over } \mathscr{X}}{\mathrm{Ho}\big(\mathrm{SmthStacks}_\infty^{/\mathscr{X}}\big)} \underset{\underset{\infty\text{-sheaf of local sections}}{\mathbb{R}\mathrm{Et}}}{\overset{\mathbb{D}\mathrm{i}_{\mathscr{X}}}{\underset{\longrightarrow}{\overset{\longleftarrow}{\bot}}}} \underset{\infty\text{-sheaves over } \mathscr{X}}{\mathrm{Ho}\big(\mathbf{\acute{E}t}_{\mathscr{X}}\big)} \underset{\underset{\text{global sections}}{\mathbb{R}\Gamma_{\mathscr{X}}}}{\overset{\Delta_{\mathscr{X}}}{\underset{\longrightarrow}{\overset{\longleftarrow}{\bot}}}} \mathrm{Ho}\big(\Delta\mathrm{Sets}_{\mathrm{Qu}}\big). \tag{11.8}$$

Definition 11.4 (Non-abelian ∞-sheaf hypercohomology over ∞-stacks). Given $\mathscr{X} \in \mathrm{Ho}(\mathrm{SmthStacks}_\infty)$ (Def. 1.25) and $\mathscr{A} \in \mathrm{Ho}\big(\mathbf{\acute{E}t}_{\mathscr{X}}\big)$ (Prop. 11.3) we say that the set of connected components of the *derived global sections* (11.8) of $\mathscr{A}$ over $\mathscr{X}$

$$H(\mathscr{X}, \mathscr{A}) \;:=\; \pi_0\big(\mathbb{R}\Gamma_{\mathscr{X}}(\mathscr{A})\big)$$

is the *non-abelian ∞-sheaf hypercohomology* of $\mathscr{X}$ with coefficients in $\mathscr{A}$.

Lemma 11.1 (Twisted differential non-abelian cohomology as non-abelian ∞-sheaf hyper-cohomology). *Given a differential twist* τ_{diff} (11.3) *on some* $\mathscr{X} \in \mathrm{Ho}(\mathrm{SmthStacks}_\infty)$ (1.83) *consider the object*

$$\underline{A}_{\tau_{\mathrm{diff}}} := \mathbb{R}\mathrm{LcllCnstnt}_{\mathscr{X}}\big(\mathbb{R}\tau^*_{\mathrm{diff}}(A /\!/ G)_{\mathrm{diff}}\big) \;\in\; \mathrm{Ho}\big(\mathbf{\acute{E}t}_{\mathscr{X}}\big) \tag{11.9}$$

in the étale ∞-topos over $\mathscr{X}$ Prop. 11.3. The non-abelian ∞-sheaf hypercohomology (Def. 11.4) of $\underline{A}_{\tau_{\mathrm{diff}}}$ over $\mathscr{X}$ coincides with the τ_{diff}-twisted differential non-abelian cohomology of $\mathscr{X}$ (Def. 11.2):

$$\overset{\substack{\text{non-abelian}\\ \infty\text{-sheaf hypercohomology}}}{\pi_0\,\mathbb{R}\Gamma_{\mathscr{X}}\big(\underline{A}_{\tau_{\mathrm{diff}}}\big)} \;\simeq\; \overset{\substack{\text{twisted differential}\\ \text{non-abelian cohomology}}}{\widehat{H}^{\tau_{\mathrm{diff}}}\big(\mathscr{X}, A\big)}. \tag{11.10}$$

Proof. As in [Sati and Schreiber (2020c), Rem. 3.34]. □

It is useful to decompose this construction of twisted differential cohomology via ∞-sheaf hypercohomology again as a homotopy pullback of corresponding ∞-sheaves representing plain twisted cohomology and plain twisted differential forms:

Remark 11.2 (Homotopy pullback of ∞-sheaves representing twisted differential cohomology). Given a differential twist τ_{diff} (11.3) on some $\mathscr{X} \in \mathrm{Ho}(\mathrm{SmthStacks}_\infty)$ (1.83) with components $(\tau, \tau_{\mathrm{dR}}, L_{\mathbb{R}}\tau)$ (Ex. 1.12),

(i) Consider the pullback of stacks over $\mathscr{X}$ in the following diagram

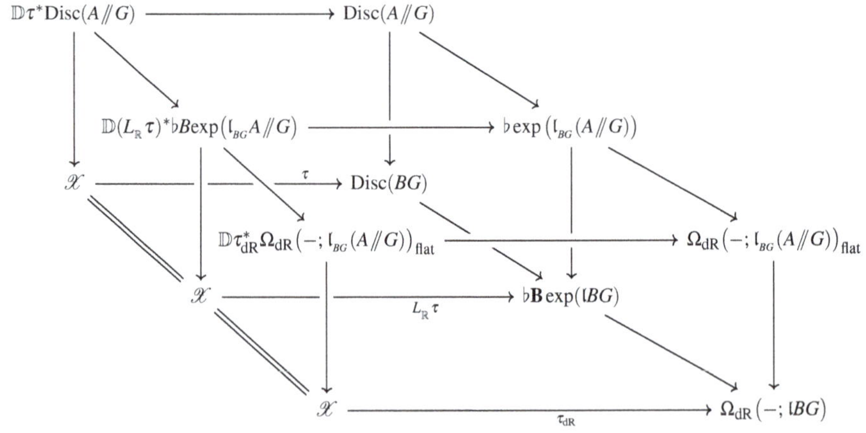

Here the right hand side is (11.1) and all front-facing squares are homotopy pullbacks (Def. 1.15).

(ii) By commutativity of homotopy limits over each other, these form a homotopy pullback square as on the right of the following diagram, which gives, under the derived right adjoint $\mathbb{R}$LcllCnstnt (11.8) a homotopy pullback diagram of ∞-sheaves of sections:

$$
\begin{array}{ccc}
\underline{A}_{\tau_{\mathrm{diff}}} & \longrightarrow & \underline{\Omega\big(-;\mathfrak{l}A\big)_{\mathrm{flat}}}_{\tau_{\mathrm{dR}}} \\
\downarrow & \text{(hpb)} & \downarrow \\
\underline{A}_{\tau} & \longrightarrow & \underline{\flat\mathfrak{B}\exp\big(\mathfrak{l}A\big)}_{L_{\mathbb{R}}\tau}
\end{array}
:=
$$

$$
\mathbb{R}\mathrm{LcllCnstst}\left(
\begin{array}{ccc}
\mathbb{R}\tau^*_{\mathrm{diff}}(A /\!\!/ G)_{\mathrm{diff}} & \longrightarrow & \mathbb{R}\tau^*_{\mathrm{dR}}\Omega_{\mathrm{dR}}\big(-;\mathfrak{l}_{BG}(A /\!\!/ G)\big) \\
\downarrow & \text{(hpb)} & \downarrow \\
\mathbb{R}\tau^*(A /\!\!/ G) & \longrightarrow & \mathbb{R}(L_{\mathbb{R}}\tau)^*\flat B\exp\big(\mathfrak{l}(A /\!\!/ G)\big)
\end{array}
\right) \in \mathrm{Ho}\big(\mathbf{\acute{E}t}_{\mathscr{X}}\big). \tag{11.11}
$$

Here the top left item $\underline{A}_{\tau_{\mathrm{diff}}}$ from (11.9) is the ∞-sheaf whose global sections give the τ_{diff}-twisted differential cohomology, by Lem. 11.1.

In differential enhancement of Prop. 3.5 and in twisted enhancement of Ex. 9.1, we have:

Example 11.1 (Twisted differential generalized cohomology). Let $\mathscr{X} = X$ be a smooth manifold (Ex. 1.21), R an E_∞-ring spectrum (Ex. 2.10), and let

$$
\begin{array}{ccc}
E_0 & \longrightarrow & (R_0) /\!\!/ \mathrm{GL}(1,R) \\
{\scriptstyle \mathbb{R}\tau^*\rho_R}\downarrow & \text{(hpb)} & \downarrow{\scriptstyle \rho_R} \\
X & \xrightarrow[\tau]{} & B\mathrm{GL}(1,R)
\end{array}
$$

be a twist for twisted generalized R-cohomology over X (3.21), as in Lem. 3.5.

(i) Then the corresponding homotopy pullback diagram (11.11), which exhibits, by Lem. 11.1, twisted differential non-abelian cohomology (Def. 11.2) with coefficients in E_0 as ∞-sheaf hypercohomology (Def. 11.4), is the image under $\mathbb{R}\Omega^\infty_X$ of the homotopy pullback diagram of sheaves of spectra considered in [Bunke and Nikolaus (2019), Def. 4.11], shown on the right below, for canonical/minimal differential refinement as in Ex. 9.1:

$$
\begin{array}{ccc}
\underline{R_0}_{\tau_{\mathrm{diff}}} & \longrightarrow & \underline{\Omega\big(-;\mathfrak{l}R_0\big)_{\mathrm{flat}}}_{\tau_{\mathrm{dR}}} \\
\downarrow & \text{(hpb)} & \downarrow \\
\underline{R_0}_{\tau} & \longrightarrow & \underline{\flat\mathfrak{B}\exp\big(\mathfrak{l}R_0\big)}_{L_{\mathbb{R}}\tau}
\end{array}
\simeq \mathbb{R}\Omega_X{}^{\infty}
\left(
\begin{array}{ccc}
\mathrm{Diff}(E) & \longrightarrow & H\mathscr{M}_{\leq 0} \\
\downarrow & \text{(hpb)} & \downarrow \\
\mathrm{Disc}(E) & \longrightarrow & H\mathscr{M}
\end{array}
\right)
$$

This is the twisted/parametrized analog of the relation (9.18).

(ii) Accordingly, the twisted differential generalized R-cohomology according to [Bunke and Nikolaus (2019), Def. 4.13] is subsumed by twisted differential non-abelian cohomology, via Lem. 11.1.

In differential enhancement of Prop. 10.1 and in twisted generalization of Ex. 9.2, we have:

Example 11.2 (Twisted Chern character in twisted differential K-theory). Consider again the local coefficient bundle

$$\begin{array}{ccc} \mathrm{KU}_0 & \longrightarrow & \mathrm{KU}_0 /\!\!/ B\mathrm{U}(1) \\ & & \big\downarrow \rho \\ & & B^2\mathrm{U}(1) \end{array}$$

for complex topological K-theory (Ex. 3.4). By Ex. 11.1, the twisted differential non-abelian cohomology theory (Def. 11.2) induced from these local coefficients is twisted differential K-theory, as discussed in [Carey *et al.* (2009)] for torsion twists (review in [Bunke and Schick (2012), §7]). By the diagram (0.16) of cohomology operations on twisted differential cohomology, one may regard the corresponding twisted curvature map (11.6)

$$\widehat{K}^{\tau_{\mathrm{diff}}}(\mathscr{X}) \xrightarrow{\left(F^{\tau_{\mathrm{dR}}}_{\mathrm{KU}_0}\right)_*} \Omega^{\tau_{\mathrm{dR}}}_{\mathrm{dR}}\big(\mathscr{X};\, \mathfrak{l}\mathrm{KU}_0\big)_{\mathrm{flat}}$$

(with values in flat $\tau_{\mathrm{dR}} \simeq H_3$-twisted differential forms, by Ex. 6.6) as an incarnation of the Chern character map on twisted differential K-theory. Unwinding this abstract construction produces the perspective taken in [Carey *et al.* (2009), p. 2][Park (2018)] for torsion twists, and in [Bunke and Nikolaus (2019), p. 6] for general twists.

However, in the spirit of the Cheeger-Simons homomorphism (9), any lift of a cohomology operation (here: rationalization) to differential cohomology should be enhanced all the way to a secondary cohomology operation (Def. 9.7, now to be generalized to a *twisted* secondary cohomology operation, Def. 11.6 below) whose codomain is itself a (twisted) differential cohomology theory. The twisted Chern character enhanced to a secondary cohomology operation this way is Ex. 11.3 below, following the perspective taken in [Grady and Sati (2019c), §3.2][Grady and Sati (2019b), §2.3].

Secondary twisted non-abelian cohomology operations. We introduce the twisted generalization of secondary non-abelian cohomology operations (Def. 11.6 below). This requires the following twisted analog of the technical condition in Def. 9.6:

Definition 11.5 (Twisted absolute minimal model). For

$$
\begin{array}{ccc}
A_1 /\!\!/ G_1 & \xrightarrow{\;c_t\;} & A_2 /\!\!/ G_2 \\
\downarrow{\scriptstyle \rho_1} & & \downarrow{\scriptstyle \rho_2} \\
BG_1 & \xrightarrow[\;c_b\;]{} & BG_2
\end{array}
\quad \in \; \Delta\mathrm{Sets}
$$

a transformation (3.29) between local coefficient bundles (3.2), and for $\mathfrak{c}_b$ an absolute minimal model (Def. 9.6) of the map c_b between spaces of twists, hence with induced transformation (9.37)

$$
\begin{array}{ccccc}
\mathrm{Disc}(BG_1) & & \xrightarrow{\;\mathrm{Disc}(c_b)\;} & & \mathrm{Disc}(BG_2) \\
& \searrow^{\mathbf{ch}_{BG_1}} & & & & \searrow^{\mathbf{ch}_{BG_2}} \\
& & \flat B\exp(\mathfrak{l}BG_1) & \xrightarrow[\;(\mathfrak{c}_b)_*\;]{} & & & \flat B\exp(\mathfrak{l}BG_1)
\end{array}
$$

between the differential character maps (Def. 9.2) on the spaces of twists, we say that a corresponding *twisted absolute minimal model* is a lift of $\mathfrak{c}_b$ to a morphism

$$
\mathfrak{l}_{BG_1}(A_1 /\!\!/ G_1) \overset{\mathfrak{c}_t}{\dashrightarrow} \mathfrak{l}_{BG_1}(A_1 /\!\!/ G_1) \tag{11.12}
$$

between the relative rational Whitehead L_∞-algebras of the local coefficient bundles (Prop. 5.16) which

(i) yields a transformation

$$
\begin{array}{ccccc}
\mathrm{Disc}(A_1 /\!\!/ G_1) & & \xrightarrow{\;\mathrm{Disc}(c_t)\;} & & \mathrm{Disc}(A_2 /\!\!/ G_2) \\
& \searrow^{\mathbf{ch}^{BG_1}_{A_1 /\!\!/ G_1}} & & & & \searrow^{\mathbf{ch}^{BG_2}_{A_2 /\!\!/ G_2}} \\
& & \flat B\exp(\mathfrak{l}_{BG_1}(A_1 /\!\!/ G_1)) & \overset{}{\underset{(\mathfrak{c}_t)_*}{\dashrightarrow}} & & & \flat B\exp(\mathfrak{l}_{BG_2}(A_2 /\!\!/ G_2))
\end{array}
$$

of the twisted differential characters (11.1) (thus being an "absolute minimal model for c_t relative to c_b"),

(ii) is compatible with the transformation of the differential characters on the twisting space, in that the following cube commutes:

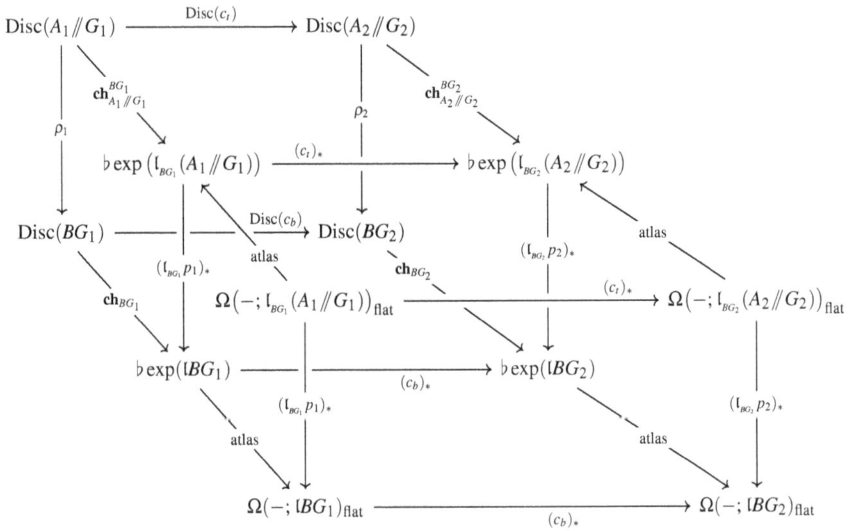
$\mathrm{Disc}(A_1 /\!\!/ G_1)$
$\mathrm{Disc}(c_t)$
$\mathrm{Disc}(A_2 /\!\!/ G_2)$
$\mathbf{ch}^{BG_1}_{A_1 /\!\!/ G_1}$
$\mathbf{ch}^{BG_2}_{A_2 /\!\!/ G_2}$
ρ_1
ρ_2
$\flat\exp\big(\mathfrak{l}_{BG_1}(A_1 /\!\!/ G_1)\big)$
$(c_t)_*$
$\flat\exp\big(\mathfrak{l}_{BG_2}(A_2 /\!\!/ G_2)\big)$
$\mathrm{Disc}(BG_1)$
$\mathrm{Disc}(c_b)$
$\mathrm{Disc}(BG_2)$
atlas
$(\mathfrak{l}_{BG_1}p_1)_*$
$(\mathfrak{l}_{BG_2}p_2)_*$
$\mathbf{ch}_{BG_2}$
$\mathbf{ch}_{BG_1}$
$\Omega\big(-;\mathfrak{l}_{BG_1}(A_1 /\!\!/ G_1)\big)_{\mathrm{flat}}$
$(c_t)_*$
$\Omega\big(-;\mathfrak{l}_{BG_2}(A_2 /\!\!/ G_2)\big)_{\mathrm{flat}}$
$\flat\exp(\mathfrak{l}BG_1)$
$(c_b)_*$
$\flat\exp(\mathfrak{l}BG_2)$
$\Omega(-;\mathfrak{l}BG_1)_{\mathrm{flat}}$
$(c_b)_*$
$\Omega(-;\mathfrak{l}BG_2)_{\mathrm{flat}}$

(11.13)

At the level of dgc-algebras, the condition that $\mathfrak{c}_t$ (11.12) is a twisted absolute minimal model for the transformation of local coefficient bundles means equivalently that it makes the following cube commute:

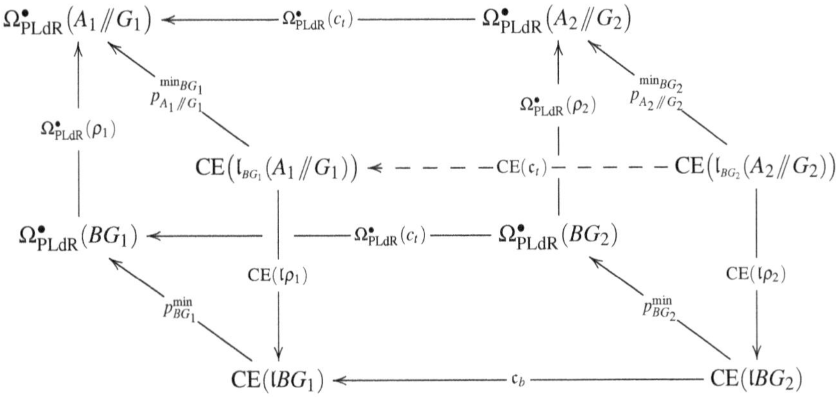
$\Omega^\bullet_{\mathrm{PLdR}}(A_1 /\!\!/ G_1)$
$\Omega^\bullet_{\mathrm{PLdR}}(c_t)$
$\Omega^\bullet_{\mathrm{PLdR}}(A_2 /\!\!/ G_2)$
$p^{\mathrm{min}BG_1}_{A_1 /\!\!/ G_1}$
$\Omega^\bullet_{\mathrm{PLdR}}(\rho_2)$
$p^{\mathrm{min}BG_2}_{A_2 /\!\!/ G_2}$
$\Omega^\bullet_{\mathrm{PLdR}}(\rho_1)$
$\mathrm{CE}\big(\mathfrak{l}_{BG_1}(A_1 /\!\!/ G_1)\big)$
$\mathrm{CE}(\mathfrak{c}_t)$
$\mathrm{CE}\big(\mathfrak{l}_{BG_2}(A_2 /\!\!/ G_2)\big)$
$\Omega^\bullet_{\mathrm{PLdR}}(BG_1)$
$\Omega^\bullet_{\mathrm{PLdR}}(c_t)$
$\Omega^\bullet_{\mathrm{PLdR}}(BG_2)$
$\mathrm{CE}(\mathfrak{l}\rho_1)$
$\mathrm{CE}(\mathfrak{l}\rho_2)$
$p^{\mathrm{min}}_{BG_1}$
$p^{\mathrm{min}}_{BG_2}$
$\mathrm{CE}(\mathfrak{l}BG_1)$
$\mathfrak{c}_b$
$\mathrm{CE}(\mathfrak{l}BG_2)$

(11.14)

In differential enhancement of Def. 3.6 and in twisted generalization of Def. 9.7, we set:

Definition 11.6 (Twisted secondary non-abelian cohomology operations). Let

$$\begin{array}{ccc} A_1 /\!\!/ G_1 & \xrightarrow{\;c_t\;} & A_2 /\!\!/ G_2 \\ {\scriptstyle \rho_1}\big\downarrow & & \big\downarrow{\scriptstyle \rho_2} \\ BG_1 & \xrightarrow{\;c_b\;} & BG_2 \end{array} \quad \in \Delta\mathrm{Sets}$$

be a transformation (3.29) between local coefficient bundles (3.2), together with an absolute minimal model $\mathfrak{c}_b$ (Def. 9.6) for the base map, and a compatible twisted absolute minimal model $\mathfrak{c}_t$ (Def. 11.5) for the total map. Then forming stage-wise homotopy pullbacks (Def. 1.15) in the required commuting cube (11.13) yields a transformation of corresponding differential coefficient bundles (11.2):

$$\begin{array}{ccc} (A_1 /\!\!/ G_1)_{\mathrm{diff}} & \xrightarrow{\;(c_t)_{\mathrm{diff}}\;} & (A_2 /\!\!/ G_2)_{\mathrm{diff}} \\ {\scriptstyle (\rho_1)_{\mathrm{diff}}}\big\downarrow & & \big\downarrow{\scriptstyle (\rho_2)_{\mathrm{diff}}} \\ (BG_1)_{\mathrm{diff}} & \xrightarrow[\;(c_b)_{\mathrm{diff}}\;]{} & (BG_2)_{\mathrm{diff}} \end{array} \quad \in \mathrm{PSh}\big(\mathrm{CartSp}, \Delta\mathrm{Sets}\big)\,. \tag{11.15}$$

This yields, in turn, a natural transformation of twisted differential non-abelian cohomology sets (Def. 11.2), hence a *twisted secondary non-abelian cohomology operation*, by pasting composition, hence by right derived base change (Ex. 1.7) along $(\rho_1)_{\mathrm{diff}}$ followed by composition with $(c_t)_{\mathrm{diff}}$ regarded as a morphism in the slice (Ex. 1.5) over $(BG_1)_{\mathrm{diff}}$:

$$\widehat{H}^{\tau_{\mathrm{diff}}}\big(\mathscr{X}; A_1\big) \xrightarrow{\;((c_t)_{\mathrm{diff}} \circ (-)) \circ ((\rho_1)_{\mathrm{diff}})_*\;} \widehat{H}^{(c_b)_{\mathrm{diff}} \circ \tau_{\mathrm{diff}}}\big(\mathscr{X}; A_2\big)\,.$$

In differential enhancement of Prop. 10.1, we have:

Example 11.3 (Twisted differential character on twisted differential K-theory). Consider the rationalization (Def. 5.2) over the actual *rational numbers* (see Rem. 5.2) of the local coefficient bundle (3.16) for degree-3 twisted complex topological K-theory (Ex. 3.4).

(i) This is captured by the diagram

$$\begin{array}{ccc} \mathrm{KU}_0 /\!\!/ B\mathrm{U}(1) & \xrightarrow{\;\eta^{\mathbb{Q}}_{\mathrm{KU}_0 /\!\!/ B\mathrm{U}(1)}\;} & L_{\mathbb{Q}}\big(\mathrm{KU}_0 /\!\!/ B\mathrm{U}(1)\big) \\ {\scriptstyle \rho}\big\downarrow & & \big\downarrow{\scriptstyle L_{\mathbb{R}}\rho} \\ B^2\mathrm{U}(1) & \xrightarrow{\;\eta^{\mathbb{Q}}_{B^2\mathrm{U}(1)}\;} & L_{\mathbb{Q}}\big(B^2\mathrm{U}(1)\big) \end{array} \tag{11.16}$$

regarded as a transformation of local coefficient bundles from twisted K-theory to twisted even-periodic rational cohomology:

$$L_{\mathbb{Q}} \mathrm{KU}_0 \simeq \Omega^\infty \Big(\underbrace{\bigoplus_k \Sigma^{2k} H\mathbb{Q}}_{=: H_{\mathrm{per}}\mathbb{Q}} \Big).$$

(ii) Since rationalization is idempotent (5.2), which here means that $L_{\mathbb{R}} \circ L_{\mathbb{Q}} \simeq L_{\mathbb{R}}$, in this situation an absolute minimal model (Def. 9.6) of the base map $c_b = \eta^{\mathbb{R}}_{B^2\mathrm{U}(1)}$ and a twisted absolute minimal model (Def. 11.5) of the total map $c_t = \eta^{\mathbb{R}}_{K_0 /\!\!/ B\mathrm{U}(1)}$ exist and are given, respectively, simply by the identity morphisms

$$\mathfrak{c}_b := \mathrm{id}_{\mathfrak{l}B^2\mathrm{U}(1)} \qquad \text{and} \qquad \mathfrak{c}_t := \mathrm{id}_{\mathfrak{l}_{B^2\mathrm{U}(1)}(K_0 /\!\!/ B\mathrm{U}(1))}.$$

(iii) Therefore, the induced twisted secondary cohomology operation Def. 11.6 exists, and is for each differential twist τ_{diff} a transformation

$$\widehat{K}^{\tau_{\mathrm{diff}}}(\mathscr{X}) \xrightarrow{\ \mathrm{ch}^{\tau_{\mathrm{diff}}}_{\mathrm{diff}} := \left(\eta^{\mathbb{R}}_{K_0 /\!\!/ B\mathrm{U}(1)}\right)_{\mathrm{diff}}\ } \widehat{H_{\mathrm{per}}\mathbb{Q}}^{L_{\mathbb{Q}}\tau_{\mathrm{diff}}}(\mathscr{X}) \tag{11.17}$$

from twisted differential K-theory to twisted differential periodic rational cohomology theory.

(iv) This is the twisted differential Chern character map on twisted differential complex K-theory as conceived in [Grady and Sati (2019c), §3.2][Grady and Sati (2019b), Prop. 4]. The analogous statement holds for the twisted differential Pontrjagin character (as in Ex. 7.3) on twisted differential real K-theory [Grady and Sati (2019d), Thm. 12].

(v) Notice that this construction is close to but more structured than the plain curvature map on twisted differential K-theory (Ex. 11.2): If we considered the transformation of local coefficients as in (11.16) but for rationalization $L_{\mathbb{R}}$ over the real numbers (Rem. 5.2), then the induced twisted secondary cohomology operation would be equivalent to the twisted curvature map. Instead, (11.17) refines the plain curvature map to a twisted secondary operation that retains information about rational periods.

Chapter 12

Twisted character on twisted differential Cohomotopy

Cohomotopical character maps. We discuss here (Ex. 12.1 below) the twisted non-abelian character map on tangentially twisted Cohomotopy (Ex. 3.8) in degree 4, and on Twistorial Cohomotopy (Ex. 3.11). We highlight the induced charge quantization (Prop. 12.1 below) and comment on the relevance to high energy physics (Rem. 12.2).

These twisted non-abelian cohomotopical character maps have been introduced and analyzed in [Fiorenza *et al.* (2020b)] and [Fiorenza *et al.* (2022)]. The general theory of non-abelian characters developed here shows how these cohomotopical characters are cousins both of generalized abelian characters such as the Chern character on twisted higher K-theory (Chapter 10), notably of the character on topological modular forms (by Ex. 7.5, and Rem. 7.4) as well as of non-abelian characters such as the Chern-Weil homomorphism (Chapter 8) and the Cheeger-Simons homomorphism (Chapter 9).

Example 12.1 (Character map on tangentially twisted Cohomotopy and on Twistorial Cohomotopy [Fiorenza *et al.* (2020b), Prop. 3.20] [Fiorenza *et al.* (2022), Prop. 3.9]**).** Let X be an 8-dimensional smooth spin manifold equipped with tangential Sp(2)-structure τ (3.33). Then the twisted non-abelian character maps (Def. V.3) on tangentially twisted Cohomotopy (Ex. 3.8) in degree 4, and on Twistorial Cohomotopy (Ex. 3.11) are of the following form (with p_1, p_2, I_8 from Ex. 8.1):

$$
\begin{array}{ccccc}
\mathbb{C}P^3 /\!\!/ \mathrm{Sp}(2) & \underset{\text{Twistorial Cohomotopy}}{\mathscr{T}^{\tau}(X) := H^{\tau}(X; \mathbb{C}P^3)} & \xrightarrow[\text{character map on Twistorial Cohomotopy}]{\mathrm{ch}^{\tau}_{\mathbb{C}P^3}} & H^{\tau_{\mathrm{dR}}}_{\mathrm{dR}}(X; \mathfrak{l}\mathbb{C}P^3) = & \left\{ \begin{array}{l|l} H_3, & dH_3 = G_4 - \tfrac{1}{4}p_1(\nabla) - F_2 \wedge F_2, \\ F_2, & dF_2 = 0, \\ 2G_7, & d\,2G_7 = -(G_4 - \tfrac{1}{4}p_1(\nabla)) \wedge (G_4 + \tfrac{1}{4}p_1(\nabla)) - \chi_8(\nabla), \\ G_4 & d\;G_4 = 0 \end{array} \right\}_{/\sim} \\
\Big\downarrow {\scriptstyle t_{\mathbb{H}} /\!\!/ \mathrm{Sp}(2)} \; \text{Borel-equivariantized twistor fibration} & & \Big\downarrow {\scriptstyle (t_{\mathbb{H}})_*} \; \text{cohomology operation along twistor fibration} & \Big\downarrow {\scriptstyle (\mathfrak{l}t_{\mathbb{H}})_*} & \Big\downarrow \begin{array}{cccc} H_3 & F_2 & 2G_7 & G_4 \\ \downarrow & \downarrow & \downarrow & \downarrow \\ 0 & 0 & 2G_7 & G_4 \end{array} \\
S^4 /\!\!/ \mathrm{Sp}(2) & \underset{\text{tang. twisted 4-Cohomotopy}}{\pi^{\tau}(X) := H^{\tau}(X; S^4)} & \xrightarrow[\text{character map in tang. twisted Cohomotopy}]{\mathrm{ch}^{\tau}_{S^4}} & H^{\tau_{\mathrm{dR}}}_{\mathrm{dR}}(X; \mathfrak{l}S^4) = & \left\{ \begin{array}{l|l} 2G_7, & d\,2G_7 = -(G_4 - \tfrac{1}{4}p_1(\nabla)) \wedge (G_4 + \tfrac{1}{4}p_1(\nabla)) - \chi_8(\nabla), \\ G_4 & d\;G_4 = 0, \end{array} \right\}_{/\sim}
\end{array}
$$

Here:

(i) The twisted non-abelian de Rham cohomology targets on the right are as shown, by Ex. 6.8. (In particular the twisted curvature forms in the first line are relative to $\mathfrak{l}S^4$.)

(ii) The vertical twisted non-abelian cohomology operation (Def. 3.6) on the left is induced from the Borel-equivariantized twistor fibration (3.35), and that on the right from its associated morphism of rational Whitehead L_∞-algebras (Prop. 5.16).

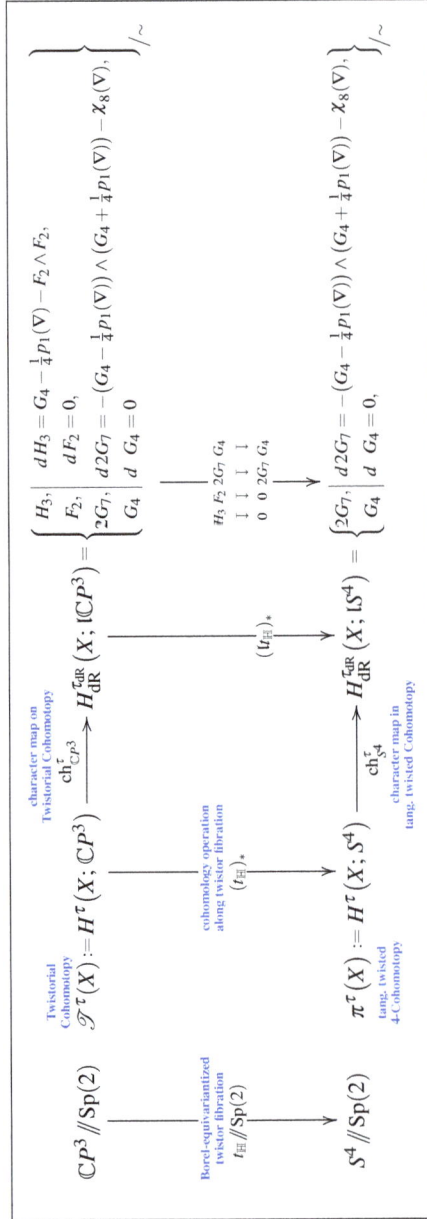

Proposition 12.1 (Charge quantization in tangentially twisted Cohomotopy [Fiorenza *et al.* (2020b), Prop. 3.13][Fiorenza *et al.* (2022), Cor. 3.11]**).** *Consider the twisted non-abelian character maps (Def. V.3) in tangentially twisted Cohomotopy and in Twistorial Cohomotopy from Ex. 12.1.*

(i) *A necessary condition for a flat* Sp(2)*-twisted* $\mathfrak{l}S^4$*-valued differential form datum* (G_4, G_7) *to lift through the tangentially twisted cohomotopical character map (i.e. to be in its image) is that the de Rham class of* G_4*, when shifted by the fourth fraction of the Pontrjagin form (Ex. 8.1), is in the image, under the de Rham homomorphism (Ex. 7.1), of an integral class:*

$$\left[G_4 - \tfrac{1}{4}p_1(\nabla)\right] \in H^4(X;\mathbb{Z}) \longrightarrow H^4_{\mathrm{dR}}(X). \tag{12.1}$$

(ii) *A necessary condition for a flat* Sp(2)*-twisted* $\mathfrak{l}\mathbb{C}P^3$*-valued differential form datum* (G_4, G_7, F_2, H_3) *to lift through the character map in Twistorial Cohomotopy is that the de Rham class of* G_4 *shifted by the fourth fraction of the Pontrjagin form (Ex. 8.1) is in the image, under the de Rham homomorphism (Ex. 7.1), of an integral class, and as such equal to the* $[F_2]$ *cup-square:*

$$\left[G_4 - \tfrac{1}{4}p_1(\nabla)\right] = \left[F_2 \wedge F_2\right] \in H^4_{\mathrm{dR}}(X). \tag{12.2}$$

Twisted differential Cohomotopy theory.

Definition 12.2 (Differential twists for twistorial Cohomotopy). Let X^8 be an 8-dimensional smooth spin manifold equipped with tangential Sp(2)-structure τ (3.33). By (3.12) in Ex. 3.2, by (3.31) in Ex. 3.9, and by (2.7) in Ex. 2.2, we have

$$[\tau] \in H^{\tau_{\mathrm{fr}}}\big(X; \mathrm{O}(n)/\mathrm{Sp}(2)\big) \xrightarrow{(Bi)_*} H\big(X^8; B\mathrm{Sp}(2)\big) \simeq \mathrm{Sp}(2)\mathrm{Bundles}(X)_{/\sim}. \tag{12.3}$$

This gives, in particular, the class of a smooth principal Sp(2)-bundle $P \to X$ to which the tangent bundle TX is associated. With (8.1), we may choose an Sp(2)-connection ∇ on P, and, by Prop. 9.4, this connection has a class $[\tau_{\mathrm{diff}}]$ in differential non-abelian cohomology (Def. 9.3) with coefficients in $B\mathrm{Sp}(2)$:

$$\begin{array}{ccccccc} H\big(X^8; B\mathrm{Sp}(2)\big) \simeq \mathrm{Sp}(2)\mathrm{Bundles}(X^8)_{/\sim} & \twoheadleftarrow & \mathrm{Sp}(2)\mathrm{Connections}(X^8)_{/\sim} & \dashrightarrow & \widehat{H}\big(X^8; B\mathrm{Sp}(2)\big) \\ [\tau] \longleftrightarrow [P] & \longleftarrow\!\!\dashv & [\nabla] & \longmapsto & [\tau_{\mathrm{diff}}]. \end{array}$$

Any such τ_{diff} serves as a *differential twist* (11.3) *for twistorial Cohomotopy* in the following.

In twisted generalization of Ex. 9.3, we have:

Example 12.2 (Differential twistorial Cohomotopy). Let X^8 be a spin 8-manifold equipped with tangential Sp(2)-structure τ (3.33), and with a corresponding differential twist τ_{diff} (Def. 12.2).

(i) Consider the local coefficient bundle (3.28), $S^4 \to S^4 /\!\!/ \mathrm{Sp}(2) \xrightarrow{J_{\mathbb{C}P^3}} B\mathrm{Sp}(2)$, for *tangentially twisted 4-Cohomotopy* (Ex. 3.8) pulled back along $B\mathrm{Sp}(2) \xrightarrow{\simeq} B\mathrm{Spin}(5) \to BO(5)$. This induces, via Def. 11.2, a twisted differential non-abelian cohomology theory $\widehat{\mathscr{T}}^{\,\tau_{\mathrm{diff}}}(-)$, which we call *tangentially twisted differential 4-Cohomotopy*, whose value on manifolds $\mathscr{X} = X^8 \times \mathbb{R}^k$ sits in a cohomology operation diagram (0.16) of this form:

$$
\begin{array}{ccc}
\underset{\substack{\text{differential} \\ \text{tang. twisted} \\ \text{4-Cohomotopy}}}{\widehat{\pi}^{\,\tau_{\mathrm{diff}}}(\mathscr{X})} \xrightarrow{\substack{\text{tang. twisted} \\ \text{cohomotopical} \\ \text{curvature} \\ F_{S^4}^{\tau_{\mathrm{dR}}^4}}} & \left\{ \begin{array}{l} 2G_7, \\ G_4 \end{array} \in \Omega_{\mathrm{dR}}^{\bullet}(\mathscr{X}) \left| \begin{array}{l} \overset{\text{tang. twisted cohomotopical Bianchi identities (Ex. 6.8)}}{d\,2G_7 = -\left(G_4 - \tfrac{1}{4}p_1(\nabla)\right) \wedge \left(G_4 + \tfrac{1}{4}p_1(\nabla)\right) - \chi_8(\nabla)} \\ d \ \ G_4 = 0 \end{array} \right. \right\} \\[3em]
\downarrow & \downarrow \\[1em]
\underset{\substack{\text{tang. twisted} \\ \text{4-Cohomotopy} \\ \text{(Ex. 3.8)}}}{\pi^{\tau^4}(\mathscr{X})} \xrightarrow[\substack{\text{character map} \\ \text{on tang. twisted Cohomotopy} \\ \text{(Ex. 12.1)}}]{\mathrm{ch}_{S^4}^{\tau^4}} & \underset{\substack{\text{tang. twisted} \\ \text{de Rham cohomology} \\ \text{(Def. 6.9)}}}{H_{\mathrm{dR}}^{\tau_{\mathrm{dR}}^4}\left(\mathscr{X}\,;\,lS^4\right).}
\end{array}
$$

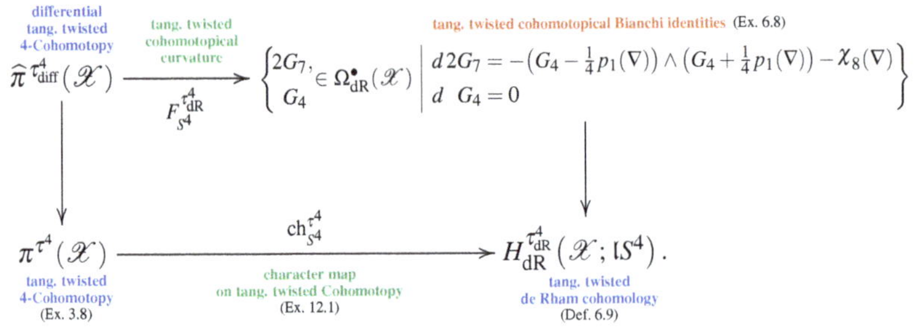

$$(12.4)$$

(ii) Consider the local coefficient bundle (3.35) $\mathbb{C}P^3 \to \mathbb{C}P^3 /\!\!/ \mathrm{Sp}(2) \xrightarrow{J_{\mathbb{C}P^3}} B\mathrm{Sp}(2)$ for *twistorial Cohomotopy* (Def. 3.11). This induces, via Def. 11.2, a twisted differential non-abelian cohomology theory $\widehat{\mathscr{T}}^{\,\tau_{\mathrm{diff}}}(-)$, which we call *differential twistorial Cohomotopy*, whose value on manifolds $\mathscr{X} = X^8 \times \mathbb{R}^k$ sits in a cohomology operation diagram (0.16) of this form:

$$
\begin{array}{ccc}
\underset{\substack{\text{differential} \\ \text{twistorial} \\ \text{Cohomotopy}}}{\widehat{\mathscr{T}}^{\,\tau_{\mathrm{diff}}}(\mathscr{X})} \xrightarrow{\substack{\text{twistorial} \\ \text{curvature} \\ F_{\mathbb{C}P^3}^{\tau_{\mathrm{dR}}}}} & \left\{ \begin{array}{l} H_3, \\ F_2, \\ 2G_7, \\ G_4 \end{array} \in \Omega_{\mathrm{dR}}^{\bullet}(\mathscr{X}) \left| \begin{array}{l} \overset{\text{twistorial Bianchi identities (Ex. 6.8)}}{d\,H_3 = G_4 - \tfrac{1}{4}p_1(\nabla) - F_2 \wedge F_2} \\ d\,F_2 = 0 \\ d\,2G_7 = -\left(G_4 - \tfrac{1}{4}p_1(\nabla)\right) \wedge \left(G_4 + \tfrac{1}{4}p_1(\nabla)\right) - \chi_8(\nabla) \\ d \ \ G_4 = 0 \end{array} \right. \right\} \\[4em]
\downarrow & \downarrow \\[1em]
\underset{\substack{\text{twistorial} \\ \text{Cohomotopy} \\ \text{(Ex. 3.11)}}}{\mathscr{T}^{\tau}(\mathscr{X})} \xrightarrow[\substack{\text{character map} \\ \text{on twistorial Cohomotopy} \\ \text{(Ex. 12.1)}}]{\mathrm{ch}_{\mathbb{C}P^3}^{\tau}} & \underset{\substack{\text{twistorial} \\ \text{de Rham cohomology} \\ \text{(Def. 6.9)}}}{H_{\mathrm{dR}}^{\tau_{\mathrm{dR}}}\left(\mathscr{X}\,;\,l\mathbb{C}P^3\right).}
\end{array}
$$

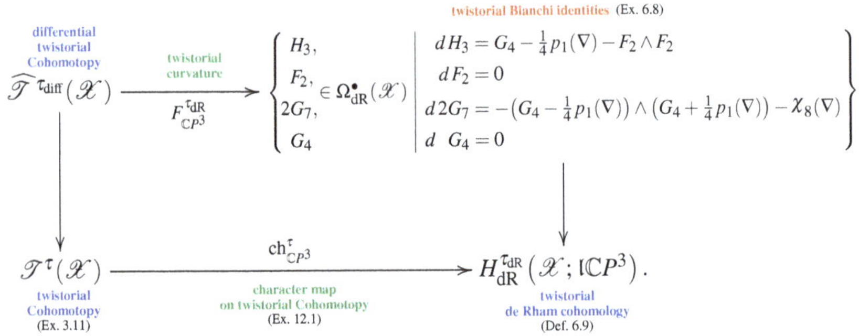

$$(12.5)$$

Proposition 12.3 (Twisted secondary cohomology operation induced by twistor fibration). *The defining twisted non-abelian cohomology operation (3.36) from twistorial Cohomotopy (Ex. 3.11) to tangentially twisted 4-Cohomotopy (Ex. 3.8), induced by the $\mathrm{Sp}(2)$-equivariantized twistor fibration $t_{\mathbb{H}} /\!\!/ \mathrm{Sp}(2)$ (3.35) refines to a twisted secondary cohomology operation (Def. 11.6) from differential twistorial Cohomotopy to differential* *twisted Cohomotopy (Ex. 12.2):*

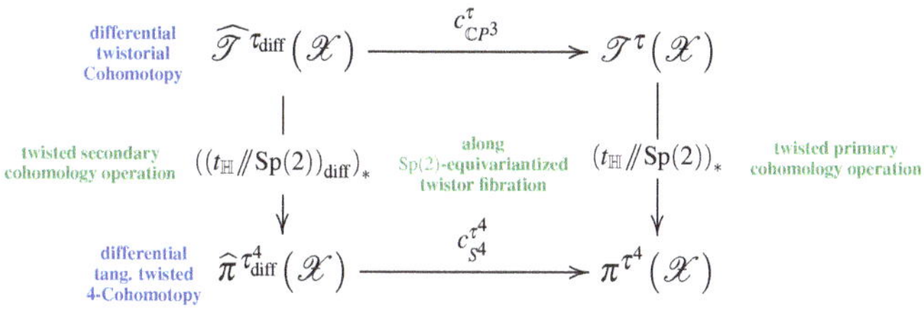

Proof. By Def. 11.6 we need to show that we have a twisted absolute minimal model (Def. 11.5) for the Sp(2)-equivariantized twistor fibration (3.35). By (11.14) this means that we can find a morphism

$$\mathfrak{l}_{B\mathrm{Sp}(2)}\big(\mathbb{C}P^3 /\!\!/ \mathrm{Sp}(2)\big) \overset{\mathfrak{t}_{\mathbb{H}} /\!\!/ \mathfrak{l}\mathrm{Sp}(2)}{\dashrightarrow} \mathfrak{l}_{B\mathrm{Sp}(2)}\big(S^4 /\!\!/ \mathrm{Sp}(2)\big) \tag{12.6}$$

between the relative Whitehead L_∞-algebras (Prop. 5.16) of the two local coefficient bundles, which makes the following cube of transformations of derived PL-de Rham adjunction units commute:

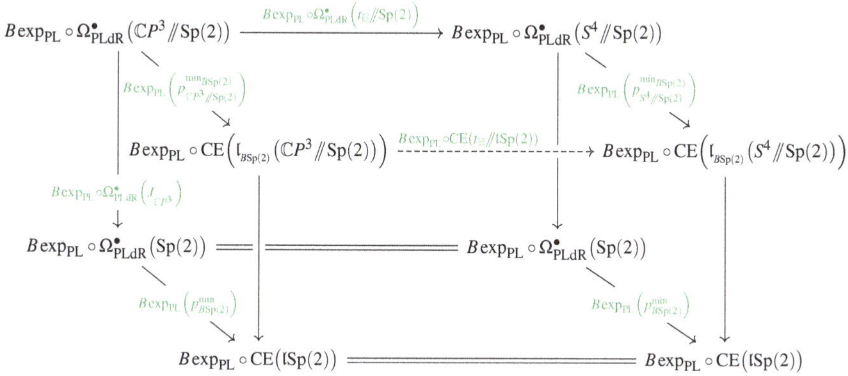

But, from Ex. 6.8, we see that the total object of the relative Whitehead L_∞-algebra of $\mathbb{C}P^3 /\!\!/ \mathrm{Sp}(2)$, relative to $\mathfrak{l}B\mathrm{Sp}(2)$, coincides with that relative to $\mathfrak{l}_{B\mathrm{Sp}(2)}\big(S^4 /\!\!/ \mathrm{Sp}(2)\big)$. Therefore, we may take the twisted absolute minimal model (12.6) to be equal to top arrow in Ex. 6.8. This makes the front square commute by construction, and it being a relative minimal model for $t_{\mathbb{H}} /\!\!/ \mathrm{Sp}(2)$ implies by Prop. 4.24 that there is an essentially unique top left morphism such that the top square commutes. □

Remark 12.1 (Lifting against the twisted differential twistor fibration). In terms of differential moduli ∞-stacks (11.1), the result of Prop. 12.3 with Ex. 12.2 says that lifting a twisted differential Cohomotopy cocycle $\widehat{C}_3$ with 4-flux density G_4 against the twisted differential refinement (11.15) of the equivariantized twistor fibration (3.35) to a differential twistorial Cohomotopy cocycle $(\widehat{C}_3, \widehat{C}_2, \widehat{C}_1)$ involves, on twisted curvature forms (11.6) the appearance of a 2-flux density F_2 and of a 3-form H_3 such that $dH_3 = G_4 - \frac{1}{4}p_1(\nabla) - F_2 \wedge F_2$.

$$
\begin{array}{ccc}
& \overset{\substack{\text{lift of C-field through}\\\text{twistor fibration}}}{} & \\
\mathbb{R}^{1,1} \times X^8 \dashrightarrow \underset{(\widehat{C}_3, \widehat{B}_2, \widehat{A}_1)}{\dashrightarrow} & \big(\mathbb{C}P^3 /\!\!/ \mathrm{Sp}(2)\big)_{\text{diff}} \\
\Big\downarrow & & \Big\downarrow {\scriptstyle (t_{\mathbb{H}} /\!\!/ \mathrm{Sp}(2))_{\text{diff}}} \overset{\substack{\text{twisted differential}\\\text{twistor fibration}}}{} \\
\mathbb{R}^{2,1} \times X^8 \xrightarrow[\text{C-field}]{\widehat{C}_3} & \big(S^4 /\!\!/ \mathrm{Sp}(2)\big)_{\text{diff}}
\end{array}
$$

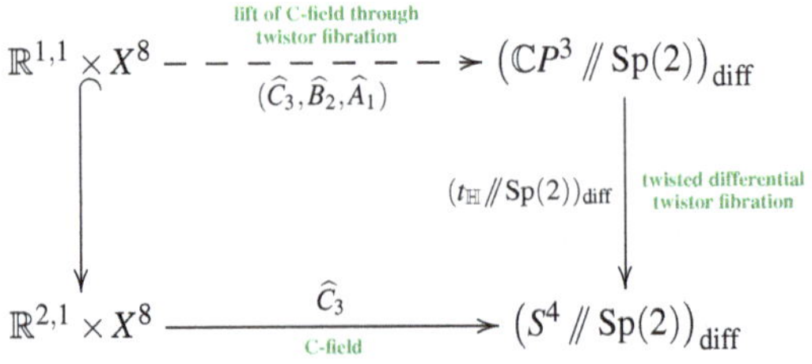

Remark 12.2 (M-theory fields and Hypothesis H). In summary, we have found:

(i) A cocycle $\widehat{C}_3$ in tangentially twisted differential 4-Cohomotopy (Ex. 12.2) has as curvature/character forms (11.6):

    (a) a closed 4-form G_4, hence a 4-flux density,

    (b) a non-closed 7-form G_7,

satisfying the following Bianchi identities (Ex. 12.1) and integrality conditions (Prop. 12.1):

$$
\underset{\substack{\text{differential}\\\text{tang. twisted 4-Cohomotopy}}}{\widehat{\pi}\,{}^{\tau_{\text{diff}}^4}(\mathcal{X})} \xrightarrow{\underset{\text{(non-abelian character form representative)}}{\text{curvature}}} \underset{\substack{\text{flat twisted cohomotopical}\\\text{differential forms}}}{\Omega_{\text{dR}}^{\tau_{\text{dR}}}(X;\mathrm{l}S^4)_{\text{flat}}}
$$

$$
(\widehat{C}_3) \longmapsto \left(\begin{array}{c|c} \begin{array}{c} G_4, \\ 2G_7 \end{array} & \begin{array}{l} \overset{\text{shifted C-field flux quantization}}{\big[G_4 - \frac{1}{4}p_1(\omega)\big]} \quad \in H^4(X;\mathbb{Z}) \\ d\ G_4 = 0 \\ d\,2G_7 = -\big(G_4 - \frac{1}{4}p_1(\omega)\big) \wedge \big(G_4 + \frac{1}{4}p_1(\omega)\big) - 24I_8(\omega) \\ \underset{\text{C-field tadpole cancellation \& M5 Hopf WZ term level quantization}}{} \end{array} \end{array} \right),
$$

$$(12.7)$$

where the characteristic forms p_1, p_2 and I_8 are from Ex. 8.1.

(ii) Lifting this cocycle through the twisted differential twistor fibration (Prop. 12.3) to a cocycle $(\widehat{C}_3, \widehat{B}_2, \widehat{A}_1)$ in differential twistorial Cohomotopy (Ex. 12.2) involves (Rem. 12.1) adjoining to the 4-flux density G_4:

    (c) a closed 2-form curvature F_2, hence a 2-flux density,

    (d) a non-closed 3-form H_3,

such that these curvature/character forms satisfy the following Bianchi identities (Ex. 12.1) and integrality conditions (Prop. 12.1):

$$
\begin{array}{ccc}
\underset{\text{twistorial Cohomotopy}}{\overset{\text{differential}}{}} \widehat{\mathcal{T}}^{\tau_{\mathrm{diff}}}(\mathscr{X}) & \xrightarrow[\text{(non-abelian character form representative)}]{\text{curvature}} & \underset{}{\overset{\text{flat twistorial differential forms}}{}} \Omega_{\mathrm{dR}}^{\tau_{\mathrm{dR}}}\big(X; \mathbb{C}P^3\big)_{\mathrm{flat}} \\
\big(\widehat{C}_3, \widehat{B}_2, \widehat{A}_1\big) & \longmapsto & \left(\begin{array}{l|l}
H_3, & d H_3 = G_4 - \tfrac{1}{4}p_1(\omega) - F_2 \wedge F_2, \\
 & \text{Hořava-Witten Green-Schwarz mechanism} \\
 & \big[G_4 - \tfrac{1}{4}p_1(\omega)\big] = \big[F_2 \wedge F_2\big] \ \in H^4(X;\mathbb{Z}) \\
G_4, F_2, & d\ G_4 = 0, \ \ d F_2 = 0, \\
2G_7 & d\, 2G_7 = -\big(G_4 - \tfrac{1}{4}p_1(\omega)\big) \wedge \big(G_4 + \tfrac{1}{4}p_1(\omega)\big) - 24 I_8(\omega) \\
 & \text{C-field tadpole cancellation \& M5 Hopf WZ term level quantization}
\end{array}\right)
\end{array}
\tag{12.8}
$$

(iii) With these cohomotopical curvature/character forms interpreted as flux densities, this is the Bianchi identities and charge quantization expected in M-theory on the supergravity C-field ($\widehat{C}_3$), the heterotic B-field ($\widehat{B}_2$) and the heterotic $S\big(\mathrm{U}(1)^2\big) \subset E_8$ gauge field ($\widehat{A}_1$), with the following prominent features:

(a) The charge quantization:

(1)	$\big[G_4 - \tfrac{1}{4}p_1(\nabla)\big] \qquad \in H^4(\mathscr{X};\mathbb{Z})$	is expected for the C-field in the M-theory bulk ([Witten (1997b), (1.2)] [Witten (1997a), (1.2)])
(2)	$\big[G_4 - \tfrac{1}{4}p_1(\nabla)\big] = [F_2 \wedge F_2] \in H^4(\mathscr{X};\mathbb{Z})$	is expected on heterotic boundaries ([Hořava and Witten (1996), (1.13)], review in [Fiorenza *et al.* (2022), §1])

(12.9)

(b) The quadratic functions:

(1)	$G_4 \mapsto \big(G_4 - \tfrac{1}{4}p_1(\omega)\big) \wedge \big(G_4 + \tfrac{1}{4}p_1(\omega)\big) + 24 I_8(\omega)$	constitute the Hopf Wess-Zumino term or Page charge ([Aharony (1996), p. 11] [Intriligator (2000)], see [Fiorenza *et al.* (2021b)][Sati and Schreiber (2021b)])
(2)	$F_2 \mapsto F_2 \wedge F_2$	constitute the 2nd Chern class of a $\mathrm{U}(1) \subset \mathrm{SU}(2) \subset E_8$-bundle [Anderson *et al.* (2011)] [Anderson *et al.* (2012)] [Fiorenza *et al.* (2022), (7)]

(12.10)

(iv) These are necessary, not yet sufficient constraints on cohomotopical lifts. Further constraints follow by Postnikov tower analysis [Grady and Sati (2021a)] and coincide with further expected conditions in M-theory (see [Fiorenza *et al.* (2020b), Table 1]).

All this suggests the *Hypothesis H* [Fiorenza *et al.* (2020b)][Fiorenza *et al.* (2021b)][Sati and Schreiber (2020b)][Sati and Schreiber (2022)][Braunack-Mayer *et al.* (2019)][Sati and Schreiber (2021b)][Fiorenza *et al.* (2022)][Fiorenza *et al.* (2021c)][Sati and Schreiber (2021a)], following [Sati (2018), §2.5], that the elusive cohomology theory which controls M-theory in analogy to how K-theory controls string theory is: (twisted, equivariant, differential) non-abelian Cohomotopy theory.

Cohomotopical character into K-theory. We may regard the *secondary non-abelian Hurewicz/Boardman homomorphism* (Ex. 9.5) from differential 4-Cohomotopy (Ex. 9.3) to differential K-theory (Ex. 9.2), as a non-abelian but K-theory valued character, lifting the target of the cohomotopical character (Ex. 12.1) from rational cohomology to K-theory (compare [Burton *et al.* (2021), Fig. 1]):

$$\underset{\text{Cohomotopy}}{\overset{\text{differential}}{\widehat{\pi}^4(\mathscr{X})}} \xrightarrow[\substack{\text{differential non-abelian}\\ \text{Boardman homomorphism}\\ \text{(Ex. 9.5)}}]{(e_{\mathrm{KU}})^4_{\mathrm{diff}}} \underset{\text{K-theory}}{\overset{\text{differential}}{\widehat{\mathrm{KU}}^0(\mathscr{X})}} \xrightarrow[\substack{\text{differential}\\ \text{Chern character}\\ \text{(Ex. (11.3))}}]{\mathrm{ch}_{\mathrm{diff}}} \underset{\text{rational cohomology}}{\overset{\text{differential}}{\widehat{H}_{\mathrm{per}}\mathbb{Q}^0(\mathscr{X})}} \tag{12.11}$$

$$\xleftarrow{\quad\text{charge-quantization in M-theory}\quad} \qquad \xleftarrow{\quad\text{charge-quantization in string theory}\quad}$$

(i) Lifting through this differential Boardman homomorphism induces *secondary charge quantization conditions on* K-theory, analogous to (0.2) but invisible even in generalized cohomology, instead now coming from non-abelian cohomology theory.

(ii) In the plain version (12.11) (i.e. disregarding twisting and equivariant enhancement) the effect of $(e_{\mathrm{KU}})^4_{\mathrm{diff}}$ on curvature forms (9.13) is (by Ex. 9.5) to forget the quadratic function (12.10) on G_4 and to inject what remains as the 4-form curvature component F_4 in differential K-theory:

$$\begin{array}{ccc}
\underset{\text{Cohomotopy}}{\overset{\text{differential}}{\widehat{\pi}^4(\mathscr{X})}} & \xrightarrow[(e_{\mathrm{KU}})^4_{\mathrm{diff}}]{\substack{\text{secondary non-abelian}\\ \text{Boardman homomorphism}}} & \underset{\text{K-theory}}{\overset{\text{differential}}{\widehat{\mathrm{KU}}^0(\mathscr{X})}} \\
\Big\downarrow (F_{S^4})_* & \text{curvatures/flux densities} & \Big\downarrow (F_{\mathrm{KU}_0})_* \\
\left\{ \begin{array}{c|l} 2G_7, & d\,2G_7 = -G_4 \wedge G_4 \\ G_4 & d\;\;G_4 = 0 \end{array} \right\} & \xrightarrow[\substack{G_4 \mapsto F_4 \\ G_7 \mapsto 0}]{} & \big\{ (F_{2k}) \,\big|\, d\,F_{2k} = 0 \big\}.
\end{array} \tag{12.12}$$

This is a 'cohomotopical enhancement' of the reduction in [Diaconescu *et al.* (2002)] of E_8 bundles in M-theory to the K-theory of type IIA string theory, now characterized by higher Postnikov stages of the Boardman homomorphism [Grady and Sati (2021a)]. The remaining RR-flux components in $\{F_{2k}\}$ are also found in the cohomotopical character, through cohomological double dimensional reduction formulated in parametrized homotopy theory: this is discussed in detail in [Braunack-Mayer *et al.* (2019)].

(iii) The twisted generalization of the non-abelian Boardman homomorphism in (12.11) and (12.12) is more subtle, since the degree-3 twist of K-theory does not arise from the tangential twist of Cohomotopy, but arises, together with the further RR-flux components, from S^1-equivariantization/double dimensional reduction of Cohomotopy [Fiorenza *et al.* (2017)][Braunack-Mayer *et al.* (2019)], reproducing the reduction of E_8 bundles from M-theory to type IIA in [Mathai and Sati (2004)].

Outlook: Equivariant enhancement. The twisted non-abelian character theory presented here enhances further to *proper* (i.e. Bredon-style not Borel-style) *equivariant* non-abelian cohomology on orbi-orientifolds, by combining it with the techniques developed in [Huerta *et al.* (2019)][Sati and Schreiber (2020c)] (essentially: parametrizing the construction here over the orbit category of the equivariance group). The resulting *character map in equivariant non-abelian cohomology* is discussed in [Sati and Schreiber (2020a), §2,3]. For example, the equivariant enhancement of the cohomotopical character into K-theory (12.11), lifting the RR-fields in equivariant K-theory through the equivariantized enhancement of the Boardman homomorphism on the left of (12.11), enforces [Sati and Schreiber (2020b)][Burton *et al.* (2021)] "tadpole cancellation" conditions expected in string theory at orbifold singularities.

Bibliography

Adamo, T. (2015). Gravity with a cosmological constant from rational curves, *J. High Energy Phys.* **2015**, 98, doi:10.1007/JHEP11(2015)098, https://doi.org/10.1007/JHEP11(2015)098.

Adams, J. F. (1962). Vector fields on spheres, *Bull. Amer. Math. Soc.* **68**, pp. 39–41, doi:10.1090/S0002-9904-1962-10693-4, https://doi.org/10.1090/S0002-9904-1962-10693-4.

Adams, J. F. (1974). *Stable homotopy and generalised homology*, Chicago Lectures in Mathematics (University of Chicago Press, Chicago, Ill.-London).

Adams, J. F. (1978). *Infinite loop spaces*, Annals of Mathematics Studies, No. 90 (Princeton University Press, Princeton, N.J.; University of Tokyo Press, Tokyo), ISBN 0-691-08207-3; 0-691-08206-5.

Addington, N. (2007). Fiber bundles and nonabelian cohomology, http://pages.uoregon.edu/adding/notes/gstc2007.pdf.

Aguilar, M., Gitler, S., and Prieto, C. (2002). *Algebraic topology from a homotopical viewpoint*, Universitext (Springer-Verlag, New York), ISBN 0-387-95450-3, doi:10.1007/b97586, https://doi.org/10.1007/b97586, translated from the Spanish by Stephen Bruce Sontz.

Aharony, O. (1996). String theory dualities from M-theory, *Nuclear Phys. B* **476**, 3, pp. 470–483, doi:10.1016/0550-3213(96)00321-5, https://doi.org/10.1016/0550-3213(96)00321-5.

Alvarez, O. (1985). Topological quantization and cohomology, *Comm. Math. Phys.* **100**, 2, pp. 279–309, http://projecteuclid.org/euclid.cmp/1103943448.

Anderson, L. B., Gray, J., Lukas, A., and Palti, E. (2011). Two hundred heterotic standard models on smooth calabi-yau threefolds, *Phys. Rev. D* **84**, p. 106005, doi:10.1103/PhysRevD.84.106005, https://link.aps.org/doi/10.1103/PhysRevD.84.106005.

Anderson, L. B., Gray, J., Lukas, A., and Palti, E. (2012). Heterotic line bundle standard models, *J. High Energy Phys.* **2012**, 113, doi:10.1007/JHEP06(2012)113, https://doi.org/10.1007/JHEP06(2012)113.

Ando, M., Blumberg, A. J., and Gepner, D. (2010). Twists of K-theory and TMF, in *Superstrings, geometry, topology, and C^*-algebras*, *Proc. Sympos. Pure Math.*, Vol. 81 (Amer. Math. Soc., Providence, RI), pp. 27–63, doi:10.1090/pspum/081/2681757, https://doi.org/10.1090/pspum/081/2681757.

Ando, M., Blumberg, A. J., Gepner, D., Hopkins, M., and Rezk, C. (2008). Units of ring spectra and Thom spectra, https://arxiv.org/abs/0810.4535.

Ando, M., Blumberg, A. J., Gepner, D., Hopkins, M. J., and Rezk, C. (2014a). An ∞-categorical approach to R-line bundles, R-module Thom spectra, and twisted R-homology, *J. Topol.* **7**, 3, pp. 869–893, doi:10.1112/jtopol/jtt035, https://doi.org/10.1112/jtopol/jtt035.

Ando, M., Blumberg, A. J., Gepner, D., Hopkins, M. J., and Rezk, C. (2014b). Units of ring spectra, orientations and Thom spectra via rigid infinite loop space theory, *J. Topol.* **7**, 4, pp. 1077–1117, doi:10.1112/jtopol/jtu009, https://doi.org/10.1112/jtopol/jtu009.

Ando, M., Hopkins, M., and Rezk, C. (2010). Multiplicative orientations of KO-theory and the spectrum of topological modular forms, https://faculty.math.illinois.edu/~mando/papers/koandtmf.pdf.

Ando, M., Hopkins, M. J., and Strickland, N. P. (2001). Elliptic spectra, the Witten genus and the theorem of the cube, *Invent. Math.* **146**, 3, pp. 595–687, doi:10.1007/s002220100175, https://doi.org/10.1007/s002220100175.

Artin, M. and Mazur, B. (1977). Formal groups arising from algebraic varieties, *Ann. Sci. École Norm. Sup. (4)* **10**, 1, pp. 87–131, http://www.numdam.org/item?id=ASENS_1977_4_10_1_87_0.

Ashmore, A., Dumitru, S., and Ovrut, B. A. (2021a). Explicit soft supersymmetry breaking in the heterotic M-theory $B-L$ MSSM, *J. High Energy Phys.* **2021**, 33, doi:10.1007/jhep08(2021)033, https://doi.org/10.1007/jhep08(2021)033.

Ashmore, A., Dumitru, S., and Ovrut, B. A. (2021b). Line bundle hidden sectors for strongly coupled heterotic standard models, *Fortschr. Phys.* **69**, 7, pp. Paper No. 2100052, 33, doi:10.1002/prop.202100052, https://doi.org/10.1002/prop.202100052.

Atiyah, M., Dunajski, M., and Mason, L. J. (2017). Twistor theory at fifty: from contour integrals to twistor strings, *Proc. A.* **473**, 2206, pp. 20170530, 33, doi:10.1098/rspa.2017.0530, https://doi.org/10.1098/rspa.2017.0530.

Atiyah, M. and Hitchin, N. (1988). *The geometry and dynamics of magnetic monopoles*, M. B. Porter Lectures (Princeton University Press, Princeton, NJ), ISBN 0-691-08480-7, doi:10.1515/9781400859306, https://doi.org/10.1515/9781400859306.

Atiyah, M. and Segal, G. (2004). Twisted K-theory, *Ukr. Mat. Visn.* **1**, 3, pp. 287–330.

Atiyah, M. and Segal, G. (2006). Twisted K-theory and cohomology, in *Inspired by S. S. Chern*, *Nankai Tracts Math.*, Vol. 11 (World Sci. Publ., Hackensack, NJ), pp. 5–43, doi:10.1142/9789812772688_0002, https://doi.org/10.1142/9789812772688_0002.

Atiyah, M. F. (1967). *K-theory* (W. A. Benjamin, Inc., New York-Amsterdam), lecture notes by D. W. Anderson.

Atiyah, M. F. (1984). Instantons in two and four dimensions, *Comm. Math. Phys.* **93**, 4, pp. 437–451, http://projecteuclid.org/euclid.cmp/1103941176.

Atiyah, M. F. and Hirzebruch, F. (1959). Riemann-Roch theorems for differentiable manifolds, *Bull. Amer. Math. Soc.* **65**, pp. 276–281, doi:10.1090/S0002-9904-1959-10344-X.

Atiyah, M. F. and Hirzebruch, F. (1961). Vector bundles and homogeneous spaces, in *Proc. Sympos. Pure Math., Vol. III* (American Mathematical Society, Providence, R.I.), pp. 7–38.

Ausoni, C. (2010). On the algebraic K-theory of the complex K-theory spectrum, *Invent. Math.* **180**, 3, pp. 611–668, doi:10.1007/s00222-010-0239-x, `https://doi.org/10.1007/s00222-010-0239-x`.

Ausoni, C. and Rognes, J. (2002). Algebraic K-theory of topological K-theory, *Acta Math.* **188**, 1, pp. 1–39, doi:10.1007/BF02392794, `https://doi.org/10.1007/BF02392794`.

Ausoni, C. and Rognes, J. (2012). Rational algebraic K-theory of topological K-theory, *Geom. Topol.* **16**, 4, pp. 2037–2065, doi:10.2140/gt.2012.16.2037, `https://doi.org/10.2140/gt.2012.16.2037`.

Baas, N. A., Dundas, B. r. I., Richter, B., and Rognes, J. (2011). Stable bundles over rig categories, *J. Topol.* **4**, 3, pp. 623–640, doi:10.1112/jtopol/jtr016, `https://doi.org/10.1112/jtopol/jtr016`.

Baas, N. A., Dundas, B. r. I., and Rognes, J. (2004). Two-vector bundles and forms of elliptic cohomology, in *Topology, geometry and quantum field theory*, *London Math. Soc. Lecture Note Ser.*, Vol. 308 (Cambridge Univ. Press, Cambridge), pp. 18–45, doi:10.1017/CBO9780511526398.005, `https://doi.org/10.1017/CBO9780511526398.005`.

Baez, J. C. and Hoffnung, A. E. (2011). Convenient categories of smooth spaces, *Trans. Amer. Math. Soc.* **363**, 11, pp. 5789–5825, doi:10.1090/S0002-9947-2011-05107-X, `https://doi.org/10.1090/S0002-9947-2011-05107-X`.

Baez, J. C. and Stevenson, D. (2009). The classifying space of a topological 2-group, in *Algebraic topology*, *Abel Symp.*, Vol. 4 (Springer, Berlin), pp. 1–31, doi:10.1007/978-3-642-01200-6_1, `https://doi.org/10.1007/978-3-642-01200-6_1`.

Baez, J. C., Stevenson, D., Crans, A. S., and Schreiber, U. (2007). From loop groups to 2-groups, *Homology Homotopy Appl.* **9**, 2, pp. 101–135, `http://projecteuclid.org/euclid.hha/1201127333`.

Baker, A. and Richter, B. (eds.) (2004). *Structured ring spectra*, *London Mathematical Society Lecture Note Series*, Vol. 315 (Cambridge University Press, Cambridge), ISBN 0-521-60305-6, doi:10.1017/CBO9780511529955, `https://doi.org/10.1017/CBO9780511529955`.

Battye, R. A., Manton, N. S., and Sutcliffe, P. M. (2010). *Skyrmions and Nuclei*, chap. 1 (World Scientific Publishing Co. Pte. Ltd., Hackensack, NJ), pp. 3–40.

Bauer, T. (2014). *Bousfield localization and the Hasse square*, *Mathematical Surveys and Monographs*, Vol. 201, chap. 6 (American Mathematical Society, Providence, RI), ISBN 978-1-4704-1884-7, pp. 79–88, doi:10.1090/surv/201, `https://doi.org/10.1090/surv/201`.

Belchí, F., Buijs, U., Moreno-Fernández, J. M., and Murillo, A. (2017). Higher order Whitehead products and L_∞ structures on the homology of a DGL, *Linear Algebra Appl.* **520**, pp. 16–31, doi:10.1016/j.laa.2017.01.008, `https://doi.org/10.1016/j.laa.2017.01.008`.

Berglund, A. (2015). Rational homotopy theory of mapping spaces via Lie theory for L_∞-algebras, *Homology Homotopy Appl.* **17**, 2, pp. 343–369, doi:10.4310/HHA.2015.v17.n2.a16, `https://doi.org/10.4310/HHA.2015.v17.n2.a16`.

Bertram, A. (2004). Stable maps and Gromov-Witten invariants, in *Intersection theory and moduli*, ICTP Lect. Notes, XIX (Abdus Salam Int. Cent. Theoret. Phys., Trieste), pp. 1–40.

Berwick-Evans, D. (2013). Perturbative sigma models, elliptic cohomology and the Witten genus, `https://arxiv.org/abs/1311.6836`.

Beĭlinson, A. A. (1984). Higher regulators and values of L-functions, in *Current problems in mathematics, Vol. 24*, Itogi Nauki i Tekhniki (Akad. Nauk SSSR, Vsesoyuz. Inst. Nauchn. i Tekhn. Inform., Moscow), pp. 181–238.

Blumberg, A. J., Gepner, D., and Tabuada, G. (2013). A universal characterization of higher algebraic K-theory, *Geom. Topol.* **17**, 2, pp. 733–838, doi:10.2140/gt.2013.17.733, `https://doi.org/10.2140/gt.2013.17.733`.

Boardman, J. M., Johnson, D. C., and Wilson, W. S. (1995). Unstable operations in generalized cohomology, in *Handbook of algebraic topology* (North-Holland, Amsterdam), pp. 687–828, doi:10.1016/B978-044481779-2/50016-X, `https://doi.org/10.1016/B978-044481779-2/50016-X`.

Borceux, F. (1994). *Handbook of categorical algebra. 2, Encyclopedia of Mathematics and its Applications*, Vol. 51 (Cambridge University Press, Cambridge), ISBN 0-521-44179-X, categories and structures.

Borel, A. and Hirzebruch, F. (1958). Characteristic classes and homogeneous spaces. I, *Amer. J. Math.* **80**, pp. 458–538, doi:10.2307/2372795, `https://doi.org/10.2307/2372795`.

Borsuk, K. (1936). Sur les groupes des classes de transformations continues, *C.R. Acad. Sci. Paris* **202**, pp. 1400–1403, doi:10.24033/asens.603, `https://doi.org/10.24033/asens.603`.

Bott, R. (1973). On the Chern-Weil homomorphism and the continuous cohomology of Lie-groups, *Advances in Math.* **11**, pp. 289–303, doi:10.1016/0001-8708(73)90012-1, `https://doi.org/10.1016/0001-8708(73)90012-1`.

Bott, R. and Tu, L. W. (1982). *Differential forms in algebraic topology*, *Graduate Texts in Mathematics*, Vol. 82 (Springer-Verlag, New York-Berlin), ISBN 0-387-90613-4.

Bousfield, A. K. (1979). The localization of spectra with respect to homology, *Topology* **18**, 4, pp. 257–281, doi:10.1016/0040-9383(79)90018-1, `https://doi.org/10.1016/0040-9383(79)90018-1`.

Bousfield, A. K. and Friedlander, E. M. (1978). Homotopy theory of Γ-spaces, spectra, and bisimplicial sets, in *Geometric applications of homotopy theory (Proc. Conf., Evanston, Ill., 1977), II, Lecture Notes in Math.*, Vol. 658 (Springer, Berlin), pp. 80–130.

Bousfield, A. K. and Gugenheim, V. K. A. M. (1976). On PL de Rham theory and rational homotopy type, *Mem. Amer. Math. Soc.* **8**, 179, pp. ix+94, doi:10.1090/memo/0179, `https://doi.org/10.1090/memo/0179`.

Bousfield, A. K. and Kan, D. M. (1972a). The core of a ring, *J. Pure Appl. Algebra* **2**, pp. 73–81, doi:10.1016/0022-4049(72)90023-0, `https://doi.org/10.1016/0022-4049(72)90023-0`.

Bousfield, A. K. and Kan, D. M. (1972b). *Homotopy limits, completions and localizations*, Lecture Notes in Mathematics, Vol. 304 (Springer-Verlag, Berlin-New York).

Bouwknegt, P., Carey, A. L., Mathai, V., Murray, M. K., and Stevenson, D. (2002). Twisted K-theory and K-theory of bundle gerbes, *Comm. Math. Phys.* **228**, 1, pp. 17–45, doi:10.1007/s002200200646, `https://doi.org/10.1007/s002200200646`.

Braunack-Mayer, V., Sati, H., and Schreiber, U. (2019). Gauge enhancement of super M-branes via parametrized stable homotopy theory, *Comm. Math. Phys.* **371**, 1, pp. 197–265, doi:10.1007/s00220-019-03441-4, `https://doi.org/10.1007/s00220-019-03441-4`.

Breen, L. (1990). Bitorseurs et cohomologie non abélienne, in *The Grothendieck Festschrift, Vol. I, Progr. Math.*, Vol. 86 (Birkhäuser Boston, Boston, MA), pp. 401–476.

Breen, L. (2010). *Notes on 1- and 2-gerbes, The IMA Volumes in Mathematics and its Applications*, Vol. 152, chap. 5 (Springer, New York), ISBN 978-1-4419-1523-8, pp. xiv+268, doi:10.1007/978-1-4419-1524-5, `https://doi.org/10.1007/978-1-4419-1524-5`.

Bressler, P., Gorokhovsky, A., Nest, R., and Tsygan, B. (2008). Chern character for twisted complexes, in *Geometry and dynamics of groups and spaces, Progr. Math.*, Vol. 265 (Birkhäuser, Basel), pp. 309–324, doi:10.1007/978-3-7643-8608-5_5, `https://doi.org/10.1007/978-3-7643-8608-5_5`.

Brown, E. H., Jr. and Szczarba, R. H. (1989). Continuous cohomology and real homotopy type, *Trans. Amer. Math. Soc.* **311**, 1, pp. 57–106, doi:10.2307/2001017, `https://doi.org/10.2307/2001017`.

Brown, E. H., Jr. and Szczarba, R. H. (1995). Real and rational homotopy theory, in *Handbook of algebraic topology* (North-Holland, Amsterdam), pp. 867–915, doi:10.1016/B978-044481779-2/50018-3, `https://doi.org/10.1016/B978-044481779-2/50018-3`.

Brown, K. S. (1973). Abstract homotopy theory and generalized sheaf cohomology, *Trans. Amer. Math. Soc.* **186**, pp. 419–458, doi:10.2307/1996573, `https://doi.org/10.2307/1996573`.

Brylinski, J.-L. (1993). *Loop spaces, characteristic classes and geometric quantization, Progress in Mathematics*, Vol. 107 (Birkhäuser Boston, Inc., Boston, MA), ISBN 0-8176-3644-7, doi:10.1007/978-0-8176-4731-5, `https://doi.org/10.1007/978-0-8176-4731-5`.

Buhné, L. (2011). Properties of integral Morava K-theory and the asserted application to the Diaconescu-Moore-Witten anomaly, `https://inspirehep.net/literature/899170`.

Buhštaber, V. M. (1970). The Chern-Dold character in cobordisms. I, *Mat. Sb. (N.S.)* **83** **(125)**, pp. 575–595.

Buijs, U., Félix, Y., and Murillo, A. (2011). L_∞ models of based mapping spaces, *J. Math. Soc. Japan* **63**, 2, pp. 503–524, `http://projecteuclid.org/euclid.jmsj/1303737796`.

Bullejos, M., Faro, E., and García-Muñoz, M. A. (2003). Homotopy colimits and cohomology with local coefficients, *Cah. Topol. Géom. Différ. Catég.* **44**, 1, pp. 63–80.

Bunke, U. (2013). Differential cohomology, `https://arxiv.org/abs/1208.3961`.

Bunke, U. and Gepner, D. (2012). Differential function spectra, the differential Becker-Gottlieb transfer, and applications to differential algebraic K-theory, https://arxiv.org/abs/1306.0247.

Bunke, U. and Nikolaus, T. (2019). Twisted differential cohomology, *Algebr. Geom. Topol.* **19**, 4, pp. 1631–1710, doi:10.2140/agt.2019.19.1631, https://doi.org/10.2140/agt.2019.19.1631.

Bunke, U., Nikolaus, T., and Völkl, M. (2016). Differential cohomology theories as sheaves of spectra, *J. Homotopy Relat. Struct.* **11**, 1, pp. 1–66, doi:10.1007/s40062-014-0092-5, https://doi.org/10.1007/s40062-014-0092-5.

Bunke, U. and Schick, T. (2009). Smooth K-theory, *Astérisque* **328**, pp. 45–135 (2010).

Bunke, U. and Schick, T. (2012). Differential K-theory: a survey, in *Global differential geometry, Springer Proc. Math.*, Vol. 17 (Springer, Heidelberg), pp. 303–357, doi:10.1007/978-3-642-22842-1_11, https://doi.org/10.1007/978-3-642-22842-1_11.

Burton, S., Sati, H., and Schreiber, U. (2021). Lift of fractional D-brane charge to equivariant cohomotopy theory, *J. Geom. Phys.* **161**, pp. Paper No. 104034, 20, doi:10.1016/j.geomphys.2020.104034, https://doi.org/10.1016/j.geomphys.2020.104034.

Cachazo, F. and Skinner, D. (2013). Gravity from rational curves in twistor space, *Phys. Rev. Lett.* **110**, p. 161301, doi:10.1103/PhysRevLett.110.161301, https://link.aps.org/doi/10.1103/PhysRevLett.110.161301.

Čadek, M. and Vanžura, J. (1998). Almost quaternionic structures on eight-manifolds, *Osaka J. Math.* **35**, 1, pp. 165–190, http://projecteuclid.org/euclid.ojm/1200787905.

Candelas, P., Horowitz, G. T., Strominger, A., and Witten, E. (1985). Vacuum configurations for superstrings, *Nuclear Phys. B* **258**, 1, pp. 46–74, doi:10.1016/0550-3213(85)90602-9, https://doi.org/10.1016/0550-3213(85)90602-9.

Carey, A. L., Mickelsson, J., and Wang, B.-L. (2009). Differential twisted K-theory and applications, *J. Geom. Phys.* **59**, 5, pp. 632–653, doi:10.1016/j.geomphys.2009.02.002, https://doi.org/10.1016/j.geomphys.2009.02.002.

Carey, A. L., Murray, M. K., and Wang, B. L. (1997). Higher bundle gerbes and cohomology classes in gauge theories, *J. Geom. Phys.* **21**, 2, pp. 183–197, doi:10.1016/S0393-0440(96)00014-9, https://doi.org/10.1016/S0393-0440(96)00014-9.

Cartan, H. (1950). Cohomologie réelle d'un espace fibré principal différentiable. i: notions d'algèbre différentielle, algèbre de weil d'un groupe de lie, in *Séminaire Henri Cartan, vol. 2 (1949-1950), Talk no. 19* (W. A. Benjamin), http://www.numdam.org/item/SHC_1949-1950__2__A18_0/.

Cartan, H. (1951). Notions d'algèbre différentielle; application aux groupes de Lie et aux variétés où opère un groupe de Lie, in *Colloque de topologie (espaces fibrés), Bruxelles, 1950* (Georges Thone, Liège; Masson & Cie, Paris), pp. 15–27.

Castellani, L., D'Auria, R., and Fré, P. (1991). *Supergravity and superstrings. A geometric perspective* (World Scientific Publishing Co., Inc., Teaneck, NJ), ISBN 981-02-0673-9, doi:10.1142/0224, https://doi.org/10.1142/0224, mathematical foundations.

Cavalcanti, G. (2005). *New aspects of the dd^c-lemma*, Ph.D. thesis, Oxford, `https://arxiv.org/abs/math/0501406`.

Cheeger, J. and Simons, J. (1985). Differential characters and geometric invariants, in *Geometry and topology (College Park, Md., 1983/84)*, *Lecture Notes in Math.*, Vol. 1167 (Springer, Berlin), pp. 50–80, doi:10.1007/BFb0075216, `https://doi.org/10.1007/BFb0075216`.

Chen, Y. and Yang, S. (2019). On the blow-up formula of twisted de Rham cohomology, *Ann. Global Anal. Geom.* **56**, 2, pp. 277–290, doi:10.1007/s10455-019-09667-8, `https://doi.org/10.1007/s10455-019-09667-8`.

Chern, S.-S. (1951). *Topics in differential geometry* (Institute for Advanced Study (IAS), Princeton, N.J.).

Chern, S.-S. (1952). Differential geometry of fiber bundles, in *Proceedings of the International Congress of Mathematicians, Cambridge, Mass., 1950, vol. 2* (Amer. Math. Soc., Providence, R.I.), pp. 397–411.

Chern, S. S. and Simons, J. (1974). Characteristic forms and geometric invariants, *Ann. of Math. (2)* **99**, pp. 48–69, doi:10.2307/1971013, `https://doi.org/10.2307/1971013`.

Chuang, J. and Lazarev, A. (2013). Combinatorics and formal geometry of the Maurer-Cartan equation, *Lett. Math. Phys.* **103**, 1, pp. 79–112, doi:10.1007/s11005-012-0586-1, `https://doi.org/10.1007/s11005-012-0586-1`.

Cisinski, D.-C. (2019). *Higher categories and homotopical algebra*, *Cambridge Studies in Advanced Mathematics*, Vol. 180 (Cambridge University Press, Cambridge), ISBN 978-1-108-47320-0, doi:10.1017/9781108588737, `https://doi.org/10.1017/9781108588737`.

Cohen, F. R., Cohen, R. L., Mann, B. M., and Milgram, R. J. (1991). The topology of rational functions and divisors of surfaces, *Acta Math.* **166**, 3-4, pp. 163–221, doi:10.1007/BF02398886, `https://doi.org/10.1007/BF02398886`.

Córdova, C., Dumitrescu, T. T., and Intriligator, K. (2021). 2-group global symmetries and anomalies in six-dimensional quantum field theories, *J. High Energy Phys.* **2021**, 252, doi:10.1007/jhep04(2021)252, `https://doi.org/10.1007/jhep04(2021)252`.

Crans, S. E. (1998). Generalized centers of braided and sylleptic monoidal 2-categories, *Adv. Math.* **136**, 2, pp. 183–223, doi:10.1006/aima.1998.1720, `https://doi.org/10.1006/aima.1998.1720`.

Cruickshank, J. (2003). Twisted homotopy theory and the geometric equivariant 1-stem, *Topology Appl.* **129**, 3, pp. 251–271, doi:10.1016/S0166-8641(02)00183-9, `https://doi.org/10.1016/S0166-8641(02)00183-9`.

Curtis, E. B. (1971). Simplicial homotopy theory, *Advances in Math.* **6**, pp. 107–209 (1971), doi:10.1016/0001-8708(71)90015-6, `https://doi.org/10.1016/0001-8708(71)90015-6`.

Dadarlat, M. and Pennig, U. (2015). Unit spectra of K-theory from strongly self-absorbing C^*-algebras, *Algebr. Geom. Topol.* **15**, 1, pp. 137–168, doi:10.2140/agt.2015.15.137, `https://doi.org/10.2140/agt.2015.15.137`.

D'Auria, R. and Fré, P. (1982). Geometric supergravity in $D = 11$ and its hidden supergroup, *Nuclear Phys. B* **201**, 1, pp. 101–140, doi:10.1016/0550-3213(82)90376-5, `https://doi.org/10.1016/0550-3213(82)90376-5`.

Day, B. and Street, R. (1997). Monoidal bicategories and Hopf algebroids, *Adv. Math.* **129**, 1, pp. 99–157, doi:10.1006/aima.1997.1649, https://doi.org/10.1006/aima.1997.1649.

Deflorin, G. (2019). The Homotopy Hypothesis, http://user.math.uzh.ch/cattaneo/deflorin.pdf.

Del Zotto, M. and Ohmori, K. (2021). 2-group symmetries of 6D little string theories and T-duality, *Ann. Henri Poincaré* **22**, 7, pp. 2451–2474, doi:10.1007/s00023-021-01018-3, https://doi.org/10.1007/s00023-021-01018-3.

Deligne, P. (1971). Théorie de Hodge. II, *Inst. Hautes Études Sci. Publ. Math.* **40**, pp. 5–57, http://www.numdam.org/item?id=PMIHES_1971__40__5_0.

Deligne, P., Griffiths, P., Morgan, J., and Sullivan, D. (1975). Real homotopy theory of Kähler manifolds, *Invent. Math.* **29**, 3, pp. 245–274, doi:10.1007/BF01389853, https://doi.org/10.1007/BF01389853.

Diaconescu, D.-E., Moore, G., and Witten, E. (2002). E_8 gauge theory, and a derivation of K-theory from M-theory, *Adv. Theor. Math. Phys.* **6**, 6, pp. 1031–1134 (2003), doi:10.4310/ATMP.2002.v6.n6.a2, https://doi.org/10.4310/ATMP.2002.v6.n6.a2.

Dirac, P. (1931). Quantised singularities in the electromagnetic field, *Proc. R. Soc. Lond. A* **133**, pp. 60–72, doi:10.1098/rspa.1931.0130, http://doi.org/10.1098/rspa.1931.0130.

Dold, A. (1958). Homology of symmetric products and other functors of complexes, *Ann. of Math. (2)* **68**, pp. 54–80, doi:10.2307/1970043, https://doi.org/10.2307/1970043.

Dold, A. (1972). London Mathematical Society Lecture Note Series, No. 4 (Cambridge University Press, London-New York), pp. 166–177.

Donaldson, S. K. (1984). Nahm's equations and the classification of monopoles, *Comm. Math. Phys.* **96**, 3, pp. 387–407, http://projecteuclid.org/euclid.cmp/1103941858.

Donovan, P. and Karoubi, M. (1970). Graded Brauer groups and K-theory with local coefficients, *Inst. Hautes Études Sci. Publ. Math.* **38**, pp. 5–25, http://www.numdam.org/item?id=PMIHES_1970__38__5_0.

Doubek, M., Markl, M., and Zima, P. (2007). Deformation theory (lecture notes), *Arch. Math. (Brno)* **43**, 5, pp. 333–371.

Douglas, C. L., Francis, J., Henriques, A. G., and Hill, M. A. (eds.) (2014). *Topological modular forms*, *Mathematical Surveys and Monographs*, Vol. 201 (American Mathematical Society, Providence, RI), ISBN 978-1-4704-1884-7, doi:10.1090/surv/201, https://doi.org/10.1090/surv/201.

Douglas, C. L. and Henriques, A. G. (2011). Topological modular forms and conformal nets, in *Mathematical foundations of quantum field theory and perturbative string theory*, *Proc. Sympos. Pure Math.*, Vol. 83 (Amer. Math. Soc., Providence, RI), pp. 341–354, doi:10.1090/pspum/083/2742433, https://doi.org/10.1090/pspum/083/2742433.

Dror, E., Dwyer, W. G., and Kan, D. M. (1980). Equivariant maps which are self homotopy equivalences, *Proc. Amer. Math. Soc.* **80**, 4, pp. 670–672, doi:10.2307/2043448, https://doi.org/10.2307/2043448.

Dugger, D. (1998). Sheaves and homotopy theory, `https://ncatlab.org/nlab/files/DuggerSheavesAndHomotopyTheory.pdf`.

Dugger, D. (2001). Universal homotopy theories, *Adv. Math.* **164**, 1, pp. 144–176, doi:10.1006/aima.2001.2014, `https://doi.org/10.1006/aima.2001.2014`.

Dugger, D. (2003). Notes on Delta-generated spaces, `https://pages.uoregon.edu/ddugger/delta.html`.

Dungan, G. (2010). Review of model categories, `https://ncatlab.org/nlab/files/DunganModelCategories.pdf`.

Dwyer, W. G. and Kan, D. M. (1984). An obstruction theory for diagrams of simplicial sets, *Nederl. Akad. Wetensch. Indag. Math.* **46**, 2, pp. 139–146.

Eckmann, B. and Hilton, P. J. (1961/62). Structure maps in group theory, *Fund. Math.* **50**, pp. 207–221, doi:10.4064/fm-50-2-207-221, `https://doi.org/10.4064/fm-50-2-207-221`.

Eilenberg, S. (1940). Cohomology and continuous mappings, *Ann. of Math. (2)* **41**, pp. 231–251, doi:10.2307/1968828, `https://doi.org/10.2307/1968828`.

Eilenberg, S. and MacLane, S. (1953). On the groups $H(\Pi, n)$. I, *Ann. of Math. (2)* **58**, pp. 55–106, doi:10.2307/1969820, `https://doi.org/10.2307/1969820`.

Eilenberg, S. and MacLane, S. (1954). On the groups $H(\Pi, n)$. II. Methods of computation, *Ann. of Math. (2)* **60**, pp. 49–139, doi:10.2307/1969702, `https://doi.org/10.2307/1969702`.

Eilenberg, S. and MacLane, S. (1954). On the groups $H(\Pi, n)$. III. operations and obstructions, *Ann. of Math. (2)* **60**, pp. 513–557, doi:10.2307/1969849, `https://doi.org/10.2307/1969849`.

Elmendorf, A. D., Kriz, I., Mandell, M. A., and May, J. P. (1997). *Rings, modules, and algebras in stable homotopy theory*, *Mathematical Surveys and Monographs*, Vol. 47 (American Mathematical Society, Providence, RI), ISBN 0-8218-0638-6, doi:10.1090/surv/047, `https://doi.org/10.1090/surv/047`, with an appendix by M. Cole.

Erdal, M. A. and Güçlükan İlhan, A. (2019). A model structure via orbit spaces for equivariant homotopy, *J. Homotopy Relat. Struct.* **14**, 4, pp. 1131–1141, doi:10.1007/s40062-019-00241-4, `https://doi.org/10.1007/s40062-019-00241-4`.

Esnault, H. (2009). Algebraic differential characters of flat connections with nilpotent residues, in *Algebraic topology*, *Abel Symp.*, Vol. 4 (Springer, Berlin), pp. 83–94, doi:10.1007/978-3-642-01200-6_5, `https://doi.org/10.1007/978-3-642-01200-6_5`.

Esnault, H. and Viehweg, E. (1988). Deligne-Beĭlinson cohomology, in *Beĭlinson's conjectures on special values of L-functions*, *Perspect. Math.*, Vol. 4 (Academic Press, Boston, MA), pp. 43–91.

Espinoza, J. and Uribe, B. (2014). Topological properties of the unitary group, *JP J. Geom. Topol.* **16**, 1, pp. 45–55.

Evslin, J. (2006). What Does(n't) K-theory Classify? `https://arxiv.org/abs/hep-th/0610328`.

Félix, Y. and Halperin, S. (2017). Rational homotopy theory via Sullivan models: a survey, *ICCM Not.* **5**, 2, pp. 14–36, doi:10.4310/ICCM.2017.v5.n2.a3, `https://doi.org/10.4310/ICCM.2017.v5.n2.a3`.

Félix, Y., Halperin, S., and Thomas, J.-C. (2001). *Rational homotopy theory*, *Graduate Texts in Mathematics*, Vol. 205 (Springer-Verlag, New York), ISBN 0-387-95068-0, doi:10.1007/978-1-4613-0105-9, `https://doi.org/10.1007/978-1-4613-0105-9`.

Félix, Y., Halperin, S., and Thomas, J.-C. (2015). *Rational homotopy theory. II* (World Scientific Publishing Co. Pte. Ltd., Hackensack, NJ), ISBN 978-981-4651-42-4, doi:10.1142/9473, `https://doi.org/10.1142/9473`.

Félix, Y., Oprea, J., and Tanré, D. (2008). *Algebraic models in geometry*, *Oxford Graduate Texts in Mathematics*, Vol. 17 (Oxford University Press, Oxford), ISBN 978-0-19-920651-3.

Fiorenza, D., Rogers, C. L., and Schreiber, U. (2014a). L_∞-algebras of local observables from higher prequantum bundles, *Homology Homotopy Appl.* **16**, 2, pp. 107–142, doi:10.4310/HHA.2014.v16.n2.a6, `https://doi.org/10.4310/HHA.2014.v16.n2.a6`.

Fiorenza, D., Sati, H., and Schreiber, U. (2013). Extended higher cup-product Chern-Simons theories, *J. Geom. Phys.* **74**, pp. 130–163, doi:10.1016/j.geomphys.2013.07.011, `https://doi.org/10.1016/j.geomphys.2013.07.011`.

Fiorenza, D., Sati, H., and Schreiber, U. (2014b). Multiple M5-branes, string 2-connections, and 7d nonabelian Chern-Simons theory, *Adv. Theor. Math. Phys.* **18**, 2, pp. 229–321, `http://projecteuclid.org/euclid.atmp/1414414836`.

Fiorenza, D., Sati, H., and Schreiber, U. (2015a). The E_8 moduli 3-stack of the C-field in M-theory, *Comm. Math. Phys.* **333**, 1, pp. 117–151, doi:10.1007/s00220-014-2228-1, `https://doi.org/10.1007/s00220-014-2228-1`.

Fiorenza, D., Sati, H., and Schreiber, U. (2015b). A higher stacky perspective on Chern-Simons theory, in *Mathematical aspects of quantum field theories*, Math. Phys. Stud. (Springer, Cham), pp. 153–211.

Fiorenza, D., Sati, H., and Schreiber, U. (2015c). Super-Lie n-algebra extensions, higher WZW models and super-p-branes with tensor multiplet fields, *Int. J. Geom. Methods Mod. Phys.* **12**, 2, pp. 1550018, 35, doi:10.1142/S0219887815500188, `https://doi.org/10.1142/S0219887815500188`.

Fiorenza, D., Sati, H., and Schreiber, U. (2015d). The Wess-Zumino-Witten term of the M5-brane and differential cohomotopy, *J. Math. Phys.* **56**, 10, pp. 102301, 10, doi:10.1063/1.4932618, `https://doi.org/10.1063/1.4932618`.

Fiorenza, D., Sati, H., and Schreiber, U. (2017). Rational sphere valued supercocycles in M-theory and type IIA string theory, *J. Geom. Phys.* **114**, pp. 91–108, doi:10.1016/j.geomphys.2016.11.024, `https://doi.org/10.1016/j.geomphys.2016.11.024`.

Fiorenza, D., Sati, H., and Schreiber, U. (2018). T-duality from super Lie n-algebra cocycles for super p-branes, *Adv. Theor. Math. Phys.* **22**, 5, pp. 1209–1270, doi:10.4310/ATMP.2018.v22.n5.a3, `https://doi.org/10.4310/ATMP.2018.v22.n5.a3`.

Fiorenza, D., Sati, H., and Schreiber, U. (2019). The rational higher structure of M-theory, *Fortschr. Phys.* **67**, 8-9, Special issue: Proceedings of the LMS/EPSRC Durham Symposium on Higher Structures in M-Theory, pp. 1910017, 28, doi:10.1002/prop.201910017, `https://doi.org/10.1002/prop.201910017`.

Fiorenza, D., Sati, H., and Schreiber, U. (2020a). Higher T-duality of super M-branes, *Adv. Theor. Math. Phys.* **24**, 3, pp. 621–708, doi:10.4310/ATMP.2020.v24.n3.a3, `https://doi.org/10.4310/ATMP.2020.v24.n3.a3`.

Fiorenza, D., Sati, H., and Schreiber, U. (2020b). Twisted cohomotopy implies M-theory anomaly cancellation on 8-manifolds, *Comm. Math. Phys.* **377**, 3, pp. 1961–2025, doi:10.1007/s00220-020-03707-2, `https://doi.org/10.1007/s00220-020-03707-2`.

Fiorenza, D., Sati, H., and Schreiber, U. (2021a). Super-exceptional embedding construction of the heterotic M5: emergence of SU(2)-flavor sector, *J. Geom. Phys.* **170**, pp. Paper No. 104349, 33, doi:10.1016/j.geomphys.2021.104349, `https://doi.org/10.1016/j.geomphys.2021.104349`.

Fiorenza, D., Sati, H., and Schreiber, U. (2021b). Twisted cohomotopy implies level quantization of the full 6d Wess-Zumino term of the M5-brane, *Comm. Math. Phys.* **384**, 1, pp. 403–432, doi:10.1007/s00220-021-03951-0, `https://doi.org/10.1007/s00220-021-03951-0`.

Fiorenza, D., Sati, H., and Schreiber, U. (2021c). Twisted cohomotopy implies twisted string structure on M5-branes, *J. Math. Phys.* **62**, 4, pp. Paper No. 042301, 16, doi: 10.1063/5.0037786, `https://doi.org/10.1063/5.0037786`.

Fiorenza, D., Sati, H., and Schreiber, U. (2022). Twistorial cohomotopy implies Green-Schwarz anomaly cancellation, *Rev. Math. Phys.* **34**, 5, pp. Paper No. 2250013, 42, doi:10.1142/S0129055X22500131, `https://doi.org/10.1142/S0129055X22500131`.

Fiorenza, D., Schreiber, U., and Stasheff, J. (2012). Čech cocycles for differential characteristic classes: an ∞-Lie theoretic construction, *Adv. Theor. Math. Phys.* **16**, 1, pp. 149–250, `http://projecteuclid.org/euclid.atmp/1358950853`.

Frankel, T. (1997). *The geometry of physics* (Cambridge University Press, Cambridge), ISBN 0-521-38334-X, an introduction.

Freed, D. S. (2000). Dirac charge quantization and generalized differential cohomology, in *Surveys in differential geometry*, *Surv. Differ. Geom.*, Vol. 7 (Int. Press, Somerville, MA), pp. 129–194, doi:10.4310/SDG.2002.v7.n1.a6, `https://doi.org/10.4310/SDG.2002.v7.n1.a6`.

Freed, D. S. (2002). Classical Chern-Simons theory. II, *Houston J. Math.* **28**, 2, pp. 293–310, special issue for S. S. Chern.

Freed, D. S. and Hopkins, M. (2000). On Ramond-Ramond fields and K-theory, *J. High Energy Phys.* **2000**, 5, pp. Paper 44, 14, doi:10.1088/1126-6708/2000/05/044, `https://doi.org/10.1088/1126-6708/2000/05/044`.

Freed, D. S. and Hopkins, M. J. (2013). Chern-Weil forms and abstract homotopy theory, *Bull. Amer. Math. Soc. (N.S.)* **50**, 3, pp. 431–468, doi:10.1090/S0273-0979-2013-01415-0.

Freed, D. S., Hopkins, M. J., and Teleman, C. (2008). Twisted equivariant K-theory with complex coefficients, *J. Topol.* **1**, 1, pp. 16–44, doi:10.1112/jtopol/jtm001, `https://doi.org/10.1112/jtopol/jtm001`.

Freed, D. S., Moore, G. W., and Segal, G. (2007). The uncertainty of fluxes, *Comm. Math. Phys.* **271**, 1, pp. 247–274, doi:10.1007/s00220-006-0181-3, `https://doi.org/10.1007/s00220-006-0181-3`.

Friedman, G. (2012). Survey article: An elementary illustrated introduction to simplicial sets, *Rocky Mountain J. Math.* **42**, 2, pp. 353–423, doi:10.1216/RMJ-2012-42-2-353, `https://doi.org/10.1216/RMJ-2012-42-2-353`.

Gaiotto, D. and Johnson-Freyd, T. (2019). Mock modularity and a secondary elliptic genus, `https://arxiv.org/abs/1904.05788`.

Gaiotto, D. and Johnson-Freyd, T. (2022). Holomorphic SCFTs with small index, *Canad. J. Math.* **74**, 2, pp. 573–601, doi:10.4153/S0008414X2100002X, https://doi.org/10.4153/S0008414X2100002X.

Gaiotto, D., Johnson-Freyd, T., and Witten, E. (2021). A note on some minimally supersymmetric models in two dimensions, in *Integrability, quantization, and geometry II. Quantum theories and algebraic geometry*, *Proc. Sympos. Pure Math.*, Vol. 103 (Amer. Math. Soc., Providence, RI), pp. 203–221.

Gajer, P. (1997). Geometry of Deligne cohomology, *Invent. Math.* **127**, 1, pp. 155–207, doi:10.1007/s002220050118, https://doi.org/10.1007/s002220050118.

Galatius, S., Tillmann, U., Madsen, I., and Weiss, M. (2009). The homotopy type of the cobordism category, *Acta Math.* **202**, 2, pp. 195–239, doi:10.1007/s11511-009-0036-9, https://doi.org/10.1007/s11511-009-0036-9.

Garzon, A. R. and Miranda, J. G. (1997). Homotopy theory for (braided) CAT-groups, *Cahiers Topologie Géom. Différentielle Catég.* **38**, 2, pp. 99–139.

Garzón, A. R. and Miranda, J. G. (2000). Serre homotopy theory in subcategories of simplicial groups, *J. Pure Appl. Algebra* **147**, 2, pp. 107–123, doi:10.1016/S0022-4049(98)00143-1, https://doi.org/10.1016/S0022-4049(98)00143-1.

Gawędzki, K., Suszek, R. R., and Waldorf, K. (2011). Bundle gerbes for orientifold sigma models, *Adv. Theor. Math. Phys.* **15**, 3, pp. 621–687, http://projecteuclid.org/euclid.atmp/1339374265.

Gelfand, S. I. and Manin, Y. I. (1996). *Methods of homological algebra* (Springer-Verlag, Berlin), ISBN 3-540-54746-0, doi:10.1007/978-3-662-03220-6, https://doi.org/10.1007/978-3-662-03220-6, translated from the 1988 Russian original.

Getzler, E. (2009). Lie theory for nilpotent L_∞-algebras, *Ann. of Math. (2)* **170**, 1, pp. 271–301, doi:10.4007/annals.2009.170.271, https://doi.org/10.4007/annals.2009.170.271.

Giraud, J. (1971). *Cohomologie non abélienne*, Die Grundlehren der mathematischen Wissenschaften, Band 179 (Springer-Verlag, Berlin-New York).

Glenn, P. G. (1982). Realization of cohomology classes in arbitrary exact categories, *J. Pure Appl. Algebra* **25**, 1, pp. 33–105, doi:10.1016/0022-4049(82)90094-9, https://doi.org/10.1016/0022-4049(82)90094-9.

Goerss, P. and Schemmerhorn, K. (2007). Model categories and simplicial methods, in *Interactions between homotopy theory and algebra*, *Contemp. Math.*, Vol. 436 (Amer. Math. Soc., Providence, RI), pp. 3–49, doi:10.1090/conm/436/08403, https://doi.org/10.1090/conm/436/08403.

Goerss, P. G. and Jardine, J. F. (1999). *Simplicial homotopy theory*, *Progress in Mathematics*, Vol. 174 (Birkhäuser Verlag, Basel), ISBN 3-7643-6064-X, doi:10.1007/978-3-0348-8707-6, https://doi.org/10.1007/978-3-0348-8707-6.

Gomi, K. and Terashima, Y. (2000). A fiber integration formula for the smooth Deligne cohomology, *Internat. Math. Res. Notices* **2000**, 13, pp. 699–708, doi:10.1155/S1073792800000386, https://doi.org/10.1155/S1073792800000386.

Gomi, K. and Terashima, Y. (2010). Chern-Weil construction for twisted K-theory, *Comm. Math. Phys.* **299**, 1, pp. 225–254, doi:10.1007/s00220-010-1080-1, https://doi.org/10.1007/s00220-010-1080-1.

Grady, D. and Sati, H. (2017). Spectral sequences in smooth generalized cohomology, *Algebr. Geom. Topol.* **17**, 4, pp. 2357–2412, doi:10.2140/agt.2017.17.2357, https://doi.org/10.2140/agt.2017.17.2357.

Grady, D. and Sati, H. (2018a). Massey products in differential cohomology via stacks, *J. Homotopy Relat. Struct.* **13**, 1, pp. 169–223, doi:10.1007/s40062-017-0178-y, https://doi.org/10.1007/s40062-017-0178-y.

Grady, D. and Sati, H. (2018b). Primary operations in differential cohomology, *Adv. Math.* **335**, pp. 519–562, doi:10.1016/j.aim.2018.07.019, https://doi.org/10.1016/j.aim.2018.07.019.

Grady, D. and Sati, H. (2018c). Twisted smooth Deligne cohomology, *Ann. Global Anal. Geom.* **53**, 3, pp. 445–466, doi:10.1007/s10455-017-9583-z, https://doi.org/10.1007/s10455-017-9583-z.

Grady, D. and Sati, H. (2019a). Higher-twisted periodic smooth Deligne cohomology, *Homology Homotopy Appl.* **21**, 1, pp. 129–159, doi:10.4310/HHA.2019.v21.n1.a7, https://doi.org/10.4310/HHA.2019.v21.n1.a7.

Grady, D. and Sati, H. (2019b). Ramond-Ramond fields and twisted differential K-theory, https://arxiv.org/abs/1903.08843.

Grady, D. and Sati, H. (2019c). Twisted differential generalized cohomology theories and their Atiyah-Hirzebruch spectral sequence, *Algebr. Geom. Topol.* **19**, 6, pp. 2899–2960, doi:10.2140/agt.2019.19.2899, https://doi.org/10.2140/agt.2019.19.2899.

Grady, D. and Sati, H. (2019d). Twisted differential KO-theory, https://arxiv.org/abs/1905.09085.

Grady, D. and Sati, H. (2021a). Differential cohomotopy versus differential cohomology for M-theory and differential lifts of Postnikov towers, *J. Geom. Phys.* **165**, pp. Paper No. 104203, 24, doi:10.1016/j.geomphys.2021.104203, https://doi.org/10.1016/j.geomphys.2021.104203.

Grady, D. and Sati, H. (2021b). Differential KO-theory: constructions, computations, and applications, *Adv. Math.* **384**, pp. Paper No. 107671, 117, doi:10.1016/j.aim.2021.107671, https://doi.org/10.1016/j.aim.2021.107671.

Green, M. B. and Schwarz, J. H. (1984). Anomaly cancellations in supersymmetric $D = 10$ gauge theory and superstring theory, *Phys. Lett. B* **149**, 1-3, pp. 117–122, doi:10.1016/0370-2693(84)91565-X, https://doi.org/10.1016/0370-2693(84)91565-X.

Greub, W., Halperin, S., and Vanstone, R. (1973). *Connections, curvature, and cohomology. Vol. II: Lie groups, principal bundles, and characteristic classes*, Pure and Applied Mathematics, Vol. 47-II (Academic Press [Harcourt Brace Jovanovich, Publishers], New York-London).

Griffiths, P. and Morgan, J. (2013). *Rational homotopy theory and differential forms*, *Progress in Mathematics*, Vol. 16, 2nd edn. (Springer, New York), ISBN 978-1-4614-8467-7, 978-1-4614-8468-4, doi:10.1007/978-1-4614-8468-4, https://doi.org/10.1007/978-1-4614-8468-4.

Guerra, J. M. G. (2008). *Models of twisted K-theory*, Ph.D. thesis, University of Michigan, https://deepblue.lib.umich.edu/handle/2027.42/60768.

Gukov, S., Pei, D., Putrov, P., and Vafa, C. (2021). 4-manifolds and topological modular forms, *J. High Energy Phys.* **2021**, 85, doi:10.1007/jhep05(2021)084, `https://doi.org/10.1007/jhep05(2021)084`.

Gurski, N. and Osorno, A. M. (2013). Infinite loop spaces, and coherence for symmetric monoidal bicategories, *Adv. Math.* **246**, pp. 1–32, doi:10.1016/j.aim.2013.06.028, `https://doi.org/10.1016/j.aim.2013.06.028`.

Hekmati, P., Murray, M. K., Szabo, R. J., and Vozzo, R. F. (2019). Real bundle gerbes, orientifolds and twisted *KR*-homology, *Adv. Theor. Math. Phys.* **23**, 8, pp. 2093–2159, doi:10.4310/atmp.2019.v23.n8.a5, `https://doi.org/10.4310/atmp.2019.v23.n8.a5`.

Henriques, A. (2008). Integrating L_∞-algebras, *Compos. Math.* **144**, 4, pp. 1017–1045, doi:10.1112/S0010437X07003405, `https://doi.org/10.1112/S0010437X07003405`.

Hess, K. (2007). Rational homotopy theory: a brief introduction, in *Interactions between homotopy theory and algebra*, *Contemp. Math.*, Vol. 436 (Amer. Math. Soc., Providence, RI), pp. 175–202, doi:10.1090/conm/436/08409, `https://doi.org/10.1090/conm/436/08409`.

Hilton, P. (1971). *General cohomology theory and K-theory*, *London Mathematical Society Lecture Note Series*, Vol. 1 (Cambridge University Press, London-New York), course given at the University of São Paulo in the summer of 1968 under the auspices of the Instituto de Pesquisas Matemáticas, Universidade de São Paulo.

Hilton, P. (1982). Nilpotency in group theory and topology, *Publ. Sec. Mat. Univ. Autònoma Barcelona* **26**, 3, pp. 47–78, workshop on algebraic topology (Barcelona, 1982).

Hilton, P. J. (1955). On the homotopy groups of the union of spheres, *J. London Math. Soc.* **30**, pp. 154–172, doi:10.1112/jlms/s1-30.2.154, `https://doi.org/10.1112/jlms/s1-30.2.154`.

Hinich, V. (2001). DG coalgebras as formal stacks, *J. Pure Appl. Algebra* **162**, 2-3, pp. 209–250, doi:10.1016/S0022-4049(00)00121-3, `https://doi.org/10.1016/S0022-4049(00)00121-3`.

Hirashima, Y. (1979). A note on cohomology with local coefficients, *Osaka Math. J.* **16**, 1, pp. 219–231, `http://projecteuclid.org/euclid.ojm/1200771839`.

Hirschhorn, P. S. (2003). *Model categories and their localizations*, *Mathematical Surveys and Monographs*, Vol. 99 (American Mathematical Society, Providence, RI), ISBN 0-8218-3279-4, doi:10.1090/surv/099, `https://doi.org/10.1090/surv/099`.

Hirschhorn, P. S. (2019). The Quillen model category of topological spaces, *Expo. Math.* **37**, 1, pp. 2–24, doi:10.1016/j.exmath.2017.10.004, `https://doi.org/10.1016/j.exmath.2017.10.004`.

Hirzebruch, F. (1956). *Neue topologische Methoden in der algebraischen Geometrie*, Ergebnisse der Mathematik und ihrer Grenzgebiete, (N.F.), Heft 9 (Springer-Verlag, Berlin-Göttingen-Heidelberg).

Hollander, S. (2008). A homotopy theory for stacks, *Israel J. Math.* **163**, pp. 93–124, doi:10.1007/s11856-008-0006-5.

Hopkins, M. J. (1995). Topological modular forms, the Witten genus, and the theorem of the cube, in *Proceedings of the International Congress of Mathematicians, Vol. 1, 2 (Zürich, 1994)* (Birkhäuser, Basel), pp. 554–565.

Hopkins, M. J. (2002). Algebraic topology and modular forms, in *Proceedings of the International Congress of Mathematicians, Vol. I (Beijing, 2002)* (Higher Ed. Press, Beijing), pp. 291–317.

Hopkins, M. J. and Singer, I. M. (2005). Quadratic functions in geometry, topology, and M-theory, *J. Differential Geom.* **70**, 3, pp. 329–452, `http://projecteuclid.org/euclid.jdg/1143642908`.

Houghton, C. J., Manton, N. S., and Sutcliffe, P. M. (1998). Rational maps, monopoles and skyrmions, *Nuclear Phys. B* **510**, 3, pp. 507–537, doi:10.1016/S0550-3213(97)00619-6, `https://doi.org/10.1016/S0550-3213(97)00619-6`.

Hovey, M. (1999). *Model categories*, *Mathematical Surveys and Monographs*, Vol. 63 (American Mathematical Society, Providence, RI), ISBN 0-8218-1359-5.

Hořava, P. and Witten, E. (1996). Eleven-dimensional supergravity on a manifold with boundary, *Nuclear Phys. B* **475**, 1-2, pp. 94–114, doi:10.1016/0550-3213(96)00308-2, `https://doi.org/10.1016/0550-3213(96)00308-2`.

Huerta, J., HishamSati, and Schreiber, U. (2019). Real ADE-equivariant (co)homotopy and super M-branes, *Comm. Math. Phys.* **371**, 2, pp. 425–524, doi:10.1007/s00220-019-03442-3, `https://doi.org/10.1007/s00220-019-03442-3`.

Iglesias-Zemmour, P. (1985). *Fibrations difféologiques et Homotopie*, Ph.D. thesis, University of Provence, `http://math.huji.ac.il/~piz/documents/TheseEtatPI.pdf`.

Iglesias-Zemmour, P. (2013). *Diffeology*, *Mathematical Surveys and Monographs*, Vol. 185 (American Mathematical Society, Providence, RI), ISBN 978-0-8218-9131-5, doi:10.1090/surv/185, `https://doi.org/10.1090/surv/185`.

Igusa, K. (2008). Pontrjagin classes and higher torsion of sphere bundles, in *Groups of diffeomorphisms*, *Adv. Stud. Pure Math.*, Vol. 52 (Math. Soc. Japan, Tokyo), pp. 21–29, doi:10.2969/aspm/05210021, `https://doi.org/10.2969/aspm/05210021`.

Imaoka, M. and Kuwana, K. (1999). Stably extendible vector bundles over the quaternionic projective spaces, *Hiroshima Math. J.* **29**, 2, pp. 273–279, `http://projecteuclid.org/euclid.hmj/1206125008`.

Intriligator, K. (2000). Anomaly matching and a Hopf-Wess-Zumino term in six-dimensional, $\mathcal{N} = (2,0)$ field theories, *Nuclear Phys. B* **581**, 1-2, pp. 257–273, doi:10.1016/S0550-3213(00)00148-6, `https://doi.org/10.1016/S0550-3213(00)00148-6`.

Ioannidou, T. and Sutcliffe, P. M. (1999). Monopoles and harmonic maps, *J. Math. Phys.* **40**, 11, pp. 5440–5455, doi:10.1063/1.533038, `https://doi.org/10.1063/1.533038`.

Iyer, J. N. and Simpson, C. (2007). Regulators of canonical extensions are torsion: the smooth divisor case, `https://arxiv.org/abs/0707.0372`.

Jardine, J. F. (1987). Simplicial presheaves, *J. Pure Appl. Algebra* **47**, 1, pp. 35–87, doi:10.1016/0022-4049(87)90100-9.

Jardine, J. F. (2001). Stacks and the homotopy theory of simplicial sheaves, *Homology Homotopy Appl.* **3**, 2, pp. 361–384, doi:10.4310/hha.2001.v3.n2.a5, `https://doi.org/10.4310/hha.2001.v3.n2.a5`, equivariant stable homotopy theory and related areas (Stanford, CA, 2000).

Jardine, J. F. (2003). Presheaves of chain complexes, *K-Theory* **30**, 4, pp. 365–420, doi:10.1023/B:KTHE.0000021707.27987.c9, special issue in honor of Hyman Bass on his seventieth birthday. Part IV.

Jardine, J. F. (2015). *Local homotopy theory*, Springer Monographs in Mathematics (Springer, New York), ISBN 978-1-4939-2299-4; 978-1-4939-2300-7, doi:10.1007/978-1-4939-2300-7, `https://doi.org/10.1007/978-1-4939-2300-7`.

Jardine, J. F. and Luo, Z. (2006). Higher principal bundles, *Math. Proc. Cambridge Philos. Soc.* **140**, 2, pp. 221–243, doi:10.1017/S0305004105008911, `https://doi.org/10.1017/S0305004105008911`.

Jarvis, S. (1998). Euclidean monopoles and rational maps, *Proc. London Math. Soc. (3)* **77**, 1, pp. 170–192, doi:10.1112/S0024611598000434, `https://doi.org/10.1112/S0024611598000434`.

Jarvis, S. (2000). A rational map for Euclidean monopoles via radial scattering, *J. Reine Angew. Math.* **524**, pp. 17–41, doi:10.1515/crll.2000.055, `https://doi.org/10.1515/crll.2000.055`.

Johnson, D. C. and Wilson, W. S. (1975). *BP* operations and Morava's extraordinary K-theories, *Math. Z.* **144**, 1, pp. 55–75, doi:10.1007/BF01214408, `https://doi.org/10.1007/BF01214408`.

Johnson-Freyd, T. (2020). Topological Mathieu Moonshine, `https://arxiv.org/abs/2006.02922`.

Johnstone, P. T. (2002). *Sketches of an elephant: a topos theory compendium.*, *Oxford Logic Guides*, Vol. 43 and 44 (The Clarendon Press, Oxford University Press, New York), ISBN 0-19-853425-6.

Joyal, A. (2008a). Notes on logoi, doi:https://ncatlab.org/nlab/files/JoyalOnLogoi2008.pdf.

Joyal, A. (2008b). Notes on quasi-categories, `https://ncatlab.org/nlab/files/JoyalNotesOnQuasiCategories.pdf`.

Joyal, A. (2008c). The theory of quasi-categories and its applications, `http://mat.uab.cat/~kock/crm/hocat/advanced-course/Quadern45-2.pdf`.

Kamiyama, Y. (2007). Remarks on spaces of real rational functions, *Rocky Mountain J. Math.* **37**, 1, pp. 247–257, doi:10.1216/rmjm/1181069329, `https://doi.org/10.1216/rmjm/1181069329`.

Kan, D. M. (1958). Functors involving c.s.s. complexes, *Trans. Amer. Math. Soc.* **87**, pp. 330–346, doi:10.2307/1993103, `https://doi.org/10.2307/1993103`.

Karoubi, M. (1968). Algèbres de Clifford et K-théorie, *Ann. Sci. École Norm. Sup. (4)* **1**, pp. 161–270, `http://www.numdam.org/item?id=ASENS_1968_4_1_2_161_0`.

Karoubi, M. (1987). Homologie cyclique et K-théorie, *Astérisque* **149**, p. 147.

Karoubi, M. (2012). Twisted bundles and twisted K-theory, in *Topics in noncommutative geometry*, *Clay Math. Proc.*, Vol. 16 (Amer. Math. Soc., Providence, RI), pp. 223–257.

Katz, S. (2006). *Enumerative geometry and string theory*, *Student Mathematical Library*, Vol. 32 (American Mathematical Society, Providence, RI; Institute for Advanced Study (IAS), Princeton, NJ), ISBN 0-8218-3687-0, doi:10.1090/stml/032, `https://doi.org/10.1090/stml/032`, iAS/Park City Mathematical Subseries.

Kirby, R., Melvin, P., and Teichner, P. (2012). Cohomotopy sets of 4-manifolds, in *Proceedings of the Freedman Fest*, *Geom. Topol. Monogr.*, Vol. 18 (Geom. Topol. Publ., Coventry), pp. 161–190, doi:10.2140/gtm.2012.18.161, `https://doi.org/10.2140/gtm.2012.18.161`.

Kobayashi, S. and Nomizu, K. (1963). *Foundations of differential geometry. Vol I* (Interscience Publishers (a division of John Wiley & Sons, Inc.), New York-London).

Kochman, S. O. (1996). *Bordism, stable homotopy and Adams spectral sequences, Fields Institute Monographs*, Vol. 7 (American Mathematical Society, Providence, RI), ISBN 0-8218-0600-9, doi:10.1090/fim/007, https://doi.org/10.1090/fim/007.

Kono, A. and Tamaki, D. (2006). *Generalized cohomology. Transl. from the Japanese by Dai Tamaki, Transl. Math. Monogr.*, Vol. 230 (Providence, RI: American Mathematical Society (AMS)), ISBN 0-8218-3514-9.

Kontsevich, M. (2003). Deformation quantization of Poisson manifolds, *Lett. Math. Phys.* **66**, 3, pp. 157–216, doi:10.1023/B:MATH.0000027508.00421.bf, https://doi.org/10.1023/B:MATH.0000027508.00421.bf.

Kosinski, A. A. (1993). *Differential manifolds, Pure and Applied Mathematics*, Vol. 138 (Academic Press, Inc., Boston, MA), ISBN 0-12-421850-4.

Kriz, I. and Sati, H. (2004). M-theory, type IIA superstrings, and elliptic cohomology, *Adv. Theor. Math. Phys.* **8**, 2, pp. 345–394, http://projecteuclid.org/euclid.atmp/1091543172.

Kriz, I. and Sati, H. (2005). Type II string theory and modularity, *J. High Energy Phys.* **2005**, 8, pp. 038, 30, doi:10.1088/1126-6708/2005/08/038, https://doi.org/10.1088/1126-6708/2005/08/038.

Krusch, S. (2003). Homotopy of rational maps and the quantization of skyrmions, *Ann. Physics* **304**, 2, pp. 103–127, doi:10.1016/S0003-4916(03)00014-9, https://doi.org/10.1016/S0003-4916(03)00014-9.

Krusch, S. (2006). Quantization of skyrmions, https://arxiv.org/abs/hep-th/0610176.

Kříž, I. and May, J. P. (1995). Operads, algebras, modules and motives, *Astérisque* **233**, pp. iv+145pp.

Lada, T. and Markl, M. (1995). Strongly homotopy Lie algebras, *Comm. Algebra* **23**, 6, pp. 2147–2161, doi:10.1080/00927879508825335, https://doi.org/10.1080/00927879508825335.

Lada, T. and Stasheff, J. (1993). Introduction to SH Lie algebras for physicists, *Internat. J. Theoret. Phys.* **32**, 7, pp. 1087–1103, doi:10.1007/BF00671791, https://doi.org/10.1007/BF00671791.

Lawson, T. (2022). An introduction to Bousfield localization, in *Stable categories and structured ring spectra, Math. Sci. Res. Inst. Publ.*, Vol. 69 (Cambridge Univ. Press, Cambridge), pp. 301–344.

Lazarev, A. (2013). Maurer-Cartan moduli and models for function spaces, *Adv. Math.* **235**, pp. 296–320, doi:10.1016/j.aim.2012.11.009, https://doi.org/10.1016/j.aim.2012.11.009.

Lerche, W., Nilsson, B. E. W., Schellekens, A. N., and Warner, N. P. (1988). Anomaly cancelling terms from the elliptic genus, *Nuclear Phys. B* **299**, 1, pp. 91–116, doi:10.1016/0550-3213(88)90468-3, https://doi.org/10.1016/0550-3213(88)90468-3.

Li, B. H. and Duan, H. B. (1991). Spin characteristic classes and reduced *K*Spin group of a low-dimensional complex, *Proc. Amer. Math. Soc.* **113**, 2, pp. 479–491, doi:10.2307/2048534, https://doi.org/10.2307/2048534.

Li, Z. (2016). A note on model (co)slice categories, *Chinese Ann. Math. Ser. B* **37**, 1, pp. 95–102, doi:10.1007/s11401-015-0955-z, https://doi.org/10.1007/s11401-015-0955-z.

Lind, J. A. (2016). Bundles of spectra and algebraic K-theory, *Pacific J. Math.* **285**, 2, pp. 427–452, doi:10.2140/pjm.2016.285.427, https://doi.org/10.2140/pjm.2016.285.427.

Lind, J. A., Sati, H., and Westerland, C. (2020). Twisted iterated algebraic K-theory and topological T-duality for sphere bundles, *Ann. K-Theory* **5**, 1, pp. 1–42, doi:10.2140/akt.2020.5.1, https://doi.org/10.2140/akt.2020.5.1.

Lott, J. (1994). $\mathbf{R}/\mathbf{Z}$ index theory, *Comm. Anal. Geom.* **2**, 2, pp. 279–311, doi:10.4310/CAG.1994.v2.n2.a6, https://doi.org/10.4310/CAG.1994.v2.n2.a6.

Low, Z. L. (2013). mathstackexchange, http://math.stackexchange.com/a/597990/58526.

Lupercio, E. and Uribe, B. (2004). Gerbes over orbifolds and twisted K-theory, *Comm. Math. Phys.* **245**, 3, pp. 449–489, doi:10.1007/s00220-003-1035-x, https://doi.org/10.1007/s00220-003-1035-x.

Lurie, J. (2009a). *Higher topos theory*, *Annals of Mathematics Studies*, Vol. 170 (Princeton University Press, Princeton, NJ), ISBN 978-0-691-14049-0; 0-691-14049-9, doi:10.1515/9781400830558, https://doi.org/10.1515/9781400830558.

Lurie, J. (2009b). On the classification of topological field theories, in *Current developments in mathematics, 2008* (Int. Press, Somerville, MA), pp. 129–280.

Lurie, J. (2010). Chromatic homotopy theory, http://people.math.harvard.edu/~lurie/252x.html.

Lurie, J. (2011). Dag xiii: Rational and p-adic homotopy theory, http://www.math.harvard.edu/~lurie/papers/DAG-XIII.pdf.

Lurie, J. (2014). Lecture 19: Algebraic k-theory of ring spectra, http://people.math.harvard.edu/~lurie/281notes/Lecture19-Rings.pdf.

Maakestad, H. (2017). Notes on the chern-character, *J. Gen. Lie Theory Appl.* **11**, 1, p. 6, doi:0.4172/1736-4337.1000253, ttps://www.hilarispublisher.com/open-access/notes-on-the-cherncharacter-.pdf.

Macdonald, L., Mathai, V., and Saratchandran, H. (2021). On the Chern character in higher twisted K-theory and spherical T-duality, *Comm. Math. Phys.* **385**, 1, pp. 331–368, doi:10.1007/s00220-021-04096-w, https://doi.org/10.1007/s00220-021-04096-w.

Mackaay, M. (2003). A note on the holonomy of connections in twisted bundles, *Cah. Topol. Géom. Différ. Catég.* **44**, 1, pp. 39–62.

Mackenzie, K. (1987). *Lie groupoids and Lie algebroids in differential geometry*, *London Mathematical Society Lecture Note Series*, Vol. 124 (Cambridge University Press, Cambridge), ISBN 0-521-34882-X, doi:10.1017/CBO9780511661839, https://doi.org/10.1017/CBO9780511661839.

Mackenzie, K. C. H. (2005). *General theory of Lie groupoids and Lie algebroids*, *London Mathematical Society Lecture Note Series*, Vol. 213 (Cambridge University Press, Cambridge), ISBN 978-0-521-49928-3; 0-521-49928-3, doi:10.1017/CBO9781107325883, https://doi.org/10.1017/CBO9781107325883.

Mahowald, M. E. and Ravenel, D. C. (1987). Toward a global understanding of the homotopy groups of spheres, in *The Lefschetz centennial conference, Part II (Mexico City, 1984)*, *Contemp. Math.*, Vol. 58 (Amer. Math. Soc., Providence, RI), pp. 57–74, doi:10.1016/0040-9383(93)90055-z, https://doi.org/10.1016/0040-9383(93)90055-z.

Manetti, M. (2022). *Lie methods in deformation theory*, Springer Monographs in Mathematics (Springer, Singapore), ISBN 978-981-19-1184-2; 978-981-19-1185-9, doi: 10.1007/978-981-19-1185-9, `https://doi.org/10.1007/978-981-19-1185-9`.

Manolescu, C. (2014). Triangulations of manifolds, *ICCM Not.* **2**, 2, pp. 21–23, doi:10.4310/ICCM.2014.v2.n2.a2, `https://doi.org/10.4310/ICCM.2014.v2.n2.a2`.

Manton, N. S. and Piette, B. M. A. G. (2001). Understanding skyrmions using rational maps, in *European Congress of Mathematics, Vol. I (Barcelona, 2000), Progr. Math.*, Vol. 201 (Birkhäuser, Basel), pp. 469–479.

Markl, M. (2012). *Deformation theory of algebras and their diagrams*, *CBMS Regional Conference Series in Mathematics*, Vol. 116 (Published for the Conference Board of the Mathematical Sciences, Washington, DC; by the American Mathematical Society, Providence, RI), ISBN 978-0-8218-8979-4, doi:10.1090/cbms/116, `https://doi.org/10.1090/cbms/116`.

Mathai, V. and Sati, H. (2004). Some relations between twisted K-theory and E_8 gauge theory, *J. High Energy Phys.* **2004**, 3, pp. 016, 22, doi:10.1088/1126-6708/2004/03/016, `https://doi.org/10.1088/1126-6708/2004/03/016`.

Mathai, V. and Stevenson, D. (2003). Chern character in twisted K-theory: equivariant and holomorphic cases, *Comm. Math. Phys.* **236**, 1, pp. 161–186, doi:10.1007/s00220-003-0807-7, `https://doi.org/10.1007/s00220-003-0807-7`.

Mathai, V. and Stevenson, D. (2006). On a generalized Connes-Hochschild-Kostant-Rosenberg theorem, *Adv. Math.* **200**, 2, pp. 303–335, doi:10.1016/j.aim.2004.11.006, `https://doi.org/10.1016/j.aim.2004.11.006`.

Mathai, V. and Wu, S. (2011). Analytic torsion for twisted de Rham complexes, *J. Differential Geom.* **88**, 2, pp. 297–332, `http://projecteuclid.org/euclid.jdg/1320067649`.

May, J. P. (1967). *Simplicial objects in algebraic topology*, Van Nostrand Mathematical Studies, No. 11 (D. Van Nostrand Co., Inc., Princeton, N.J.-Toronto, Ont.-London).

May, J. P. (1972). *The geometry of iterated loop spaces*, Lecture Notes in Mathematics, Vol. 271 (Springer-Verlag, Berlin-New York).

May, J. P. (1977a). *E_∞ ring spaces and E_∞ ring spectra*, Lecture Notes in Mathematics, Vol. 577 (Springer-Verlag, Berlin-New York), with contributions by Frank Quinn, Nigel Ray, and Jørgen Tornehave.

May, J. P. (1977b). Infinite loop space theory, *Bull. Amer. Math. Soc.* **83**, 4, pp. 456–494, doi:10.1090/S0002-9904-1977-14318-8, `https://doi.org/10.1090/S0002-9904-1977-14318-8`.

May, J. P. (1999). *A concise course in algebraic topology*, Chicago Lectures in Mathematics (University of Chicago Press, Chicago, IL), ISBN 0-226-51182-0; 0-226-51183-9.

May, J. P. and Ponto, K. (2012). *More concise algebraic topology*, Chicago Lectures in Mathematics (University of Chicago Press, Chicago, IL), ISBN 978-0-226-51178-8; 0-226-51178-2, localization, completion, and model categories.

May, J. P. and Sigurdsson, J. (2006). *Parametrized homotopy theory*, *Mathematical Surveys and Monographs*, Vol. 132 (American Mathematical Society, Providence, RI), ISBN 978-0-8218-3922-5; 0-8218-3922-5, doi:10.1090/surv/132, `https://doi.org/10.1090/surv/132`.

Mazur, B. and Messing, W. (1974). *Universal extensions and one dimensional crystalline cohomology*, Lecture Notes in Mathematics, Vol. 370 (Springer-Verlag, Berlin-New York).

Menichi, L. (2015). Rational homotopy—Sullivan models, in *Free loop spaces in geometry and topology*, *IRMA Lect. Math. Theor. Phys.*, Vol. 24 (Eur. Math. Soc., Zürich), pp. 111–136.

Merkulov, S. A. (2004). Operads, deformation theory and F-manifolds, in *Frobenius manifolds*, Aspects Math., E36 (Friedr. Vieweg, Wiesbaden), pp. 213–251.

Milnor, J. (1956). Construction of universal bundles. II, *Ann. of Math. (2)* **63**, pp. 430–436, doi:10.2307/1970012, https://doi.org/10.2307/1970012.

Milnor, J. W. and Stasheff, J. D. (1974). *Characteristic classes*, Annals of Mathematics Studies, No. 76 (Princeton University Press, Princeton, N. J.; University of Tokyo Press, Tokyo).

Moerdijk, I. and Mrčun, J. (2003). *Introduction to foliations and Lie groupoids*, *Cambridge Studies in Advanced Mathematics*, Vol. 91 (Cambridge University Press, Cambridge), ISBN 0-521-83197-0, doi:10.1017/CBO9780511615450, https://doi.org/10.1017/CBO9780511615450.

Mosher, R. E. and Tangora, M. C. (1968). *Cohomology operations and applications in homotopy theory* (Harper & Row, Publishers, New York-London).

Mostow, M. A. (1979). The differentiable space structures of Milnor classifying spaces, simplicial complexes, and geometric realizations, *J. Differential Geometry* **14**, 2, pp. 255–293, http://projecteuclid.org/euclid.jdg/1214434974.

Munkres, J. R. (1966). *Elementary differential topology*, revised edn., Annals of Mathematics Studies, No. 54 (Princeton University Press, Princeton, N.J.), lectures given at Massachusetts Institute of Technology, Fall, 1961.

Murray, M. K. (1996). Bundle gerbes, *J. London Math. Soc. (2)* **54**, 2, pp. 403–416, doi:10.1093/acprof:oso/9780199534920.003.0012, https://doi.org/10.1093/acprof:oso/9780199534920.003.0012.

Nakahara, M. (2003). *Geometry, topology and physics*, 2nd edn., Graduate Student Series in Physics (Institute of Physics, Bristol), ISBN 0-7503-0606-8, doi:10.1201/9781420056945, https://doi.org/10.1201/9781420056945.

Narasimhan, M. S. and Ramanan, S. (1961). Existence of universal connections, *Amer. J. Math.* **83**, pp. 563–572, doi:10.2307/2372896, https://doi.org/10.2307/2372896.

Narasimhan, M. S. and Ramanan, S. (1963). Existence of universal connections. II, *Amer. J. Math.* **85**, pp. 223–231, doi:10.2307/2373211, https://doi.org/10.2307/2373211.

Nikolaus, T., Sachse, C., and Wockel, C. (2013). A smooth model for the string group, *Int. Math. Res. Not. IMRN* **2013**, 16, pp. 3678–3721, doi:10.1093/imrn/rns154, https://doi.org/10.1093/imrn/rns154.

Nikolaus, T., Schreiber, U., and Stevenson, D. (2015a). Principal ∞-bundles: general theory, *J. Homotopy Relat. Struct.* **10**, 4, pp. 749–801, doi:10.1007/s40062-014-0083-6, https://doi.org/10.1007/s40062-014-0083-6.

Nikolaus, T., Schreiber, U., and Stevenson, D. (2015b). Principal ∞-bundles: presentations, *J. Homotopy Relat. Struct.* **10**, 3, pp. 565–622, doi:10.1007/s40062-014-0077-4, https://doi.org/10.1007/s40062-014-0077-4.

Nikolaus, T. and Waldorf, K. (2013). Four equivalent versions of nonabelian gerbes, *Pacific J. Math.* **264**, 2, pp. 355–419, doi:10.2140/pjm.2013.264.355, `https://doi.org/10.2140/pjm.2013.264.355`.

Nowak, S. (2003). Stable cohomotopy groups of compact spaces, *Fund. Math.* **180**, 2, pp. 99–137, doi:10.4064/fm180-2-1, `https://doi.org/10.4064/fm180-2-1`.

Park, B. (2018). Geometric models of twisted differential K-theory I, *J. Homotopy Relat. Struct.* **13**, 1, pp. 143–167, doi:10.1007/s40062-017-0177-z, `https://doi.org/10.1007/s40062-017-0177-z`.

Pennig, U. (2016). A non-commutative model for higher twisted K-theory, *J. Topol.* **9**, 1, pp. 27–50, doi:10.1112/jtopol/jtv033, `https://doi.org/10.1112/jtopol/jtv033`.

Peterson, F. P. (1956). Some results on cohomotopy groups, *Amer. J. Math.* **78**, pp. 243–258, doi:10.2307/2372514, `https://doi.org/10.2307/2372514`.

Pontryagin, L. S. (1959). Smooth manifolds and their applications in homotopy theory, in *American Mathematical Society Translations, Ser. 2, Vol. 11* (American Mathematical Society, Providence, R.I.), pp. 1–114.

Prezma, M. (2012). Homotopy normal maps, *Algebr. Geom. Topol.* **12**, 2, pp. 1211–1238, doi:10.2140/agt.2012.12.1211, `https://doi.org/10.2140/agt.2012.12.1211`.

Pridham, J. P. (2010). Unifying derived deformation theories, *Adv. Math.* **224**, 3, pp. 772–826, doi:10.1016/j.aim.2009.12.009, `https://doi.org/10.1016/j.aim.2009.12.009`.

Quillen, D. (1969). Rational homotopy theory, *Ann. of Math. (2)* **90**, pp. 205–295, doi:10.2307/1970725, `https://doi.org/10.2307/1970725`.

Quillen, D. G. (1967). *Homotopical algebra*, Lecture Notes in Mathematics, No. 43 (Springer-Verlag, Berlin-New York).

Ravenel, D. C. (1986). *Complex cobordism and stable homotopy groups of spheres*, *Pure and Applied Mathematics*, Vol. 121 (Academic Press, Inc., Orlando, FL), ISBN 0-12-583430-6; 0-12-583431-4.

Rezk, C. (2010). Toposes and homotopy toposes, `https://faculty.math.illinois.edu/~rezk/homotopy-topos-sketch.pdf`.

Reznikov, A. (1995). All regulators of flat bundles are torsion, *Ann. of Math. (2)* **141**, 2, pp. 373–386, doi:10.2307/2118525, `https://doi.org/10.2307/2118525`.

Reznikov, A. (1996). Rationality of secondary classes, *J. Differential Geom.* **43**, 3, pp. 674–692, `http://projecteuclid.org/euclid.jdg/1214458328`.

Rho, M. and Zahed, I. (eds.) (2016). *The multifaceted skyrmion, Second Edition* (World Scientific Publishing Co.), doi:10.1142/9710, `https://doi.org/10.1142/9710`.

Richter, B. (2020). *From categories to homotopy theory*, *Cambridge Studies in Advanced Mathematics*, Vol. 188 (Cambridge University Press, Cambridge), ISBN 978-1-108-47962-2, doi:10.1017/9781108855891, `https://doi.org/10.1017/9781108855891`.

Richter, B. (2022). Commutative ring spectra, in *Stable categories and structured ring spectra, Math. Sci. Res. Inst. Publ.*, Vol. 69 (Cambridge Univ. Press, Cambridge), pp. 249–299.

Riehl, E. (2009). A concise definition of model category, `http://www.math.jhu.edu/~eriehl/modelcat.pdf`.

Riehl, E. (2014). *Categorical homotopy theory*, *New Mathematical Monographs*, Vol. 24 (Cambridge University Press, Cambridge), ISBN 978-1-107-04845-4, doi:10.1017/CBO9781107261457, `https://doi.org/10.1017/CBO9781107261457`.

Riehl, E. and Verity, D. (2021). Elements of ∞-category theory, `https://math.jhu.edu/~eriehl/elements.pdf`.

Roberts, D. M. and Stevenson, D. (2016). Simplicial principal bundles in parametrized spaces, *New York J. Math.* **22**, pp. 405–440, `http://nyjm.albany.edu:8000/j/2016/22_405.html`.

Rogers, C. L. (2020). An explicit model for the homotopy theory of finite-type Lie n-algebras, *Algebr. Geom. Topol.* **20**, 3, pp. 1371–1429, doi:10.2140/agt.2020.20.1371, `https://doi.org/10.2140/agt.2020.20.1371`.

Rognes, J. (2014). Chromatic redshift (lecture notes), `https://arxiv.org/abs/1403.4838`.

Rohm, R. and Witten, E. (1986). The antisymmetric tensor field in superstring theory, *Ann. Physics* **170**, 2, pp. 454–489, doi:10.1016/0003-4916(86)90099-0, `https://doi.org/10.1016/0003-4916(86)90099-0`.

Roiban, R., Spradlin, M., and Volovich, A. (2004). Tree-level S matrix of Yang-Mills theory, *Phys. Rev. D (3)* **70**, 2, pp. 026009, 10, doi:10.1103/PhysRevD.70.026009, `https://doi.org/10.1103/PhysRevD.70.026009`.

Rosenberg, J. (1989). Continuous-trace algebras from the bundle theoretic point of view, *J. Austral. Math. Soc. Ser. A* **47**, 3, pp. 368–381.

Rudolph, G. and Schmidt, M. (2017). *Differential geometry and mathematical physics. Part II*, Theoretical and Mathematical Physics (Springer, Dordrecht), ISBN 978-94-024-0958-1; 978-94-024-0959-8, doi:10.1007/978-94-024-0959-8, `https://doi.org/10.1007/978-94-024-0959-8`, fibre bundles, topology and gauge fields.

Rudyak, Y. B. (1998). *On Thom spectra, orientability, and cobordism*, Springer Monographs in Mathematics (Springer-Verlag, Berlin), ISBN 3-540-62043-5, with a foreword by Haynes Miller.

Sati, H. (2008). An approach to anomalies in M-theory via KSpin, *J. Geom. Phys.* **58**, 3, pp. 387–401, doi:10.1016/j.geomphys.2007.11.010, `https://doi.org/10.1016/j.geomphys.2007.11.010`.

Sati, H. (2009). A higher twist in string theory, *J. Geom. Phys.* **59**, 3, pp. 369–373, doi:10.1016/j.geomphys.2008.11.009, `https://doi.org/10.1016/j.geomphys.2008.11.009`.

Sati, H. (2010). Geometric and topological structures related to M-branes, in *Superstrings, geometry, topology, and C^*-algebras*, *Proc. Sympos. Pure Math.*, Vol. 81 (Amer. Math. Soc., Providence, RI), pp. 181–236, doi:10.1090/pspum/081/2681765, `https://doi.org/10.1090/pspum/081/2681765`.

Sati, H. (2011). Anomalies of E_8 gauge theory on string manifolds, *Internat. J. Modern Phys. A* **26**, 13, pp. 2177–2197, doi:10.1142/S0217751X1105333X, `https://doi.org/10.1142/S0217751X1105333X`.

Sati, H. (2014). M-theory with framed corners and tertiary index invariants, *SIGMA Symmetry Integrability Geom. Methods Appl.* **10**, pp. Paper 024, 28, doi:10.3842/SIGMA.2014.024, `https://doi.org/10.3842/SIGMA.2014.024`.

Sati, H. (2015). Ninebrane structures, *Int. J. Geom. Methods Mod. Phys.* **12**, 4, pp. 1550041, 24, doi:10.1142/S0219887815500413, https://doi.org/10.1142/S0219887815500413.

Sati, H. (2018). Framed M-branes, corners, and topological invariants, *J. Math. Phys.* **59**, 6, pp. 062304, 25, doi:10.1063/1.5007185, https://doi.org/10.1063/1.5007185.

Sati, H. (2019). Six-dimensional gauge theories and (twisted) generalized cohomology, https://arxiv.org/abs/1908.08517.

Sati, H. and Schreiber, U. (2018). Higher t-duality in m-theory via local supersymmetry, *Physics Letters B* **781**, pp. 694–698, doi:https://doi.org/10.1016/j.physletb.2018.04.058, https://www.sciencedirect.com/science/article/pii/S0370269318303526.

Sati, H. and Schreiber, U. (2020a). The character map in equivariant twistorial cohomotopy implies the green-schwarz mechanism with heterotic m5-branes, https://arxiv.org/abs/2011.06533.

Sati, H. and Schreiber, U. (2020b). Equivariant cohomotopy implies orientifold tadpole cancellation, *J. Geom. Phys.* **156**, pp. 103775, 40, doi:10.1016/j.geomphys.2020.103775, https://doi.org/10.1016/j.geomphys.2020.103775.

Sati, H. and Schreiber, U. (2020c). Proper orbifold cohomology, https://arxiv.org/abs/2008.01101.

Sati, H. and Schreiber, U. (2021a). M/f-theory as mf-theory, https://arxiv.org/abs/2103.01877.

Sati, H. and Schreiber, U. (2021b). Twisted cohomotopy implies M5-brane anomaly cancellation, *Lett. Math. Phys.* **111**, 5, pp. Paper No. 120, 25, doi:10.1007/s11005-021-01452-8, https://doi.org/10.1007/s11005-021-01452-8.

Sati, H. and Schreiber, U. (2022). Differential cohomotopy implies intersecting brane observables via configuration spaces and chord diagrams, *Adv. Theor. Math. Phys.* **26**, 4, https://arxiv.org/abs/1912.10425.

Sati, H., Schreiber, U., and Stasheff, J. (2009). L_∞-algebra connections and applications to String- and Chern-Simons n-transport, in *Quantum field theory* (Birkhäuser, Basel), pp. 303–424, doi:10.1007/978-3-7643-8736-5_17, https://doi.org/10.1007/978-3-7643-8736-5_17.

Sati, H., Schreiber, U., and Stasheff, J. (2012). Twisted differential string and fivebrane structures, *Comm. Math. Phys.* **315**, 1, pp. 169–213, doi:10.1007/s00220-012-1510-3, https://doi.org/10.1007/s00220-012-1510-3.

Sati, H. and Westerland, C. (2015). Twisted Morava K-theory and E-theory, *J. Topol.* **8**, 4, pp. 887–916, doi:10.1112/jtopol/jtv020, https://doi.org/10.1112/jtopol/jtv020.

Sati, H. and Wheeler, M. (2018). Variations of rational higher tangential structures, *J. Geom. Phys.* **130**, pp. 229–248, doi:10.1016/j.geomphys.2018.04.001, https://doi.org/10.1016/j.geomphys.2018.04.001.

Schellekens, A. N. and Warner, N. P. (1986). Anomalies and modular invariance in string theory, *Phys. Lett. B* **177**, 3-4, pp. 317–323, doi:10.1016/0370-2693(86)90760-4, https://doi.org/10.1016/0370-2693(86)90760-4.

Schellekens, A. N. and Warner, N. P. (1987). Anomalies, characters and strings, *Nuclear Phys. B* **287**, 2, pp. 317–361, doi:10.1016/0550-3213(87)90108-8, https://doi.org/10.1016/0550-3213(87)90108-8.

Schlafly, R. (1980). Universal connections, *Invent. Math.* **59**, 1, pp. 59–65, doi:10.1007/BF01390314, https://doi.org/10.1007/BF01390314.

Schlank, T. M. and Yanovski, L. (2019). The ∞-categorical Eckmann-Hilton argument, *Algebr. Geom. Topol.* **19**, 6, pp. 3119–3170, doi:10.2140/agt.2019.19.3119, https://doi.org/10.2140/agt.2019.19.3119.

Schlichtkrull, C. (2004). Units of ring spectra and their traces in algebraic K-theory, *Geom. Topol.* **8**, pp. 645–673, doi:10.2140/gt.2004.8.645, https://doi.org/10.2140/gt.2004.8.645.

Schreiber, U. (2013). Differential cohomology in a cohesive infinity-topos, https://arxiv.org/abs/1310.7930.

Schreiber, U., Schweigert, C., and Waldorf, K. (2007). Unoriented WZW models and holonomy of bundle gerbes, *Comm. Math. Phys.* **274**, 1, pp. 31–64, doi:10.1007/s00220-007-0271-x, https://doi.org/10.1007/s00220-007-0271-x.

Schwänzl, R. and Vogt, R. M. (1994). Basic constructions in the K-theory of homotopy ring spaces, *Trans. Amer. Math. Soc.* **341**, 2, pp. 549–584, doi:10.2307/2154572, https://doi.org/10.2307/2154572.

Schwarz, J. H. (2012). *Gravity, unification, and the superstring*, chap. 3 (Cambridge University Press), p. 37–62, doi:10.1017/CBO9780511977725.005.

Schwede, S. and Shipley, B. (2003a). Equivalences of monoidal model categories, *Algebr. Geom. Topol.* **3**, pp. 287–334, doi:10.2140/agt.2003.3.287, https://doi.org/10.2140/agt.2003.3.287.

Schwede, S. and Shipley, B. (2003b). Stable model categories are categories of modules, *Topology* **42**, 1, pp. 103–153, doi:10.1016/S0040-9383(02)00006-X, https://doi.org/10.1016/S0040-9383(02)00006-X.

Schweigert, C. and Waldorf, K. (2011). Gerbes and Lie groups, in *Developments and trends in infinite-dimensional Lie theory*, *Progr. Math.*, Vol. 288 (Birkhäuser Boston, Boston, MA), pp. 339–364, doi:10.1007/978-0-8176-4741-4_10, https://doi.org/10.1007/978-0-8176-4741-4_10.

Segal, G. (1968). Classifying spaces and spectral sequences, *Inst. Hautes Études Sci. Publ. Math.* **34**, pp. 105–112, http://www.numdam.org/item?id=PMIHES_1968__34__105_0.

Segal, G. (1979). The topology of spaces of rational functions, *Acta Math.* **143**, 1-2, pp. 39–72, doi:10.1007/BF02392088, https://doi.org/10.1007/BF02392088.

Sharma, A. (2019). On the homotopy theory of g-spaces, *Intern. J. Math. Stat. Inv.* **7**, pp. 22–55.

Shimakawa, K., Yoshida, K., and Haraguchi, T. (2018). Homology and cohomology via enriched bifunctors, *Kyushu J. Math.* **72**, 2, pp. 239–252, doi:10.2206/kyushujm.72.239, https://doi.org/10.2206/kyushujm.72.239.

Simons, J. and Sullivan, D. (2008). Axiomatic characterization of ordinary differential cohomology, *J. Topol.* **1**, 1, pp. 45–56, doi:10.1112/jtopol/jtm006, https://doi.org/10.1112/jtopol/jtm006.

Simons, J. and Sullivan, D. (2010). Structured vector bundles define differential K-theory, in *Quanta of maths*, *Clay Math. Proc.*, Vol. 11 (Amer. Math. Soc., Providence, RI), pp. 579–599.

Simpson, C. (1997a). The Hodge filtration on nonabelian cohomology, in *Algebraic geometry—Santa Cruz 1995, Proc. Sympos. Pure Math.*, Vol. 62 (Amer. Math. Soc., Providence, RI), pp. 217–281, doi:10.1090/pspum/062.2/1492538, `https://doi.org/10.1090/pspum/062.2/1492538`.

Simpson, C. (1997b). Secondary kodaira-spencer classes and nonabelian dolbeault cohomology, `https://arxiv.org/abs/alg-geom/9712020`.

Simpson, C. (2002). Algebraic aspects of higher non-Abelian Hodge theory, in *Motives, polylogarithms and Hodge theory. Part II: Hodge theory. Papers from the International Press conference, Irvine, CA, USA, June 1998* (Somerville, MA: International Press), ISBN 1-57146-091-8, pp. 417–604.

Souriau, J. M. (1980). Groupes différentiels, Differential geometrical methods in mathematical physics, Proc. Conf. Aix-en-Provence and Salamanca 1979, Lect. Notes Math. 836, 91-128 (1980).

Souriau, J.-M. (1984). *Groupes différentiels et physique mathématique*, *Lecture Notes in Physics*, Vol. 201 (Springer), pp. 511–513, doi:10.1007/BFb0016198, `https://doi.org/10.1007/BFb0016198`.

Spanier, E. H. (1949). Borsuk's cohomotopy groups, *Ann. Math. (2)* **50**, pp. 203–245, doi:10.2307/1969362.

Spanier, E. H. (1966). Algebraic topology, McGraw-Hill Series in Higher Mathematics. New York etc.:McGraw-Hill Book Company. XIV, 528 p. (1966).

Stasheff, J. D. (1971). H-spaces and classifying spaces: Foundations and recent developments, Algebraic Topology, Proc. Sympos. Pure Math. 22, 247-272 (1971).

Steenrod, N. (1951). *The topology of fibre bundles*, *Princeton Math. Ser.*, Vol. 14 (Princeton University Press, Princeton, NJ).

Steenrod, N. E. (1943). Homology with local coefficients, *Ann. Math. (2)* **44**, pp. 610–627, doi:10.2307/1969099.

Steenrod, N. E. (1947). Products of cocycles and extensions of mappings, *Ann. Math. (2)* **48**, pp. 290–320, doi:10.2307/1969172.

Steenrod, N. E. (1962). *Cohomology operations. Lectures. Written and revised by D.B.A. Epstein.*, *Ann. Math. Stud.*, Vol. 50 (Princeton University Press, Princeton, NJ), doi:10.1515/9781400881673.

Steenrod, N. E. (1967). A convenient category of topological spaces, *Mich. Math. J.* **14**, pp. 133–152, doi:10.1307/mmj/1028999711.

Steenrod, N. E. (1968). Milgram's classifying space of a topological group, *Topology* **7**, pp. 349–368, doi:10.1016/0040-9383(68)90012-8.

Steenrod, N. E. (1972). Cohomology operations, and obstructions to extending continuous functions, *Adv. Math.* **8**, pp. 371–416, doi:10.1016/0001-8708(72)90004-7, `hdl.handle.net/2027/coo.31924001073042`.

Stel, H. G. (2010). ∞-stacks and their function algebras, `https://dspace.library.uu.nl/handle/1874/203187`.

Stevenson, D. (2004). Bundle 2-gerbes, *Proc. Lond. Math. Soc. (3)* **88**, 2, pp. 405–435, doi:10.1112/S0024611503014357.

Stevenson, D. (2012). Classifying theory for simplicial parametrized groups, `https://arxiv.org/abs/1203.2461`.

Stolz, S. and Teichner, P. (2011). Supersymmetric field theories and generalized cohomology, in U. S. H. Sati (ed.), *Mathematical foundations of quantum field theory and perturbative string theory.* (Providence, RI: American Mathematical Society (AMS)), ISBN 978-0-8218-5195-1, pp. 279–340.

Stretch, C. T. (1981). Stable cohomotopy and cobordism of Abelian groups, *Math. Proc. Camb. Philos. Soc.* **90**, pp. 273–278, doi:10.1017/S0305004100058734.

Strickland, N. (2009). The category of cgwh spaces, `http://neil-strickland.staff.shef.ac.uk/courses/homotopy/cgwh.pdf`.

Strickland, N. (2020). The model structure for chain complexes, `https://arxiv.org/abs/2001.08955`.

Sullivan, D. (1977). Infinitesimal computations in topology, *Publ. Math., Inst. Hautes Étud. Sci.* **47**, pp. 269–331, doi:10.1007/BF02684341.

Sutcliffe, P. M. (1997). BPS monopoles, *Int. J. Mod. Phys. A* **12**, 26, pp. 4663–4705, doi:10.1142/S0217751X97002504.

Taylor, L. (2012). The principal fibration sequence and the second cohomotopy set, in *Proceedings of the Freedman Fest. Based on the conference on low-dimensional manifolds and high-dimensional categories, Berkeley, CA, USA, June 6–10, 2011 and the Freedman symposium, Santa Barbara, CA, USA, April 15–17, 2011 dedicated to Mike Freedman on the occasion of his 60th birthday* (Coventry: Geometry & Topology Publications), pp. 235–251.

Teleman, C. (2004). K-theory of the moduli space of bundles on a surface and deformations of the Verlinde algebra, in *Topology, geometry and quantum field theory. Proceedings of the 2002 Oxford symposium in honour of the 60th birthday of Graeme Segal, Oxford, UK, June 24–29, 2002* (Cambridge: Cambridge University Press), ISBN 0-521-54049-6, pp. 358–378.

Thomas, E. (1962). On the cohomology groups of the classifying space for the stable spinor group, *Bol. Soc. Mat. Mex., II. Ser.* **7**, pp. 57–69.

Toën, B. (2002). Stacks and Non-abelian cohomology. Lecture at Introductory Workshop on Algebraic Stacks, Intersection Theory, and Non-Abelian Hodge Theory, `https://perso.math.univ-toulouse.fr/btoen/files/2015/02/msri2002.pdf`.

Toën, B. (2006). Affine stacks, *Sel. Math., New Ser.* **12**, 1, pp. 39–134, doi:10.1007/s00029-006-0019-z.

Toën, B. and Vezzosi, G. (2005). Homotopical algebraic geometry. I: Topos theory, *Adv. Math.* **193**, 2, pp. 257–372, doi:10.1016/j.aim.2004.05.004.

tom Dieck, T. (2008). *Algebraic topology*, EMS Textb. Math. (Zürich: European Mathematical Society (EMS)), ISBN 978-3-03719-048-7, doi:10.4171/048.

Tu, J.-L. and Xu, P. (2006). Chern character for twisted K-theory of orbifolds, *Adv. Math.* **207**, 2, pp. 455–483, doi:10.1016/j.aim.2005.12.001, `https://doi.org/10.1016/j.aim.2005.12.001`.

Vallette, B. (2020). Homotopy theory of homotopy algebras, *Ann. Inst. Fourier (Grenoble)* **70**, 2, pp. 683–738, `http://aif.cedram.org/item?id=AIF_2020__70_2_683_0`.

van Nieuwenhuizen, P. (1983). Free graded differential superalgebras, in *Group theoretical methods in physics (Istanbul, 1982)*, *Lecture Notes in Phys.*, Vol. 180 (Springer, Berlin), pp. 228–247, doi:10.1007/3-540-12291-5_29, `https://doi.org/10.1007/3-540-12291-5_29`.

Voisin, C. (2007). *Hodge theory and complex algebraic geometry. I & II*, *Cambridge Studies in Advanced Mathematics*, Vol. 76, 77, english edn. (Cambridge University Press, Cambridge), ISBN 978-0-521-71801-1, translated from the French by Leila Schneps.

Weibel, C. A. (1994). *An introduction to homological algebra*, *Cambridge Studies in Advanced Mathematics*, Vol. 38 (Cambridge University Press, Cambridge), ISBN 0-521-43500-5; 0-521-55987-1, doi:10.1017/CBO9781139644136, `https://doi.org/10.1017/CBO9781139644136`.

Weil, A. (2014). *Géométrie différentielle des espaces fibres*, Springer Collected Works in Mathematics (Springer, Heidelberg), ISBN 978-3-662-45256-1, pp. 422–436, reprint of the 2009 [MR2883738] and 1979 [MR0537937] editions.

Weinstein, A. (1996). Groupoids: unifying internal and external symmetry. A tour through some examples, *Notices Amer. Math. Soc.* **43**, 7, pp. 744–752.

Wendt, M. (2011). Classifying spaces and fibrations of simplicial sheaves, *J. Homotopy Relat. Struct.* **6**, 1, pp. 1–38.

Whitehead, G. W. (1962). Generalized homology theories, *Trans. Amer. Math. Soc.* **102**, pp. 227–283, doi:10.2307/1993676, `https://doi.org/10.2307/1993676`.

Whitney, H. (1957). *Geometric integration theory* (Princeton University Press, Princeton, N. J.).

Wierstra, F. (2017). Hopf Invariants in Real and Rational Homotopy Theory, `http://urn.kb.se/resolve?urn=urn:nbn:se:su:diva-146246`.

Witten, E. (1987). Elliptic genera and quantum field theory, *Comm. Math. Phys.* **109**, 4, pp. 525–536, `http://projecteuclid.org/euclid.cmp/1104117076`.

Witten, E. (1997a). Five-brane effective action in M-theory, *J. Geom. Phys.* **22**, 2, pp. 103–133, doi:10.1016/S0393-0440(97)80160-X, `https://doi.org/10.1016/S0393-0440(97)80160-X`.

Witten, E. (1997b). On flux quantization in M-theory and the effective action, *J. Geom. Phys.* **22**, 1, pp. 1–13, doi:10.1016/S0393-0440(96)00042-3, `https://doi.org/10.1016/S0393-0440(96)00042-3`.

Witten, E. (2004). Perturbative gauge theory as a string theory in twistor space, *Comm. Math. Phys.* **252**, 1-3, pp. 189–258, doi:10.1007/s00220-004-1187-3, `https://doi.org/10.1007/s00220-004-1187-3`.

Würgler, U. (1991). Morava K-theories: a survey, in *Algebraic topology Poznań 1989*, *Lecture Notes in Math.*, Vol. 1474 (Springer, Berlin), pp. 111–138, doi:10.1007/BFb0084741, `https://doi.org/10.1007/BFb0084741`.

Index

www.ingramcontent.com/pod-product-compliance
Lightning Source LLC
Chambersburg PA
CBHW050555190326
41458CB00007B/2049
* 9 7 8 9 8 1 1 2 7 6 6 9 9 *